Space Economics

Edited by
Joel S. Greenberg
Princeton Synergetics, Inc.
Princeton, New Jersey
and
Henry R. Hertzfeld
HRH Associates
Bethesda, Maryland

Volume 144
PROGRESS IN
ASTRONAUTICS AND AERONAUTICS

A. Richard Seebass, Editor-in-Chief
University of Colorado at Boulder
Boulder, Colorado

Published by the American Institute of Aeronautics and Astronautics,
Inc., 370 L'Enfant Promenade, SW, Washington, DC 20024-2518

Copyright © 1992 by the American Institute of Aeronautics and Astronautics, Inc. Printed in the United States of America. All rights reserved. Reproduction or translation of any part of this work beyond that permitted by Sections 107 and 108 of the U.S. Copyright Law without the permission of the copyright owner is unlawful. The code following this statement indicates the copyright owner's consent that copies of articles in this volume may be made for personal or internal use, on condition that the copier pay the per-copy fee ($2.00) plus the per-page fee ($0.50) through the Copyright Clearance Center, Inc., 21 Congress Street, Salem, Mass. 01970. This consent does not extend to other kinds of copying, for which permission requests should be addressed to the publisher. Users should employ the following code when reporting copying from this volume to the Copyright Clearance Center:

1-56347-042-X/92 $2.00+.50

Data and information appearing in this book are for informational purposes only. AIAA is not responsible for any injury or damage resulting from use or reliance, nor does AIAA warrant that use or reliance will be free from privately owned rights.

ISSN 0079-6050

Progress in Astronautics and Aeronautics

Editor-in-Chief
A. Richard Seebass
University of Colorado at Boulder

Editorial Board

Richard G. Bradley
General Dynamics

Allen E. Fuhs
Carmel, California

George J. Gleghorn
*TRW Space
and Technology Group*

Dale B. Henderson
Los Alamos National Laboratory

Carolyn L. Huntoon
NASA Johnson Space Center

Reid R. June
Boeing Military Airplane Company

John L. Junkins
Texas A&M University

John E. Keigler
*General Electric Company
Astro-Space Division*

Daniel P. Raymer
*Lockheed Aeronautical Systems
Company*

Martin Summerfield
*Princeton Combustion Research
Laboratories, Inc.*

Charles E. Treanor
*Arvin/Calspan
Advanced Technology Center*

Jeanne Godette
Series Managing Editor
AIAA

Table of Contents

Preface vii

Introduction ix

Chapter 1. Financial/Investment Considerations

Financial/Investment Analysis 3
Joel S. Greenberg, *Princeton Synergetics, Inc., Princeton, New Jersey*

Financing Space Projects 35
Jerome Simonoff, *U.S. Department of Transportation, Commercial Space Transportation Advisory Committee, Washington, DC*

Chapter 2. Cost Analysis

Standard Cost Elements for Technology Programs 45
Carissa Bryce Christensen, *Princeton Synergetics, Inc., Princeton, New Jersey*, and Carl Wagenfuehrer, *NASA Headquarters, Washington, DC*

Cost-Estimating Relationships for Space Programs 57
Humboldt C. Mandell Jr., *NASA Johnson Space Center, Houston, Texas*

Cost-Estimating Relationships: A DCAA Perspective 89
Michael Thibault, *Defense Contract Audit Agency, Alexandria, Virginia*

Cost Estimating for Technology Programs 97
George A. Hazelrigg, *National Science Foundation, Washington, DC*

Space System Life Cycle Cost and Availability Analysis 117
Joel S. Greenberg, *Princeton Synergetics, Inc., Princeton, New Jersey*

Chapter 3. Benefit/Cost and Cost Effectiveness Models

Measuring Returns to Space Research and Development 151
Henry R. Hertzfeld, *HRH Associates, Bethesda, Maryland*

Measuring and Managing Spinoffs: The Case of the Spinoffs
   Generated by ESA programs 171
L. Bach, P. Cohendet, G. Lambert, and M. J. Ledoux, *Louis Pasteur University of Strasbourg, France*

Economics and Regulation of Space Activities 207
Carissa Bryce Christensen, *Princeton Synergetics, Inc., Princeton, New Jersey*

Chapter 4. Economics of the Marketplace

Remote Sensing 233
David Moore, *Congressional Budget Office, Washington, DC*

Product and Service Pricing: Launch Vehicles 263
Eric Gabler, *Silver Spring, Maryland*

Product/Service Pricing: Support Facilities (Space Facilities) 293
Chester M. Lee, *Spacehab, Washington, DC*

Space Insurance 307
Daniel E. Cassidy, *Technology Programmatics International, Arlington, Virginia*

Chapter 5. Relationship of Economics to Major Issues

Commercial Development of Space: Government/Industry
   Relationship 323
Joel S. Greenberg, *Princeton Synergetics, Inc., Princeton, New Jersey*

Economics of Strategic Planning 357
Joseph Fuller and Kevin Lacobie, *Futron Corporation, Bethesda, Maryland*

Remote Sensing: The Inconsistency of U.S. Space Policy 369
Simon P. Worden, *Space Defense Initiative Organization, Washington, DC,* and Jordan S. Katz, *Comprehensive Technologies, Inc., International, Arlington, Virginia*

Engineering Design and Decision Making 381
George A. Hazelrigg, *National Science Foundation, Washington, DC*

Resource Allocation in the Congress 403
Terry Dawson, *U.S. House of Representatives, Washington, DC*

Economic Issues Facing the United States in International
   Space Activities 417
Henry R. Hertzfeld, *HRH Associates, Bethesda, Maryland*

Author Index for Volume 144 437

List of Series Volumes 438

Preface

Merging the principles of different disciplines is never an easy task. Space activities are the playing field of scientists and engineers. Economic considerations tend to take a back seat to the goals of advanced design, exploration, and discovery. However, sooner or later the economic realities of allocating human, physical, and financial resources in a manner that will meet the objectives of any given activity hits the technicians. Often, scientists and engineers do not rigorously analyze economic issues until late in the planning stages. When that occurs, economics and economic analysis (and, by association, economists and financial analysts) earn reputations as hindrances to progress and to helping space activities meet their goals.

This negative connotation to economics need not occur. The purpose of this book is to expose scientists and engineers active in space projects to the many different and useful ways that economic analysis and methodology can help get the job done. Whether it be through an understanding of cost-estimating procedures or through a better insight into the use of economics in strategic planning and marketing, the space professional will find that the use of a formal and structured economic analysis early in the design of a program will make later decisions easier and more informed.

The chapters in this book will also help guide noneconomists to the use of the appropriate and proper economic tools to answer specific questions. One would, for example, not use a cost-benefit model to establish a new space business venture. Nor would one use a strategic business plan to design a deep space exploration program. However, a government official would use a cost-effectiveness model to help decide the optimal combination of resources to meet well-defined space transportation needs. Also, an entrepreneur who identifies a new and potentially profitable space business activity would need to perform market research, evaluate pricing options, and complete a business plan.

Two principles should be kept in mind. The first is that although economics is an analytic tool, it is based on the actions of people and society rather than on natural physical or chemical properties. The methodology and models contain many assumptions. What are even more important than the results of economic analysis are the thought process and the insights into economic and budgetary problems that are not intuitively obvious without performing the analysis. The second is that economics is not static. The results of economic analyses change in response to both changes in program and technical designs as well as to changes in external factors such as political decisions, business cycle fluctuations, interest rates, etc. The lesson for scientists and engineers is that economic analyses and predictions must be kept as current and up-to-date as new technical designs for programs and projects.

We hope that this book will be a valuable reference source for the space professional. It is intended as a guide to those who must make decisions concerning the allocation of scarce resources among and within space programs. Economic analysis can and should be an engineer's ally, not an adversary. Used properly, economic and financial modeling provide a valuable input into decision making. Used improperly, they can destroy a project as completely as an incorrect design can. This book is aimed at encouraging the knowledgeable and appropriate use of economics in situations space scientists and engineers are likely to encounter.

Henry R. Hertzfeld
Joel S. Greenberg
June 1992

Introduction

THE objectives of this book are to put economic analysis into perspective with respect to real-world decision making in the space industry and to expand the perspective of the reader with respect to the type of tools and analyses that might be brought to bear on complex business and government problems.

There are many aspects to "economics," all of which are aimed at providing information for clarifying and improving decisions concerning the commitment of resources. Operationally, economic analysis includes financial and investment analysis, cost analysis, benefit/cost and cost-effectiveness analyses, as well as traditional micro- and macroeconomic analyses. These methods, in turn, utilize many disciplines including mathematical, simulation, and econometric modeling, probability and statistics, and decision analysis. These and other aspects of space economic analysis and associated disciplines are discussed in the following pages with emphasis placed upon applications. Sufficient theory is presented as the applications are developed so that the methods and techniques described can be applied to the economic aspects of space applications beyond those chosen as examples in the articles in this book.

To appreciate the scope of economic analysis and its applications, a number of rather diverse problems are indicated below where economic and related analyses are likely to make important contributions in the decision making process:

1) Development of private sector options for the financing of space business ventures (for example, Spacehab and Geostar) and establishing the need and appropriate role for the government in helping to achieve the necessary financing.

2) Establishment of the appropriateness and likelihood of success of privatization or commercialization of government assets or provision of services previously provided by the government.

3) From the government's perspective, establishment of the likely financing requirements for commercial space ventures and the consequent need for and the form of government assistance.

4) Estimation of the cost for development and production of products including new satellites and launch vehicles.

5) Establishment of more realistic cost estimates and program plans for long-term multiphase technology programs (such as the Space Exploration initiative or the nuclear propulsion program).

6) Development of pricing policies for products and/or services that will maintain or improve competitive positions and achieve desired financial performance objectives (such as achieving a specified return on investment).

7) Establishment of the value of a new space transportation system and selecting the best cost and performance alternatives.

8) Assessment of the value of technology transfer alternatives.

9) Assisting in the formulation of efficient technology programs comprising multiple research and development projects, each having multiple benefit attributes and resource requirements, the sum of which exceed budgetary constraints.

10) Setting of facility and third-party insurance requirements for commercial space launches from government launch facilities.

11) Establishment of the likely impacts of the use of nonmarket economy goods and services on the U.S. economy and on specific organizations.

12) Establishment of the likely impacts on U.S. industry of making excess government assets available for use.

13) Establishment of the economic impacts of government regulations and policies such as setting of insurance requirements for space launches, charging launch license fees, and establishing space debris remediation policies.

14) Establishment of the appropriate tradeoffs between reliability, performance, and maintenance and sparing strategies aimed at minimizing life cycle cost of space missions while achieving system performance constraints.

15) Assessment of the economic value of government investments in technology programs on the U.S. economy and the role of such investments in the creation of jobs.

The papers in this book discuss a number of the above areas. Because of the finite space available, some topics are addressed indirectly, whereas others are discussed only by inference.

Many space businesses are unique in that they require large amounts of up-front capital funding, have long periods before payback, involve risks greater than those of other businesses (technological, market, and government intervention), and are subject to many external influences. Most financial analyses are oriented toward the short run and toward well-defined products.

Chapter 1, Financial/Investment Considerations, discusses these differences and develops the basic concepts of financial analysis and its inclusion in the business plan. This includes cash flow and present value analysis. It introduces the subjects of uncertainty and risk in decision making and illustrates how uncertainty and risk can be quantitatively taken into account. This is done through the use of a communication satellite business financial risk analysis model. Several applications of this model are presented, including assessing the value of technology programs, impacts of insurance pricing on financial performance, the selection of a space launch service taking into account the multiple attribute nature of transportation alternatives, and the potential financial impacts of alternative space debris remediation policies. This chapter also presents a discussion of the capital markets and their structure and the attributes that a business must have in order to be successful in obtaining required financing. Also discussed are sources of and conditions necessary for debt financing. An example of a communcation satellite business is used to illustrate how total financing for a business can be arranged.

Corporate and government cost analysis is oriented toward applying historical factors in production to new products. Most often, the cost-estimation procedure is designed for products that are expected to have large production runs such as major defense equipment. Space projects are typified as being low production with prototype equipment. They are also state of the art and often require significant amounts of new technological development. Therefore, there are times when hardware-oriented cost-estimation models are of limited use in evaluating new space research and development programs. Chapter 2, Cost Analysis, contains several papers that cover a number of subjects relating to cost analysis. The first problem addressed is that of comparability of cost estimates made by different organizations for elements of an overall research and development program or mission. A standard cost element structure is developed for technology programs. Next, the development and utilization of cost estimating relationships (CERs) for space systems is described. The applicability of CERs is discussed and their limitations described. It is appropriate to utilize CERs when the physical attributes and performance of the element being costed are known and there is little doubt that the element can be built. But what about costing and planning of the SEI or similar programs? These are long-term multiphase programs that require the development of new technology. The resulting performance attributes cannot be predicted with certainty, nor can schedule. And there is the possibility that certain performance goals will not be achieved. A paper is presented that describes this problem in detail and presents a cost estimating and program planning methodology appropriate for such activities.

Chapter 2 also addresses the interface and potential problems between industry cost estimating and the Defense Contract Audit Agency's (DCAA) responsibility for cost verification. Finally, Chapter 2 concludes with a detailed discussion of life cycle costing methods and life cycle cost and availability analysis as applied to space missions. A stochastic simulation model is presented that can be used to assess the impacts of launch vehicle reliability and scheduling delays, satellite subsystem reliability, sensor criticality, and maintenance and sparing strategies on mission life cycle cost and sensor availability statistics.

A number of different tools of economic analysis are used in the assessment of economic impacts of space-related research and development. Various studies have applied microeconmic and macroeconomic techniques, depending on the specific problem addressed. The measurement of the impact of particular technologies on specific industrial or geographic sectors usually involves a microeconomic approach to analysis. Benefit/cost analysis as well as surveys of the areas affected are the primary tools used. When the measurement of alternative methods of achieving a desired performance specification are required, cost-effectiveness models are employed. And when the impacts of all space research and development efforts are analyzed for their long-term effect on the economy (as measured in GNP, employment, international trade balances, etc.), a production function, macroeconomic analysis is often performed.

Chapter 3 includes papers that summarize the advantages and disadvantages of using these various econometric tools, along with a report on the various studies conducted by major space agencies in the United States and in Europe on the impacts of their contracting, their research and development efforts, and the spinoff technologies that have resulted from the state-of-the-art technology development required by the unique conditions of operating in space.

The various methods of measuring economic returns to space-related research and development are reviewed. A number of NASA-sponsored studies have shown very positive returns to investments in space. A number of different approaches have been taken, including microeconomic analysis of specific technologies as well as macroeconomic modeling of long-term productivity gains. Because of the wide variety of simplifying assumptions behind these economic models, none of the studies is conclusive. However, taken together, they tell a very dramatic picture of the robust nature of government research and development in stimulating the economy. Also, when put into the perspective of the fact that the U.S. government investment in space has been made on defense and prestige motives rather than on economic stimulation criteria, the measured returns are all that much more impressive.

Chapter 3 also explores the relationship between economics and regulation of space activities. Federal agencies are required to assess the economic impacts of regulations they impose. Such an assessment is designed to serve as an aid to decision making—the agency's responsibility is to achieve its objectives in the most cost-effective manner possible and to ensure that its regulations are in the public interest. When economic benefits exceed economic costs, the public interest is served. The new industry behind space launch operations presents a challenge for performing economic analysis. Federal regulations for economic impact analysis are reviewed. Examples described in the chapter are drawn from the activities of the U.S. Department of Transportation, which oversees and regulates commercial space transportation activities. These examples include launch insurance requirements and the levying of user fees.

The competitive free market is the intellectual and operational backbone of the economic system. Supply and demand determine the price of goods and services. Space business activity at the present time is characterized by a hybrid system where government investments and government intervention in pricing and production intervene in the market. This results because of the historical role of government involvement in defense activities which are closely tied to space research and development efforts, the fact that space is a new industry and government facilities represent the only effective way to get to and from space, and the existence of foreign nations with different economic structures that may include government owned and operated production facilities and/or nonmarket pricing mechanisms.

In spite of these space and corporate institutions that are outside of the competitive market pricing system, many companies that provide space products do operate in a free market environment. Chapter 4 provides an overview of the operation of the pricing and market system as it involves various space activities, particularly the remote sensing, space infrastructure, and launch vehicle industries. In addition, services to the space sector in the form of capital and insurance, and private launch facilities provide examples of the interaction of government ownership facilities and government regulation working in tandem with free market supply and demand forces.

The Land Remote Sensing Commercialization Act of 1984 began a 10-year experiment with the objective of transforming the Landsat system from a research project into the productive base for a new private industry. Seven years into this experiment in space commercialization, its premise—that a new private industry can be created by the transfer of federal assets and with limited subsidies—is under challenge. While some progress has been made in privatizing Landsat, the prospects for a fully private system by the end of this decade are dim. Chapter 4 reviews the history of the Landsat program, including the structure, scale, cost, and economic market conditions of remote sensing products. Because remote sensing is closely connected to various government functions, regulatory initiatives as well as policy options are also topics within this discussion.

The second paper in Chapter 4 utilizes economic theory and publicly available information to provide an overview of the economics of the commercial space launch industry and its pricing practices. A brief history is provided of the commercial launch services industry, as well as a description of the competitive environment of the international launch market and the entities that participate in the market. The cost structure of the commercial launch industry which highlights the extraordinary importance of government space launch programs to the competitiveness of the launch providers is a central focus of this paper. Economists often use a pricing behavior model to analyze oligopolistic market structures. A similar model that explains the various terms and conditions that affect the price and value of launch services is presented in this paper. The paper concludes with a discussion of the role of the government in maintaining free and fair markets for launch products and services.

Whereas the launch industry is concerned primarily with pricing decisions in an internationally competitive environment, pricing of space facilities to be provided by Spacehab are more concerned with user affordability and impact of price on demand. Spacehab, Inc. will, in essence, provide additional facilities on board the Space Shuttle Orbiter for which it will charge a fee. The mechanism for setting the fee structure starts with the specifics of the business venture and attributes of the services provided by Spacehab. The paper discusses the various advantages and disadvantages of multiattribute pricing policies and describes the rationale leading to the current pricing policy.

Although commercial space activities are different from normal business activities, in many ways they are the same. Space decisions involve risk, financing, and analysis of returns on investments and of the alternative business uses of the funds. One of the important services the financial industry provides to the companies involved in private activities in space is insurance. Because space is a very risky environment, launch and operations insurance is a virtual necessity for most firms with valuable payloads to send into space. Over time the availability of insurance at a reasonable price has varied with the history of launch successes. The role of the insurance industry and of governments is therefore reviewed in relation to the assessment of technological and financial risk.

Whereas Chapter 4 focuses primarily on the pricing system for goods and services, Chapter 5 extends the analysis to include other forms of the relationship of economics to major political and social issues. U.S. government policy is often a seemingly uncoordinated mixture of short-term necessity, annual budget worries, and concern about U.S. prestige and technical leadership. Space research and development efforts are expensive and require long gestation periods before successful missions are realized. International competitiveness is becoming an increasingly important consideration as the rest of the world grows rapidly in its space and economic capabilities. This chapter addresses many important considerations of economic analysis that go beyond traditional economic theory. Often this requires a mix of business and financial planning, long-term thinking policy decisions, and coordinated decision making. These goals may be hard to reach in today's government policy environment. The papers in this concluding chapter of the book explore the role of economic analysis in merging the often conflicting roles of economic analysis in society and in government decision making. More questions are posed than answered. The search for answers to these problems will be the subject of many stimulating and lively debates during the 1990s as the space industries leave an "infant industry" status and assume an important economic role for the twenty-first century.

As the U.S. government has become active in promoting the commerical development of space, it has, in essense, begun to play the role of a venture capitalist and/or investment banker providing its scarce resources (the use of funds, infrastructure, and personnel) to help initiate commercial ventures that are in the public interest. The various tools that the government can use to "encourage" commercial endeavors are reviewed and their relationships to private sector investment decision variables are discussed. Since government actions effect private sector investment decisions, investment attitudes are described quantitatively in terms of the likelihood of investment in terms of expected return on investment, risk as measured as the variability of ROI, magnitude of investment and payback period. A concept is developed for assessing the impact of government programs through the investment likelihood function. A number of examples are presented.

Strategic planning in the business community has developed as a separate and important adjunct to decision making during the past 20 years. A firm must set its goals and then organize its resources to best meet those goals. Whether they be increasing market share, making a higher profit, or being a leader in technology development with a goal to long-term market penetration, firms often undergo a formal and complex review of the economics and politics involved in obtaining the end results. Strategic planning in the space sector of the economy is different because of the unique characteristics of the aerospace industry. These differences focus on the involvement of the government in contracting, in research and development, and in procurement of space systems. In recent years, the government itself (at the agency level) has begun to embrace the concept of strategic planning for its own programs. Different economic factors are at play in government planning than in business planning. The economics and the concept of strategic planning in these various applications are reviewed.

Remote sensing policy well illustrates the inconsistency that has arisen in the overall planning of U.S. space policy. Specifically, the lessons learned from the LANDSAT program are related to similar issues that are present in the current U.S. Global Change Research Program. The paper finds that NASA and other remote sensing agencies should embrace a strategy based on the following attributes: the 10-year goal should be that all civil remote sensing systems should be commercially developed and operated; the U.S. government should assume the role of "anchor tenant"; and the U.S. government should demonstrate (jointly with potential commerical operators), as a research program, a distributed "lightsat" remote sensing system by 1995. The argument is made that in the long term, new technology development and validation should remain a U.S. Government function.

Engineering design can be thought of as a process of decision making instead of the more traditional approach as a process of problem solving. Space engineers are continuously making engineering decisions that have far-reaching consequences that are infrequently considered in their decision. Within Chapter 5, a paper is presented that develops a methodology which places cost analysis into a context within the process of engineering decision making.

No anthology of the economic aspects of space investments would be complete without a review of the government's process of allocating resources to space. A paper in Chapter 5 therefore traces the steps involved in this process followed by the U.S. Congress every year to decide on the budget for space. From the requirement of annual appropriations for long-term research and development programs, to the way Congress is organized with space budgets competing in Committee with veterans' benefits and housing programs, to the whims of pork barrel politics, space programs face careful but sometimes illogical oversight and comparisons. However, in spite of the many roadblocks and problems it faces, space has fared well in Congress.

Commercial space activities are on the increase. The changes in the former Soviet Union have made the space race between the United States and the Soviet Union for military and technological leadership that dominated the period of the 1950s–1980s seem almost like an anachronism. The final paper reviews the past and suggests issues and trends that will dominate space activities in the future. The role of economics in U.S. space policy, ranging from business decisions to the allocation of government resources, is changing and becoming a more immediate and more important force in policy formation.

Chapter 1. Financial/Investment Considerations

Financial/Investment Analysis

Joel S. Greenberg*
Princeton Synergetics, Inc., Princeton, New Jersey 08540

Introduction

AN investment decision involves the commitment of funds with the hope of future benefit. Many investment decisions are concerned with transforming an idea or emerging technology into a business venture. This transformation process requires the use of capital and labor resources to develop and provide products/services that satisfy market needs in such a manner that market-determined rates of return are provided on employed capital. In order to obtain necessary capital (from investment banks, venture capital firms, corporations, and other parties) it is necessary to prepare a business plan. A major objective of the business plan is to demonstrate to the investment community (including corporate management) the need for and the magnitude and timing of resources to be employed, products/services to be developed and provided, markets to be satisfied, and the financial performance that will result from the planned business venture. The business plan serves as the justification for the capital and other forms of investment. It also, upon commitment of resources, serves as both the initial plan for conducting business and the means for measuring performance.

The business plan normally consists of several supporting plans and analyses, the specific and details of which are a function of the anticipated source of capital. These may include (as necessary to obtain financing): a market analysis and resulting sales forecast; a competitive analysis including rationale for achieving forecasted market shares; a detailed product/service cost analysis; a financial analysis that develops pro-forma income, cash flow, and balance sheet statements for the considered planning horizon; a market/sales plan that identifies the distribution channels, sales force (and

---

Copyright © 1992 by Joel S. Greenberg. Published by the American Institute of Aeronautics and Astronautics, Inc. with permission.
*President.

training), and maintenance requirements; an advertising and promotion plan; a staffing plan for obtaining required personnel; and a management plan including identification of required and available skills (including identification of key individuals). These are all mutually supporting and feed the financial analysis.

The role of the financial analysis is to integrate all of the supporting analyses and plans into a framework that is reasonably standardized and readily understandable by potential investors. It provides the raison d'etre for the investment decision in the form of anticipated financial performance measures, such as annual profit (loss); annual and peak cash requirements; payback period; net present value of cash flow; various rates of return, such as return on sales, return on assets, and return on investment; and quantitative measures of risk (i.e., the possible variability associated with key financial performance measures).[1]

Financial Analysis

Financial analysis is concerned with integrating the results of other analyses, forecasts, and plans and establishing their financial consequences. (For a good general reference on financial analysis, see Ref. 2.) The financial consequences are usually expressed in terms of after-tax profit (revenue less cost and less taxes), cash flow (after-tax profit less capital expenditures, plus depreciation and less changes in other balance sheet items), indebtedness (negative of cumulative cash flow to any point in time with the maximum indebtedness being indicative of the maximum funding requirement), payback period (the time at which indebtedness becomes negative, i.e., the time when net cash inflow is equal to net cash outflow), return on sales (ratio of after-tax profit to net revenue), return on assets (ratio of after-tax profit to value of assets less accumulated depreciation), net present value of cash flow (the discounted—at the firm's cost of capital— value of the cash flow stream suitably adjusted so as to eliminate financing costs; not eliminating financing costs from the present value computation would amount to double counting), and discounted return on investment (the discount rate that results in a net present value of cash flow equal to zero). These parameters are described in more detail in the following paragraphs.

Before-tax profit is the difference between revenues and expenses. After-tax profit (ATP) adjusts this amount to account for federal and other taxes and takes into account carry-forward losses and available tax credits. ATP is given in simplified form by

$$\text{ATP}_t = [1 - \text{tax rate}/100] * [\Sigma \text{revenues}_t - \Sigma \text{expenses}_t - \text{depreciation}_t]$$

where the subscript t indicates the time period (i.e., years).

Capital expenditures are not explicitly included in the profit computation, but occur indirectly (and in any one year only partially) through the depreciation expense. Cash flow (CF), on the other hand, reflects the flow of funds through the business. CF is given by

$$\text{CF}_t = \text{ATP}_t + \text{depreciation}_t + \text{change in payables}_t$$
$$- \text{change in inventory}_t - \text{change in receivables}_t - \text{capital expenditures}_t$$

The cash flow computation includes the magnitude and timing of the inflow and outflow of funds. It includes such measures as after-tax profit, depreciation, increase in payables (i.e., due but not yet paid), decrease in inventories, and decrease in receivables (i.e., due but not yet received) as cash inflows (i.e., sources of funds), and such measures as losses (negative values of ATP), capital expenditures, decrease in payables, increase in inventories, and increase in receivables as cash outflows (i.e., uses of funds). Cash flow (which includes profit and loss as a component), not profit, is the most important determinant of the value of a venture. Profit is an accounting artifact; cash flow is a basic measure. A profitable business venture may fail because of cash flow problems. The significance of profit cannot be overlooked, however, since it is a key consideration when evaluating the availability of funds from the financial community. (Stock prices are usually measured in terms of price-earnings ratios.)

Indebtedness is defined as the negative of the cumulative cash flow to any point in time:

$$\text{Indebtedness}_T = -\sum_{t=1}^{T} \text{cash flow}_t$$

where T is the point in time at which indebtedness is to be measured.

Figure 1 illustrates typical profit, cash flow, and indebtedness patterns. Figures 2 and 3 illustrate typical pro-forma profit and cash flow statement formats, for a "fixed satellite service" communication satellite business venture. [Not normally provided, but indicated in Figs. 2 and 3, are quantitative risk measures. This is discussed in following paragraphs. In Fig. 3, *Present Value "A"* represents the present value component resulting from the cash flow during the planning horizon (15 years in this case) and *Present Value "B"* represents the present value component resulting from the cash flow after the explicit planning horizon (i.e., referred to as infinite horizon discounting).] The cash flow and profit streams normally start off as net cash outflows and losses, respectively, due to research and development (R & D) expenditures, engineering efforts, and initial operating or startup

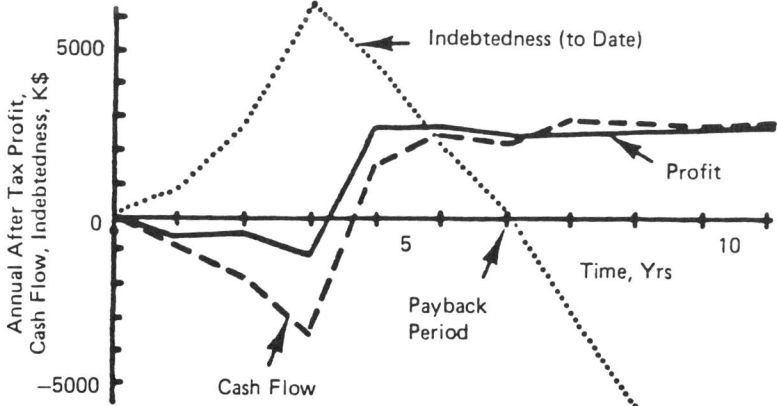

Fig. 1 Typical after-tax profit, cash flow, and indebtedness projections.

PROFORMA INCOME STATEMENT ($ THOUSANDS)

	YEAR				
	1	2	3	4	5
PROTECTED	0.	0.	0.	9373.	62770.
PROTECTED/PREEMPT.	0.	0.	0.	0.	0.
UNPROTECTED/NON-PREEMPT.	0.	0.	0.	0.	0.
PREEMPTIBLE	0.	0.	0.	1277.	3255.
TOTAL REVENUE	0.	0.	0.	10651.	66025.
	0.*	0.*	0.*	4895.*	17871.*
LAUNCH OPERATIONS	0.	0.	0.	1816.	4043.
LAUNCH INSURANCE	0.	0.	0.	753.	1623.
SATELLITE	0.	0.	0.	3220.	6826.
OTHER	0.	415.	1228.	1228.	1228.
DEPRECIATION EXPENSE	0.	415.	1228.	7016.	13720.
S/C CONTROL OPERATIONS	0.	0.	0.	714.	1717.
ENGINEERING EXPENSE	1000.	1000.	1000.	1000.	1382.
RESEARCH & DEVELOPMENT	1000.	1000.	1000.	1000.	1382.
TOTAL OPERATIONS EXPENSE	2000.	2415.	3228.	9730.	18201.
	0.*	20.*	48.*	2915.*	3124.*
GROSS MARGIN ($)	-2000.	-2415.	-3228.	921.	47824.
	0.*	20.*	48.*	2130.*	16009.*
S/C NONRECURRING COST	16766.	4457.	0.	0.	0.
G & A EXPENSE	500.	500.	500.	1277.	1358.
DEBT SERVICE EXPENSE	0.	1322.	4995.	13063.	20163.
BEFORE TAX PROFIT	-19266.	-8695.	-8723.	-13420.	26302.
INCOME TAX	-6936.	-3130.	-3140.	-4831.	9469.
INVESTMENT TAX CREDIT	0.	498.	975.	5788.	6704.
AFTER TAX PROFIT	-12330.	-5066.	-4608.	-2801.	23537.
	586.*	196.*	710.*	2230.*	8327.*
RETURN ON ASSETS (%)	-4267.	-31.	-5.	-2.	13.
	0.*	34.*	1.*	2.*	5.*
RETURN ON SALES (%)	0.	0.	0.	-14.	34.
	0.*	0.*	0.*	25.*	12.*

* STANDARD DEVIATION

Fig. 2 Typical pro-forma income statement (in thousands) for a communication satellite business venture.

FINANCIAL/INVESTMENT ANALYSIS

CASH FLOW PROJECTION ($ THOUSANDS)

	YEAR				
	1	2	3	4	5
AFTER TAX PROFIT	0.	0.	0.	1.	23665.
INCREASE IN PAYABLES	1599.	1416.	3208.	540.	130.
DECREASE IN RECEIVABLES	0.	0.	0.	0.	0.
DECREASE IN CASH	0.	15.	0.	22.	200.
DEPRECIATION	0.	415.	1228.	7016.	13720.
TOTAL CASH INFLOW	1599.	1847.	4436.	7579.	37715.
LOSS	12330.	5066.	4608.	2802.	128.
DECREASE IN PAYABLES	0.	83.	0.	123.	1107.
INCREASE IN RECEIVABLES	0.	0.	0.	1779.	9248.
INCREASE IN CASH	289.	256.	580.	98.	23.
CAPITAL EXPENDITURES	0.	27047.	66483.	61946.	41229.
TOTAL CASH OUTFLOW	12619.	32453.	71671.	66747.	51734.
NET CASH FLOW	-11020.	-30606.	-67235.	-59168.	-14019.
	523.*	9223.*	11278.*	12505.*	21276.*
INDEBTEDNESS	11020.	41626.	108861.	168028.	182047.
	523.*	9247.*	19933.*	17476.*	19544.*

	1	2	3	4	5
DISCOUNT RATE (%)	10.	15.	20.	25.	40.
NET PRESENT VALUE "A"	71795.	11268.	-20471.	-36691.	-47683.
NET PRESENT VALUE "B"	185841.	60837.	23094.	9614.	980.
NET PRESENT VALUE	257636.	72105.	2623.	-27076.	-46703.
	102318.*	55922.*	35407.*	24284.*	10331.*

* STANDARD DEVIATION

Fig. 3 Typical cash flow projection (in thousands) for a communication satellite business venture.

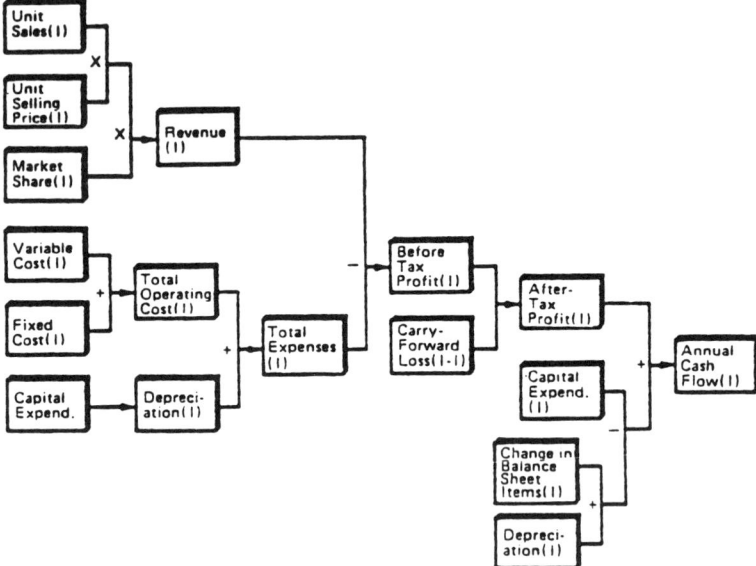

Fig. 4 Simplified cash flow structure. *I* represents time (years).

costs, which precede revenue from sales. Maximum annual net cash outflow decreases, eventually becoming a net cash inflow. The maximum funding requirement, i.e., exposure, is indicated by the peak of the indebtedness curve. When the indebtedness is positive, the total investment has not been recovered and the cumulative cash outflow exceeds the cumulative cash inflow. When the indebtedness is negative, the cumulative cash inflow exceeds the cumulative cash outflow. The indebtedness decreases to zero when sufficient cash has been generated to "pay off" the total investment. The time for this to occur is referred to as the *payback period*.

A simplified but typical cash flow computation is illustrated in Fig. 4. Revenue is the product of forecasted industry unit sales, unit selling price, and market share. Variable cost added to fixed costs yield total operating cost. Total expenses include total operating costs and depreciation, which is a function of the magnitude and timing of capital expenditures and the form of depreciation selected. Before-tax profit is revenue less total expenses, and after-tax profit is the before-tax profit plus previous year's carry-forward tax loss, multiplied by 1 minus the tax rate. Annual cash flow is the sum of after-tax profit and depreciation less capital expenditures and changes in other balance sheet items such as inventory, receivables, and payables.

As indicated, the financial analysis appears to be rather straightforward but in real life is complicated by many factors, including the need to consider many time periods, multiple products that may be interrelated, price and manufacturing cost elasticities, and alternative pricing policies ranging from specifying annual prices to establishing prices that provide desired gross margins or rates of return but are constrained by the price of competitive or substitute products. In addition, there is usually a need to consider different business structures, such as manufacturing of components, subsystems, and so forth, and manufacturer of final products selling through a distributor network or directly to users. There is also a need to consider alternative financing arrangements. Finally, there are many areas of uncertainty that contribute to the financial risk associated with a business venture that should be conveyed to potential investors through the financial analysis and business plan. The uncertainty and risk are discussed in the following paragraphs.

The payback period (the time until cumulative cash inflow equals cumulative cash outflow) criterion emphasizes exclusively the time required to recover the investment. Note that cash flows that are anticipated to occur after the time when net cash inflow equals net cash outflow are totally ignored even if they are large and persist for long periods of time. The usual argument is that shorter payback periods are preferable to longer payback periods. This results from a desire to avoid risk. Longer payback periods are equated with increased risk.

Another performance measure, somewhat less frequently used, is return on assets. It is the ratio of after-tax profit to asset value where asset value includes cash, receivables, inventory, and book value (capital expenditures less accumulated depreciation) of capital items.

Present value or discounted cash flow analysis seeks to adjust cash flows occurring in future time periods in a way that allows for their timing. The rationale behind the adjustment is that a dollar received in the future is worth less than a dollar received today, since the dollar in hand today could be put to work to create additional earnings. The adjustment process, known as discounting, establishes a present or "now" value of future cash flows. Shifting a proposed project's future cash flows to their now equivalent makes it possible to arrive at a single figure representing the value of the project. This in turn allows fair comparisons of alternative projects with different cash flow patterns over time and payback periods.

The computational approach is to reduce the cash flow occurring in a particular future period by a discount factor such that the discounted amount is the one which, if invested at the discount rate from the present to the corresponding future time, would be equal to the unadjusted value. In a sense, this process is the complement of compounding interest on a savings account. Net present value (NPV) refers to the present value of a future cash flow stream. NPV of a cash flow occurring in a given time period t is given by

$$\text{NPV}_t = \text{CF}_t/(1 + r/100)^t$$

where r is the discount rate (percent) and it is assumed that the cash flow occurs at the end of the time period. The net present value, that is, the algebraic sum of all present value contributions, is given by

$$\text{NPV} = \sum_{t=0}^{n} \text{CF}_t/(1 + r/100)^t$$

where n is the planning horizon and is finite. CF_0 implies the cash flow at the beginning of the first period.

The NPV of a project depends on the magnitude of the cash flows, their timing, and the discount rate. If cash flow is in current dollars, then the discount rate should include an inflation component. If the cash flow is in constant dollars, then the discount rate should not contain an inflation component and is the "real" discount rate. This is illustrated in Fig. 5 where a typical cash flow pattern is indicated along with its NPV as a function of discount rate. If the discount rate is large, the NPV normally becomes negative, because of the heavy discounting of future cash flows (in fact, as the discount rate becomes very large, the net present value approaches asymptotically the value of the initial cash flow component).

Central to the use of the NPV criterion is the choice of the appropriate discount rate. Although economists have generally agreed that the firm's adjusted weighted average cost of capital (taking into account the specific debt-equity situation of the firm) is the appropriate rate to be used for the private sector, controversy still exists about the appropriate rate for use in government decision making. Some have maintained that the long-term government bond rates are the most appropriate. Others have maintained that the rate should be no lower than the typical rate of return achieved by investments in the private sector. The Office of Management and Budget

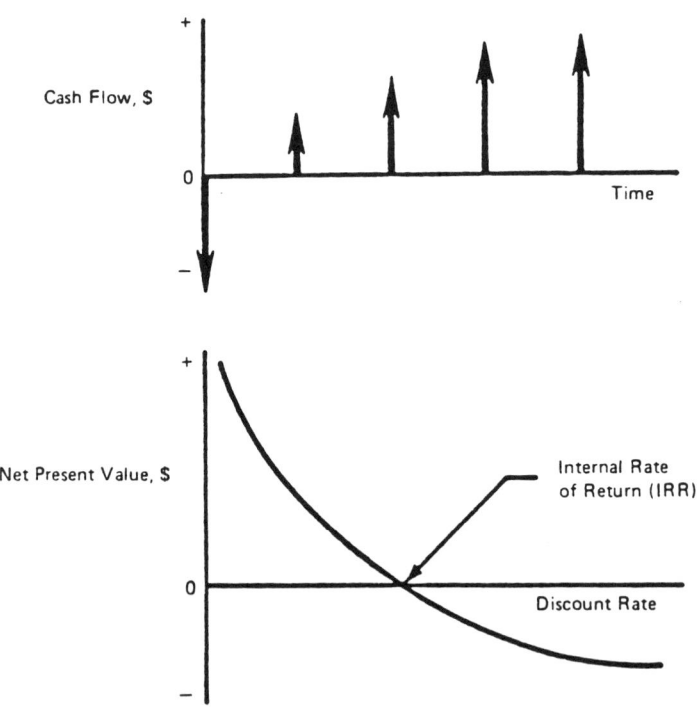

Fig. 5 Present value and internal rate of return.

(OMB) has set a rate of 10% for use in evaluating government projects on an equitable basis.[3] (This seems to be somewhat excessive when costs and benefits are expressed in terms of constant dollars. In this case a rate of perhaps 3–5% is more appropriate, i.e., the real cost of capital which does not include the inflation rate component.)

NPV is a widely used criterion, though normally not the sole criterion, in the process of evaluating and selecting investments.[1] Theoretically, in a perfect world of certainty and unconstrained budgets, all worthy projects with a NPV greater than zero should be undertaken, projects with a NPV less than zero should not be undertaken, and one can be indifferent about undertaking those projects with a NPV equal to zero. However, resources are seldom sufficient to undertake all projects passing this test, and resource rationing is said to exist. This lack of resources for all projects that an organization would like to undertake leads to the problem of choosing the set of projects or investments expected to maximize the net present value of the firm. This is commonly called the *portfolio selection problem.* Optimum solutions can be developed utilizing linear or integer programming techniques and reasonably good or near-optimum solutions can be obtained utilizing heuristic techniques.[4,5]

The discount rate at which the NPV of a proposed project would be zero is known as the internal rate of return (IRR) of the investment. This is also referred to as the discounted return on investment or just return on investment (ROI). In Fig. 6, internal rates of return for hypothetical projects A and B are indicated. The IRR, in effect, represents the maximum rate of return that might be paid for funds borrowed to make the investment. The IRR computation, which can be performed by computing the net present value at several different discount rates, and interpolating to establish that rate that yields a net present value of zero [or by solving the NPV equation for the real positive root(s) that yields NPV = 0] attempts to avoid the issue of fixing a single correct discount rate. Each project has its associated discount rate. When using IRR as an investment criterion, projects should be undertaken as long as the IRR exceeds the currently accepted threshold or "hurdle rate." The hurdle rate would be the same discount rate used in the net present value computations. Note that a determination of discount rate (the cutoff or hurdle rate) is still necessary. The hurdle rate is normally somewhat greater than the firm's cost of capital in order to compensate for risk.

An interesting situation is illustrated in Fig. 6. The net present value of two projects is shown as a function of discount rate. The cost of capital to the firm is r_0. At this cost of capital, the net present value of project A exceeds that of project B and, according to the NPV criteria, project A ranks ahead of project B. On the other hand, the internal rate of return of project B exceeds that of project A (both exceed r_0). Therefore, according to IRR criteria, project B ranks ahead of project A. In general, the mix of projects selected will depend on the criteria used for ranking. The conflict can only be resolved by considering other projects and determining which portfolio of projects maximizes the net present value of the entire portfolio of projects within specified cost constraints.

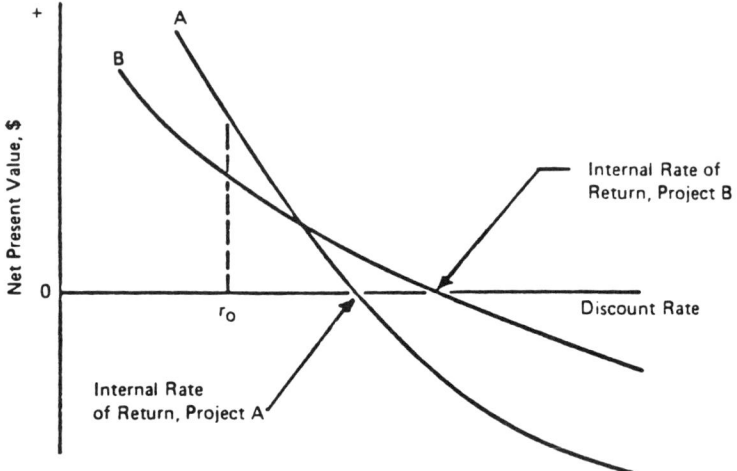

Fig. 6 Comparison of internal rate of return for hypothetical projects.

Financial Risk Analysis

Since investment decisions are concerned with the commitment of funds with the hope of future benefits, and since there is uncertainty about the exact course of future events, the specific results of an investment decision should be described in terms of the probability of achieving different outcomes.[1,6] In other words, uncertainty about the exact course of future events creates risk—fluctuations in the resulting costs, benefits, and cash flow patterns. Decision making should explicitly take into account these uncertainties. This requires the quantification of risk. The technique that makes provision for the explicit quantitative consideration of uncertainty and its effects, i.e., the risk, in the analysis of a business situation is called risk analysis and is described in the following paragraphs via examples relating to communication satellite business situations and related decisions. However, before presenting these examples, computer simulation techniques that are required for the performance of risk analysis computations are described.

Mathematical models are used to simulate the business situation being analyzed (in the following paragraphs, the DOMSAT communication satellite financial planning model is described). Monte Carlo simulation techniques are used to simulate a venture many times (perhaps 1000 or more), each time selecting the input data in a random fashion according to weighting provided by specified uncertainty profiles. The result establishes risk profiles of appropriate performance measures as well as the usual expected value financial performance measures. Figure 7 illustrates the concept of financial analysis with the explicit consideration of uncertainty and resulting risk.

Basic input data consist of deterministic and probabilistic data. Examples of deterministic data are the number of time periods to be considered, the discount rates, and tax rates. Probabilistic data consist of the probability density functions, here referred to as "uncertainty profiles," associated with the variables whose values cannot be predicted or known in advance with a reasonable degree of certainty. Typical uncertainty variables are demand for products/services, market share, expense items, and capital expenditures. Uncertainty profiles are subjective estimates, which describe the range and form (shape) of the uncertainty. For many ventures, particularly those relating to space, there is another type of variable, though possibly known with certainty, that can significantly affect risk. Variables relating to reliability are in this class. Reliability measures, such as probability of success and mean-time-to-failure, lead to variable or unpredictable timing of events (i.e., time of failure) which affects both the magnitude and timing of capital expenditures and related revenues and expenses.

These data are put into a financial simulation model that represents the real-world situation being evaluated. The illustrated model (Fig. 7) states that revenue at a given time I is equal to the product of unit sales, selling price, and market share; before-tax profit is equal to revenue less the sum of all expense items less the depreciation expense; after-tax profit is the before-tax profit multiplied by one minus the tax rate.

Risk analysis is performed by random sampling of input data (according to weighting of the uncertainty profiles), performing computations contained within the simulation model, saving the results, then repeating the process. This process is repeated many times (i.e., Monte Carlo) until a reasonable set of histograms can be developed from the saved output. These histograms are worked into the desired form to indicate the variability of performance measures, such as profit, cash flow, indebtedness, rate of return, and net present value. A convenient form of displaying performance measures is a "risk profile," which indicates the chance of a performance measure exceeding specific levels (i.e., the complement of the cumulative probability distribution).

Figure 8 illustrates a simplified cash flow structure with uncertainty and risk considerations. This is the same structure as illustrated in Fig. 4 but with certain input variables now specified as ranges of uncertainty and the

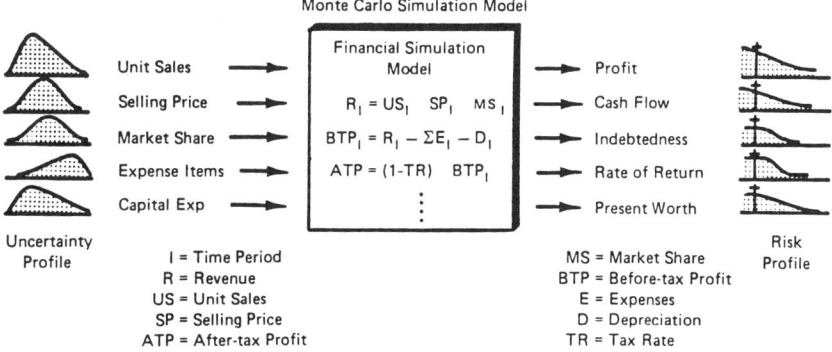

Fig. 7 **Financial analysis with the consideration of uncertainty and risk.**

form of the uncertainty (i.e., the uncertainty profile or probability density function) within the range of uncertainty. The same financial computations are performed as in Fig. 4 with the difference being that the values of the input variables are obtained by the random sampling of the associated uncertainty profiles, the results of the computations saved, repeating the random sampling of the input variables, and continuing the repetition of the process until sufficient data have been saved so that the saved data can be transformed into the risk profiles of the pertinent financial and other performance measures as indicated in Fig. 8. The important point is that the same financial computations need to be performed with or without the formal consideration of uncertainty and risk. The primary difference is in the description of the input data and the form of the developed results.

To establish risk profiles, uncertainty profiles associated with basic input parameters must be established. Informed estimates need to be made of the ranges of uncertainty of key variables and their probability distributions within the ranges. Uncertainty assessments can be made by individuals, or by an experienced group of individuals using Delphi-type techniques.

Uncertainty estimates are subjective. They express quantitatively attitudes regarding uncertainties, reflecting past experience with similar efforts, typical problems encountered in the past, and insights into problem areas that might develop. Uncertainty profiles, being subjective estimates, call for expert opinion in each area. Ordinarily, manufacturing personnel estimate the uncertainty surrounding manufacturing costs; marketing personnel estimate uncertainty surrounding the sales forecast and marketing costs; and so on through each category of input. Risk analysis forces more

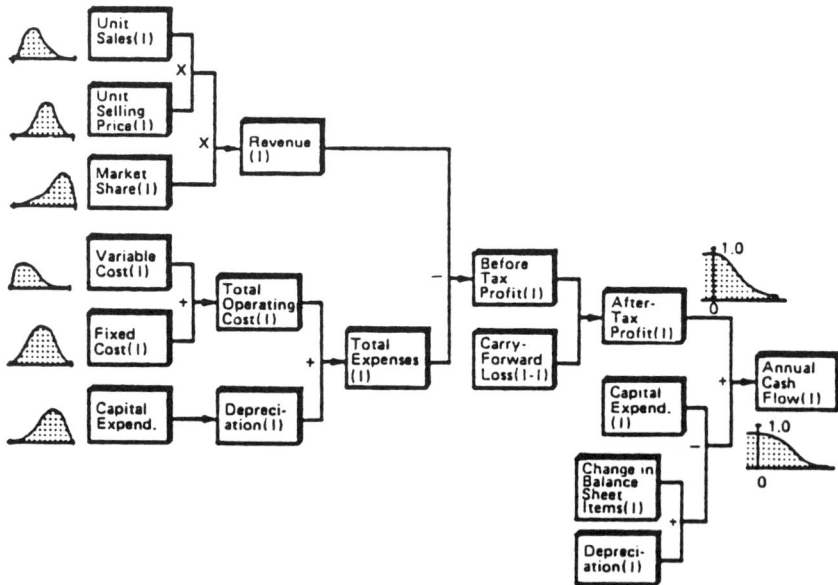

Fig. 8 Simplified cash flow structure with uncertainty and risk considerations.

detailed consideration of what can go right and what can go wrong. It results in a more carefully thought out financial analysis and resulting business plan.

A useful and frequently used procedure for establishing the shape of an uncertainty profile is as follows (refer to Fig. 9):

A) Estimate the range of uncertainty—minimum and maximum bounds (little or no chance of falling outside these bounds). Divide this range into a number of equal intervals—five have been found, through experimentation, to be useful.

B) Make a relative ranking of the likelihood of the variable falling into each of the intervals; this establishes the general shape of the uncertainty profiles (i.e., skewed left, right or central).

C) Set relative values for the chance of falling into each interval (for the Fig. 9 case, the chance of falling into interval 1 is half that of falling into interval 2).

A. Specification of range of uncertainty.

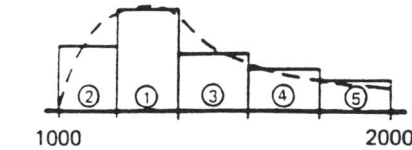

B. Qualitative ranking.

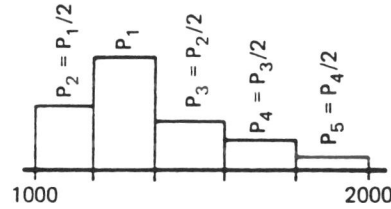

C. Establishment of relative likelihoods.

$P_1 + P_2 + P_3 + P_4 + P_5 = 1$
By Substituting from (c) Solve for P Values

D. Establishment of quantitative values.

Fig. 9 A method for establishing uncertainty profiles.

D) Having assumed the probability of falling within the range of uncertainty as 1.0, the chance of falling in each of the five intervals can be summed and set equal to unity. This equation can be solved (by substituting the relative values as obtained above in step C) for the probabilities associated with each interval.

This can become a long procedure when many uncertainty variables and/or many time periods must be dealt with in making assessments. To minimize this problem, many uncertainty profiles may be stored in the computer and pictures of these shown to the evaluators. The evaluator need specify only the minimum and maximum values and the name of the applicable uncertainty profile. If the appropriate uncertainty profile has not been stored, it can be created by the process just outlined.

As discussed previously, net present value constitutes an important indicator of financial merit of a business venture. NPV takes into account magnitude and timing of cash flow patterns and the cost of capital (i.e., the discount rate). It represents the maximum amount a firm could pay for the opportunity of making the investment (undertaking the venture) without being financially worse off.

In the world of certainty, the following investment rules apply:
1) If NPV is greater than zero, undertake the venture.
2) If NPV is less than zero, do not undertake the venture.
3) If NPV is equal to zero, the decision is immaterial.

Investment decision strategy should focus on maximizing net present value.

Uncertainty makes annual cash flow probabilistic. Therefore, NPV must be characterized by a probability distribution that represents the chance of achieving each of its possible levels. The probability of NPV exceeding specified levels can be obtained by measuring the area under the probability distribution curve for all the values greater than the specified level—the risk profile of NPV.

Figure 10 illustrates typical risk profiles of NPV. The vertical scale represents the chance or probability p of exceeding the various levels of NPV, indicated by the horizontal scale. In general, the steeper the curve, the

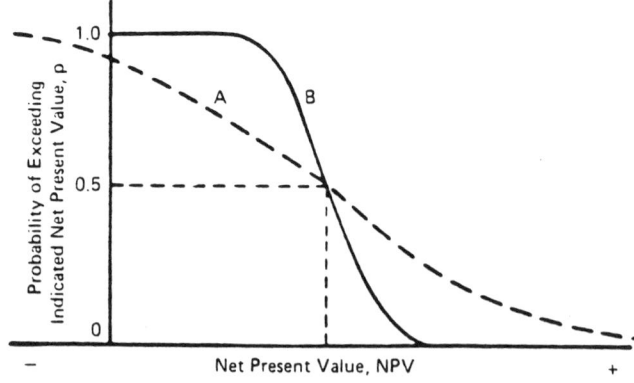

Fig. 10 Typical risk profiles of net present values.

lower the risk (i.e., variability). When comparing alternatives, it is important to compare the expected or "most likely" net present values. [The NPV probability distributions are near normal as would be expected according to the central-limit theorem, since a reasonably large number of probability distributions are usually considered in the financial computations. As a result, expected values and most likely values can be considered, for all practical purposes, as being equal. Further, the approaching normality allows the NPV probability distributions to be summarized meaningfully in terms of expected values and standard deviations (this has been done in the NPV results summarized in Fig. 3).] Comparing risk levels is equally important.

Figure 10 illustrates the NPV risk profiles for hypothetical alternatives A and B. A decision maker performing a conventional analysis without the explicit and quantitative consideration of uncertainty and resulting risk usually evaluates quantitatively only the "most likely" present value. To this uninformed decision maker, alternatives A and B "look alike," because they show equal ($p = .5$) expected and "most likely" values.

In the absence of performing a risk analysis, the decision maker will try to pick the alternative yielding maximum NPV. In risk analysis, the selection process has more information and in a sense is made more difficult. Trade-offs must be made between alternatives possessing different expected present values and associated levels of risk. When the risk dimension is added, the decision maker finds alternatives A and B in Fig. 10, for instance, quite different. Alternative A assumes greater risk (variability) than alternative B. Thus, a conservative decision maker (averse to risk) would probably select B (if there were no other unquantified pressures to select alternative A).

The decision maker will usually identify many alternatives necessitating this kind of choice; but the task will be eased somewhat by the fact that the NPV probability distributions usually take the form of the familiar bell-shaped curve. That is, it takes the normal distribution or Gaussian form. In this case the distributions can be fully characterized by their expected values or means m and standard deviation σ, and each alternative can be represented by a point on the m-σ plane. For example, Fig. 11 illustrates two alternatives (1 and 2) having the same level of risk ($\sigma_1 = \sigma_2$), but with the expected NPV of alternative 2 greater than that of alternative 1. Therefore, alternative 2 is preferable to alternative 1. Similarly, it can be argued that alternative 3 is preferable to alternative 4, that alternative 3 is preferable to alternative 2 since both have the same expected NPV but alternative 2 is riskier, and so on through the whole list of alternatives.

If this process is continued, the decision maker can chart a frontier of "best" alternatives, each choice on the chart different in both risk and expected NPV. This frontier might be called the "class of best alternatives." The "best" alternative can then be selected based on the decision maker's risk judgment. That is, the decision maker must make the trade-off between an increase in expected NPV and an accompanying increase in risk. This is discussed in more detail in the chapter entitled "Commerical Development of Space: Government/Industry Relationship."

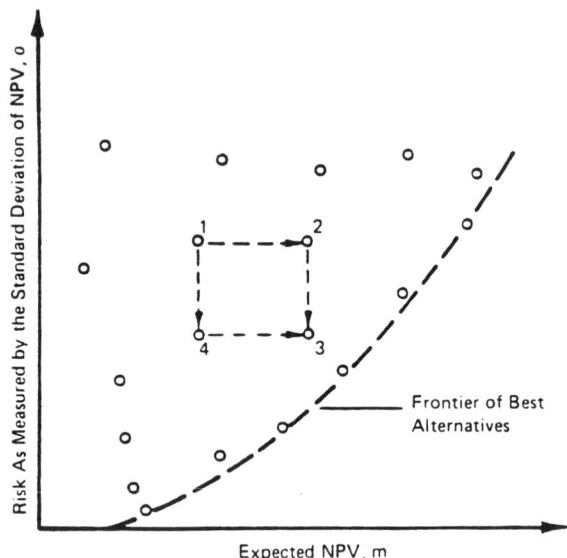

Fig. 11 Risk comparison of alternatives.

Risk analyses have been performed for many different situations, such as capital budgeting within large corporations, evaluation and planning of startup situations, evaluation of alternative technologies, and analyses of the potential impacts of government policies on private sector business ventures. Many different risk analysis models have been developed and are in current use, from those requiring the analyst to write short programs (as part of the input data stream) that tailor the model to the specific venture being analyzed, to generalized models that can simulate a wide variety of situations.[7] In the following paragraphs a general financial planning model (DOMSAT) for communication satellite business ventures is described and a number of its applications are discussed. This model explicitly considers uncertainty and unreliability and the resulting risk consequences.

Space Business Financial Planning Model (DOMSAT)

DOMSAT is a communications satellite financial planning model specifically developed to provide a means for evaluating the impacts of a broad range of policies and decisions on the financial performance of communication satellite business ventures. DOMSAT provides the means for evaluating the impacts of both government and industry policies and decisions in terms of financial performance measures such as after-tax profit, cash flow, present value of cash flow, and return on investment.

DOMSAT is a Monte Carlo simulation model that stochastically simulates the performance of a broad range of communication satellite business ventures, explicitly and quantitatively, taking into account uncertainty, unreliability, and resulting risk. The model works with data provided by the user as a range of uncertainty and the form of the uncertainty, describing the parameters of a communication satellite business. Included are data describing the reliability and performance of satellite and launch vehicle subsystems, desired launch schedule and possible launch delays, low Earth orbit (LEO) to geostationary Earth orbit (GEO) transfer time, number of transponders, sparing and grouping of transponders, services to be provided and rates for each, annual demand and elasticities, recurring and nonrecurring satellite costs, launch costs, insurance costs, depreciation, and financial and balance sheet-related data.

To establish the quantitative financial and risk measures, the model uses Monte Carlo techniques wherein the complete business scenario is simulated frequently over a period of time, each time randomly sampling from the uncertainty profiles and the reliability characteristics that are specified. The results of all of the business analyses are saved and histograms are developed of the financial performance measures. These are summarized in terms of expected values and standard deviations (typical results are illustrated in Figs. 2 and 3).

The model consists of basically two parts. The first, using the desired schedule of events, demand for communication services, satellite configuration, specified launch scenario, and reliability characteristics, establishes the specific timing and number of events and their costs. The availability of transponders (taking into account failures, sparing concepts, and services offered) is matched against demand in order to establish the schedule for replacement launches and the timing of additional capital expenditures for replacement satellites and launches. The timing and cost information is then passed to the second part of the model which performs the financial computations and establishes values of the economic performance measures.

The model may be used to evaluate the competitiveness of different launch vehicles (such as Ariane vs Atlas, taking into account differences in reliability, pricing, insurance rates, final payload placement accuracy, and schedule delays that may result from various types of failures) and the "fairness" of their pricing policies, or the effect of launch delays by establishing a base case scenario from which variations can be measured. Vehicles have many attributes and cannot be compared strictly on the basis of any one (such as price) alone. An Ariane flight, for instance, may be cheaper; however, Atlas has a higher demonstrated reliability. DOMSAT provides a way to measure and compare the implications of the various attributes by translating them into a common format (financial performance). The base case would represent the company launching on one of the vehicles. Data pertaining to the launch vehicle (such as price and reliability data) would then be changed to represent the other vehicle. The financial performance (as represented by the resulting financial statements) of the company under the two scenarios would then be compared. Similarly, other data points can be varied to formulate several scenarios for comparison to determine the effects, for example, of launch delays, insurance requirements, and so on. A number of these situations are discussed in more detail in the following paragraphs.

The level of detail of the model and the type of analyses that may be performed can be appreciated by considering the input data provided to the model. Data must be provided which characterizes the general structure of the business, the transportation system, satellite configuration, revenue and demand for communication services, and costs and expenses. The following input data may be specified to the model via a user-friendly input system:

General Data

1) Number of years in the business plan
2) Maximum number of operational satellites to be considered
3) Desired launch schedule
4) Identification of specific operations for which insurance will be taken
5) Insurance cost for each operation and associated cost spreading
6) Relaunch threshold in terms of number of operational transponders
7) Depreciation lives for transportation, satellite, and other capital expenditures
8) Interest rate
9) Tax-related data
10) Discount rates to be used in present value computations
11) Balance sheet-related data (such as typical level of receivables and payables)

Transportation-Related Data

1) Identification of initial placement and maintenance transportation scenarios as a function of time selected from the following (Fig. 12 illustrates the first of the following transportation scenarios):
   a) Direct placement of satellites using ground-based assets
   b) Placement and return of satellites using ground-based assets
   c) On-orbit repair of satellites using ground-based assets
   d) Replace/return/repair of satellites using ground-based assets
   e) Direct placement of satellites using space-based assets (i.e., space-based transfer vehicles)
   f) Placement and return of satellites using space-based assets
   g) On-orbit repair of satellites using space-based assets
   h) Replace/return/repair of satellites using space-based assets
   i) Return/replace/repair satellites on and from space-based assets
2) Probability of success of each of the major operations in the selected initial placement and maintenance scenarios (e.g., probability of success of first stage, probability of recovery of first stage given first-stage success, probability of second-stage recovery given a first-stage failure, etc.)
3) Transportation cost associated with each major operation in the selected launch and maintenance scenarios and associated cost spreading

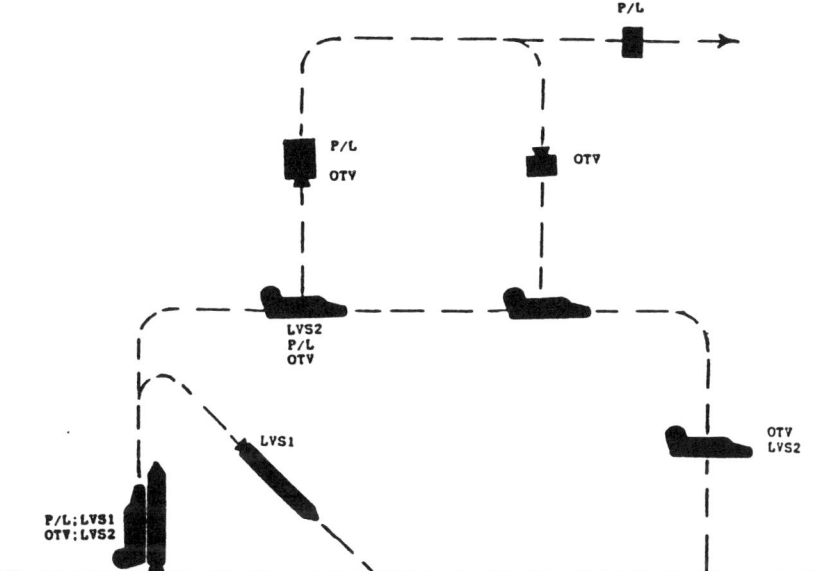

Fig. 12 Transportation scenario: direct placement using ground-based assets. Consists of reusable launch vehicle (LV) (LSVS1 may be reusable or expendable) or LV and transfer vehicle (OTV) or expendable LV and OTV for initial payload (P/L) placement and/or replacing failed P/Ls. If reusable launch vehicles are considered, then OTV and P/L checkout failures in LEO may be corrected when and if returned to Earth.

4) Possible launch delays (probability distributions of possible delays conditional upon the type of failure)

Satellite Configuration Data

1) Number of narrow-band transponder groups per satellite
2) Number of wideband transponder groups per satellite
3) Number of transponders per narrow-band group
4) Number of transponders per wideband group
5) Number of spare transponders per narrow-band group
6) Number of spare transponders per wideband group
7) Transponder reliability characteristics (i.e., mean time to failure, expected wearout life and standard deviation of wearout life)
8) Spacecraft support subsystems reliability characteristics (i.e., mean time to failure, expected wearout life, standard deviation of wearout life)

Revenue-Related Data

1) Types of communication services provided (i.e., protected, protected/preemptible, unprotected, and preemptible)

2) Rates per narrow- and wideband transponders for each type of communication service
3) Annual demand for narrow- and wideband transponders in terms of type of service and satellite location
4) Price elasticity of demand for each offered service

Cost/Expense-Related Data

1) Annual cost of satellite operations
2) Annual general and administrative (G & A) expense in terms of fixed and variable components
3) Annual R & D/engineering expense in terms of fixed and variable components
4) Satellite unit recurring cost and cost spreading
5) Satellite nonrecurring cost and cost spreading
6) Satellite unit recurring cost learning rate

Many of the above variables may be considered either as deterministic or specified as uncertainty variables (as discussed in previous paragraphs) requiring the specification of the range and form of uncertainty. For example, all costs and expenses, schedule delays, transponder pricing, and demand variables may be considered as uncertainty variables.

The DOMSAT model allows uncertainty and unreliability (initial, random, and wearout failures) to be considered explicitly and quantitatively. This is absolutely necessary when considering programs and policies that are specifically aimed at reducing uncertainty and altering reliability, both of which affect perceived risk and hence investment decisions. As already discussed, the model develops many financial performance measures, including annual after-tax profit, annual cash flow, cumulative cash flow, return on sales, return on assets, payback period, and net present value. Expected values and standard deviations are established for all of these. The net present value is established at a number of discount rates so that the return on investment can be established.

The model is implemented such that certainty conditions can be easily analyzed as well as the uncertainty situations. A user-friendly system has been developed for entering data into the model. The data are entered via Lotus 123 and the DOMSAT model is implemented in Fortran. The system has been designed for operation on the IBM PC with all DOMSAT programs residing on a single disk.

DOMSAT can be used to establish the impacts on the financial performance of communication satellite business ventures of many policies, programs, and decisions including the following:

1) Utilizing alternative space transportation system (such as Atlas, Delta, Ariane, Zenit, Long-March, etc.)
2) Achieving improved final payload placement accuracy
3) Delays resulting from launch infrastructure constraints

FINANCIAL/INVESTMENT ANALYSIS

4) Regulatory programs such as those establishing user fees and insurance requirements
5) Transportation system technology programs (e.g., low thrust transfer from LEO to GEO and improved upper stage reliability)
6) Spacecraft technology programs (e.g., on-orbit propulsion and space power)
7) Different insurance rates as compared with the self-insurance option explicitly taking into account the level of risk (insurance may be considered separately for each operation including launch to LEO, checkout in LEO, transfer from LEO to GEO, and initial payload startup)
8) Space transportation system pricing policies
9) Pricing policies for transponders and related services
10) Satellite configuration alternatives including transponder arrangements and sparing concepts
11) Alternative placement/replacement/service/repair policies utilizing ground-based facilities
12) Alternative placement/replacement/service/repair policies utilizing space-based facilities

The above capability can be used to provide quantitative measures relating to such questions as:

1) Should a new communication satellite business venture be entered into?
2) What is the value of space launch infrastructure investments?
3) Are foreign launch vehicles priced to account for their capability differences?
4) How will the availability (or lack of availability) of less costly foreign launch services affect the U.S. communication satellite business?
5) How sensitive are forecasts at required levels of satellite replacement to assumptions about the relationship between satellite design life and replacement schedule and the impact of specific technology developments?
6) What will be the financial impacts on communication satellite businesses of alternative strategies (such as moving GEO satellites to higher altitudes prior to end of life, and restricting transfer stage trajectories) aimed at reducing space debris?

A number of the above questions are considered in the following paragraphs.

Example: Ion-Thruster and GaAs Solar Cell Technology[8]

This example is concerned with assessing the value of two NASA spacecraft (S/C) technology programs in terms of their impacts on the financial performance of a fixed satellite point to multipoint service (FSS) business venture consisting of three Ku-band satellites. The considered technology programs (i.e., ion thrusters for on-orbit propulsion and gallium arsenide solar cells) have the ability of reducing overall spacecraft mass without altering the performance attributes of the spacecraft. The S/C may be designed with reduced mass with the result that transportation charges may be reduced (when a smaller launch vehicle may be used or charges are proportional to mass) leading to an increase in expected return on investment with little or no change in risk. On the other hand, the mass may be put back in a number of different ways, each of which alters S/C attributes such as on-orbit propulsion system life, number of active transponders, number of spare transponders, and so forth. This is illustrated hypothetically in Fig. 13 where the financial implications of the possible alternative S/C configurations (i.e., use of mass savings resulting from the introduction of the new technology) are indicated by plotting expected ROI of the communication satellite business venture against σ. The $m - \sigma$ plot allows the financial performance of the business venture to be summarized in terms of its expected ROI (m) and the standard deviation of ROI (σ). For example, a considerable increase in expected ROI, with an accompanying increase in risk, may result from introducing an ion propulsion system with sufficient propellant to extend satellite wearout life but with a perceived reduction in mean-time-to-failure. Note that in Fig. 13 all changes are relative to the base case, which is the business venture performance (m and σ) in the absence of the technology programs.

Determination of the best use of the mass requires the establishment of the preference curve or risk aversion attitudes. To illustrate, given the risk aversion attitudes indicated by the *dashed lines* in Fig. 13, alternative D offers the best use of the mass savings and therefore represents the maximum value of the technology program when the results of the program are used in the postulated business scenario. When alternative technology programs are to be compared, the comparison must use the maximum

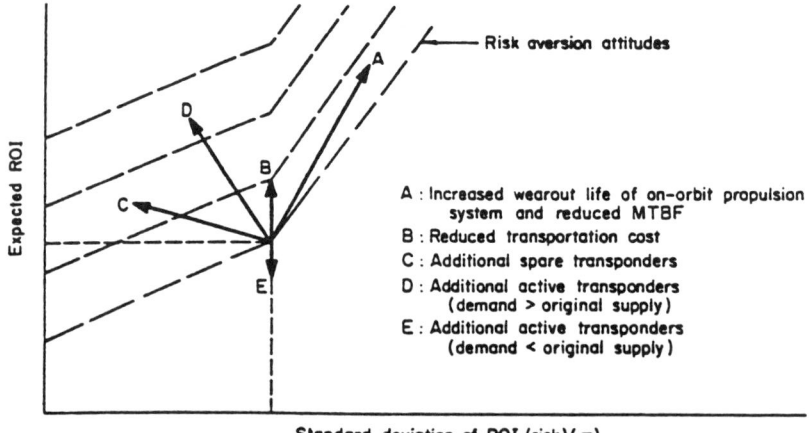

Fig. 13 Assessment of the best use of mass savings for a particular technology program.[8]

values of each of the technologies as illustrated in Fig. 14. Here T2 represents case D in Fig. 13. From Fig. 14, the choice is between technology 1 (T1) and 2 (T2), both of which offer approximately the same value. It should be noted that both the expected ROI and risk of technology 1 exceed those of technology 2.

The FSS spacecraft was reconfigured using the ion-thruster and gallium arsenide solar cell technologies assumed to result from technology programs. Mass savings on the order of 90 and 15 kg resulted from the incorporation of the ion-thruster and gallium arsenide technologies, respectively, into the FSS S/C. These mass savings allowed four additional active transponders and 2 years of life to be redesigned into the ion-thruster satellite and two additional active transponders into the gallium arsenide satellite. Extended capability was therefore designed into the satellites so that the mass at liftoff was kept approximately the same in all cases.

The business scenario data were adjusted for the new parameters: new S/C nonrecurring and recurring costs, the number of transponders, and on-orbit life. The parameter that were adjusted are summarized in Table 1 for each scenario. All other variables were held constant. The DOMSAT model was then used to reanalyze the business scenario with the new parameters and the financial results (ROI, profit, net present value, etc.) were then compared with results produced from the base case scenario.

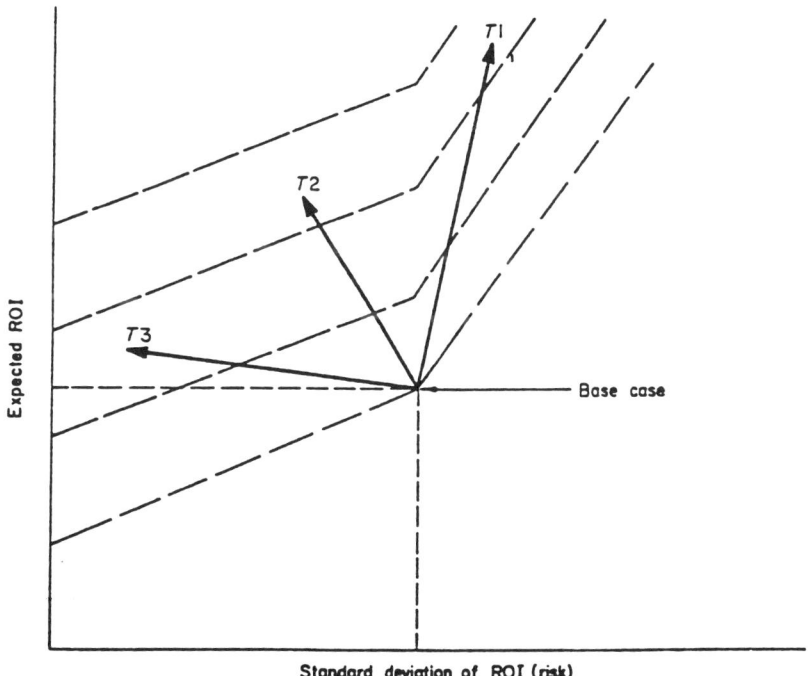

Fig. 14 Comparison of technology programs.[8]

Table 1 S/C cost and capability summary

Spacecraft FSS	Expected nonrecurring cost, M$	Expected recurring cost, M$	No. of active transponders	Wearout life, yr
Base case	20.8	38.6	16	8
Ion thruster	44.2	43.9	20	10
GA solar cell	31.0	38.8	18	8

Typical results are illustrated in Figs. 15 and 16. Figure 15 illustrates the expected profit for the base case and two technology scenarios. It is evident that increased losses are likely to be sustained in the early years in order to achieve increased profits in later years. The effect of this timing is captured in the computation of the present value of each of the scenarios. The present value risk profiles are illustrated in Fig. 16. The risk associated with each of the scenarios is essentially the same as indicated by the similarity of the slopes of all the curves. It can be seen that the use of the new technologies increases the expected value of the base case by approximately $25 million.

Example: Insurance vs No Insurance

Launch insurance rates have risen rapidly from the initial values of approximately 10% of launch and satellite costs to 25% or more of launch

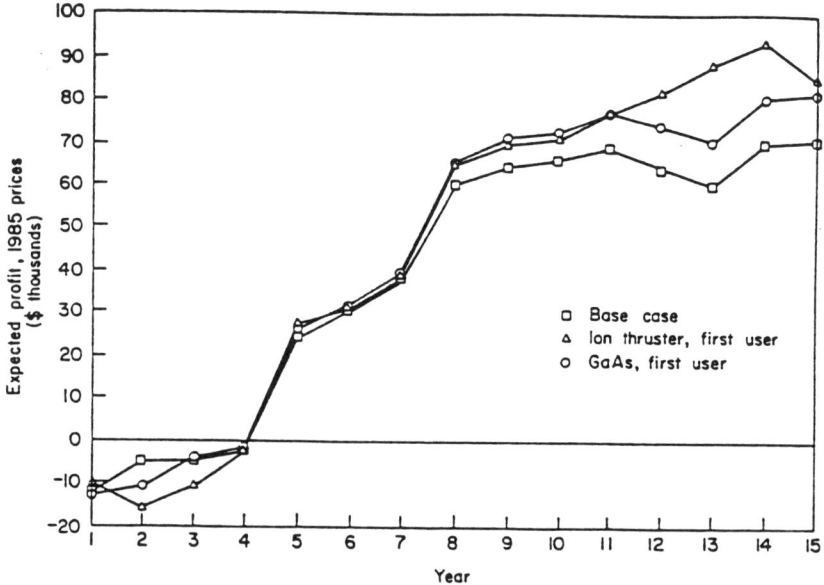

Fig. 15 FSS expected profit as a function of time.[8]

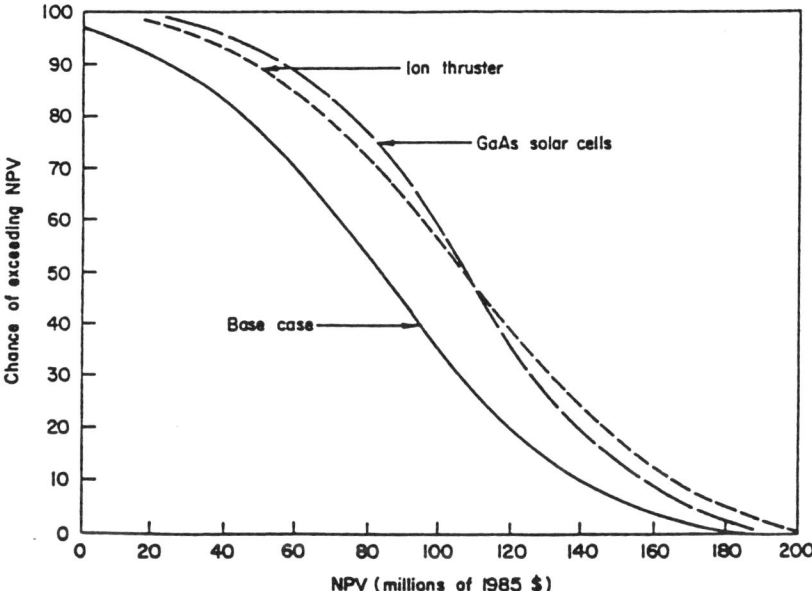

Fig. 16 FSS net present value (NPV) at 15% discount rate.[8]

and satellite costs. Insurance rates are currently on the order of 17–20%. The effect of insurance rates and the self-insurance option on the financial performance of communications satellite business ventures can be easily assessed with the DOMSAT model.

The insurance analysis, using the base case FSS business scenario, assumes that an insurance premium is paid to cover the cost of the satellite launch and the cost of the satellite.[9] The premium is specified as a percentage of these costs with the premium amount treated parametrically in the analysis. It is assumed that when insurance is taken, any loss resulting from a launch failure or a satellite failure (prior to revenue generation) will be covered by the insurance (i.e., no additional outlays will be required by the business venture in making itself whole again).

The alternative to taking insurance is not to take insurance. In this case, the business venture becomes a self-insurer and, in the event of a failure, sustains the cost of additional launches and satellite repurchases.

Figure 17 illustrates the effect of increasing insurance rates upon the financial performance of the base case FSS business venture. The indicated financial performance measure is that of return on investment. The ROI risk profiles are summarized in terms of expected value and standard deviation (i.e., the risk measure). Also indicated in Fig. 17 is a single point that represents the self-insurance option. The self-insurance option results in approximately the same expected ROI as would be obtained with a 17% insurance premium; however, there is a significant increase in the level of risk (i.e., the standard deviation of ROI).

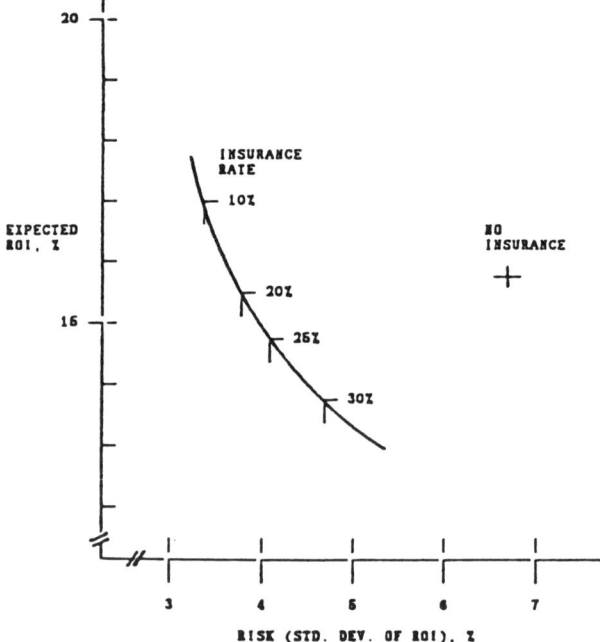

Fig. 17 Comparison of launch insurance at different rates with the self-insurance option (FSS scenario).

It is anticipated that a communication satellite business is likely to accept slightly lower expected values of ROI than would be possible through the self-insurance option, in order to avoid the increased risk associated with the self-insurance option. It is not clear exactly what this trade-off is (and it undoubtedly differs from one venture to another), but it is likely that some ventures will switch to the self-insurance option if the insurance rates increase much beyond the 20% level.

Example: Selection of a Space Transportation Service[9]

The selection of a space transportation system for a particular mission is a complex process that requires the consideration of many factors, including availability, cost, payload delivery capability, payload placement accuracy, reliability of launch operations, failure/recovery modes, and cost and availability of insurance. Because of the significant variations among many of these factors from one transportation system to the next, rarely is only one parameter such as launch cost considered when deciding upon a launch service. Many of these factors are considered implicitly or explicitly (qualitatively or quantitatively) to establish their combined effect on mission or business economics with final decisions bearing heavily on the resulting financial performance projections.

The importance of considering multiple attributes of transportation systems in transportation selection is demonstrated in Figs. 18 and 19, which were developed by utilizing the DOMSAT model and varying the multiple attributes of launch vehicles to provide the transportation services for a business venture utilizing three FSS satellites. Figure 18 illustrates the comparison of launch alternatives (at a fixed transportation price) in terms of the reliability of the launch vehicle and final payload placement accuracy. The effect of improved payload placement accuracy is manifested in an extended payload life (i.e., additional station-keeping propellant); the actual extent of the life increase is dependent on the reliability performance of all of the payload subsystems. Two curves are shown, the lower curve reflecting a satellite with an 8-year wearout life of the on-orbit propulsion system and the upper curve reflecting an increase of 5 years in the wearout life due to improved payload placement accuracy (i.e., more station-keeping propellant available as a result of more accurate satellite placement).

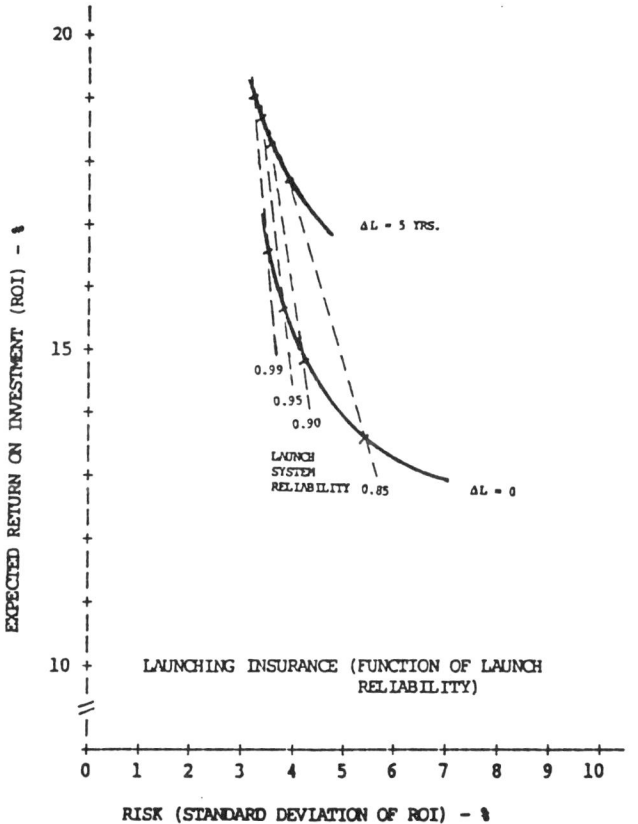

Fig. 18 Transportation system impacts on ROI statistics of a typical communication satellite business: impacts of launch system reliability and placement accuracy.[9]

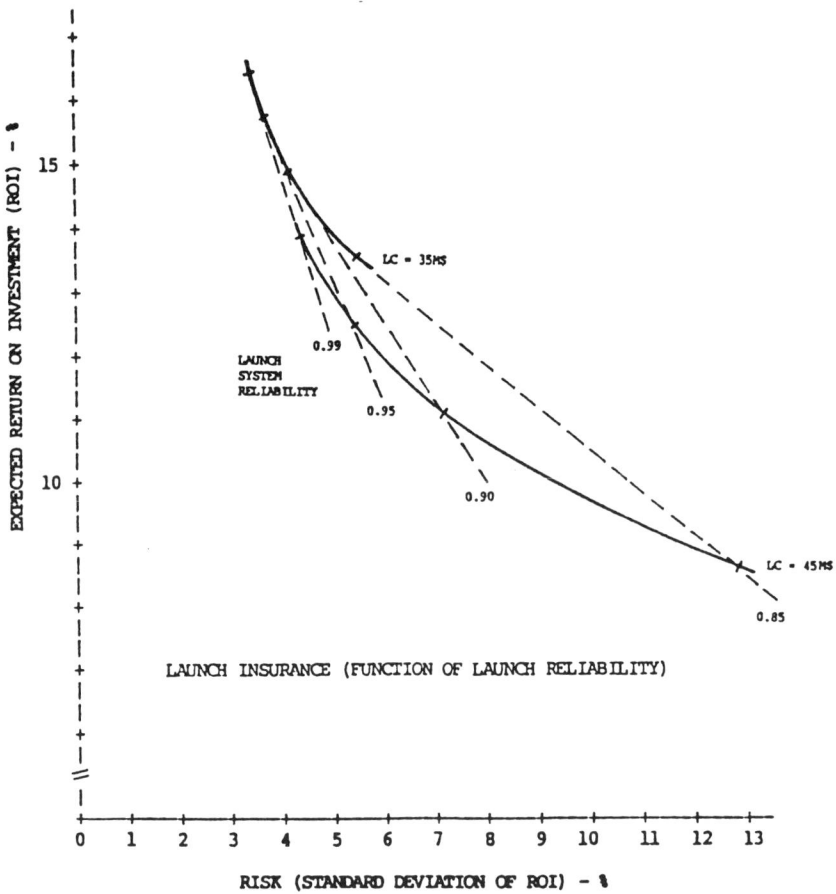

Fig. 19 Transportation system impacts on ROI statistics of a typical communication satellite business: impacts of launch system reliablity and launch cost.[9]

Figure 19 illustrates the trade-off between launch system reliability and launch cost. The significant effect of increased transportation cost in combination with low reliability is evident both on the expected ROI and the risk.

Example: Financial Impacts of Orbital Debris Remediation Policies[10]

Because of increasing concern with the orbital debris problem, serious consideration is being given to taking remedial actions, such as requiring GEO satellites to be moved to higher altitudes upon reaching the end of useful life, and placing constraints on transfer orbits. However, these remedial actions will have an impact on the financial performance of communication satellite business ventures.

The basic direct impact of both of these alternatives is a reduction in satellite performance and/or increased transportation cost. Moving satellites to higher altitudes requires mass which may be in the form of propellant and/or propellant and thrusters—in either case it amounts to shortening the on-orbit station-keeping life of the satellite or utilizing a launch vehicle with greater payload delivery capability with associated higher cost. Either choice produces financial impacts. Constraining transfer trajectories will most likely reduce the mass delivery capability of existing vehicles, which in turn will lead to reduced satellite capability—possibly less station-keeping propellant which will reduce on-orbit life, less subsystem redundancy (which will affect failure rates), etc. In any event, it seems that both alternatives will affect satellite life characteristics and/or transportation cost which will have financial consequences. These consequences have been evaluated through the use of the DOMSAT financial planning model.

The DOMSAT model allows satellite life characteristics (as might result from the above debris reduction alternatives) to be varied and the effects to be observed in terms of the financial impacts on a business. The above mentioned debris reduction impacts can be evaluated by estimating their impacts on satellite life characteristics and then evaluating the impacts of these life characteristics on business venture financial performance.

The results of an analysis of the impacts of satellite life are shown in Fig. 20, which indicates the change in expected net present value of cash flow (at a 10% discount rate) and the change in expected ROI as a function of satellite life reduction. Once specific debris remediation policies are identified (i.e., move communication satellites to a specified altitude beyond GEO), these policies can be translated into impacts on satellite life characteristics and the specific financial impacts can be determined.

Summary

An investment decision involves the commitment of funds with the hope of future rewards. Many investment decisions are concerned with transforming an idea or emerging technology into a business venture. This transformation process requires the use of capital and labor resources that

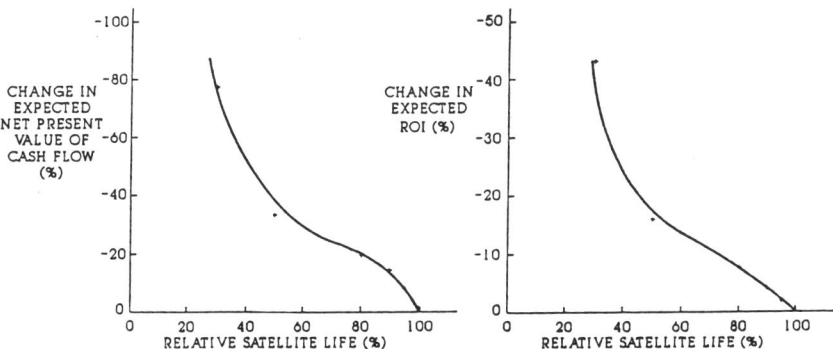

Fig. 20 Effect of satellite life reduction on the financial performance of communication satellite ventures.

normally require justification for their commitment and utilization. An important part of the justification process is the development of a business plan that describes the contemplated business and its operating environment, its short- and long-term objectives, and the intended means and methods for achieving those objectives. The business plan is aimed at demonstrating to the investment community the need for and the magnitude and timing of resources to be employed and the financial results (the rewards) that are anticipated. Thus, a major part of the business plan is concerned with financial performance as measured in terms of attributes, such as profit, cash flow, return on assets, return on investment, payback period, and net present value. It is the role of the financial/investment analysis to develop these and other financial performance measures that indicate the consequences of the plans and assumptions made in all of the other aspects of the business plan.

Space-related business ventures are normally characterized by large capital requirements, long payback periods, and high risk. The high risk results from many areas of uncertainty (beyond those of normal business risk), including government policies, availability and pricing of space-related insurance, and space transportation system reliability and availability, with all of their ramifications. It is thus important to include risk considerations in the financial analysis of space business ventures. The DOMSAT model was specifically developed to perform financial analyses of communication satellite business ventures explicitly and quantitatively, taking into account many areas of uncertainty, space transportation reliability and the consequences of failures, and satellite subsystem reliability, and developing quantitative measures of risk. The preceding paragraphs describe this model and various uses of the model such as developing the financial plans for a communication satellite business venture, analyzing the value of alternative technology programs, establishing the financial consequences of utilizing alternative launch vehicles, and establishing the financial impact of alternative orbital debris remediation policies.

The bottom line of financial analysis is the development of rational, credible, and defensible estimates of the financial impacts of alternative policies, choices, and resulting decisions.

References

[1]Greenberg, J. S., *Investment Decisions: The Influence of Risk and Other Factors*, American Management Associations, New York, 1982.

[2]Brealey, R. A., and Myers, S. C., *Principles of Corporate Finance*, McGraw-Hill, New York, 1988.

[3]Office of Management and Budget, "Discount Rates To Be Used in Evaluating Time Distributed Costs and Benefits," OMB Circular A-94, Washington, DC, March 27, 1972.

[4]Markowitz, H. M., *Portfolio Selection*, Wiley, New York, 1959.

[5]Thierauf, R. J., and Kelkamp, R. C., *Decision Making Through Operations Research*, Wiley, New York, 1975.

[6]Greenberg, J. S., "Risk Analysis," *AIAA Journal*, Nov. 1984.

[7]Greenberg, J. S., "A Financial Risk Analysis of a District Heating Business Venture," *Proceedings of the 15th Intersociety Energy Conversion Engineering Conference*, Seattle, WA, Aug. 1980.

[8]Greenberg, J. S., "Communications Satellite Business Ventures: Measuring the Impact of Technology Programmes and Related Policies," *Space Policy*, Vol. 2, No. 1, Feb. 1986.

[9]Greenberg, J. S., and Christensen, C. B., "The Selection of a Launch Vehicle," Canaveral Council of Technical Societies Twenty-Sixth Space Congress, Cocoa Beach, FL, April 1989.

[10]Greenberg, J. S., "Orbital Debris Cleanup May Be Costly," *Aerospace America*, Aug. 1991.

Financing Space Projects

Jerome Simonoff*
Consultant, Bethesda, Maryland 20814

What Is a Satellite in Financial Terms?

TO an investor, a satellite has four parts: the flying equipment, the transportation service, the insurance service, and the cost of money employed to build the satellite. In today's market, the cost of the hardware for a communication satellite is approximately equal to the transportation. The insurance adds another 25%. The cost of money is 10–15% if supported by borrowing, but considerably more in terms of an investor who may have opportunities in other places.

Confusion

Perhaps nothing is more confusing in space finance as understanding exactly what is financed. There is the space equipment itself. Also there is the business of employing space. The confusion caused by the unique properties of space systems have led to many false starts in financing, which in turn have resulted in a slowing in the development of the industry.

The key element of confusion (and some missed opportunities) is understanding what is actually being bought. The satellite and the supporting satellite control facilities are obvious. What may not be realized is that financing a satellite may imply financing a system in which the total investment overwhelms the value of the satellite. Examples are the television and especially the cable TV satellites. The cable satellites if purchased new today might require a total investment of $200–450 million. However, without these satellites, the cable TV industry would not be viable at the size that it is today, notwithstanding fiber optic systems. Therefore, a

---

Copyright © 1992 by the American Institute of Aeronautics and Astronautics, Inc. All rights reserved.
*Commercial Space Transportation Advisory Committee, U.S. Department of Transportation.

multibillion dollar investment in franchises and distribution equipment depends upon the existence of these satellites. Similarly, it is estimated that perhaps an additional $100–150 million of satellite costs may be incurred to communicate with household receivers. Today, there are estimated to be between 1 and 1.25 million "backyard dishes" installed today at an average cost of $1500–3000. This means that while the space segment if bought today would cost $300–600 million, the ground segment if purchased today represents an investment of $1.5–3.75 billion.

Investors and Lenders

Investors (the providers of equity financing) take risks associated with the commercial viability of a project or company. Consequently they seek higher returns on their investment than do lenders. Their risk concerns whether the market will purchase services, whether the management will work, whether the technology will work, and so forth. (Generally, investors will not accept leading-edge technology for space investment.)

Borrowed funds are generally used to overcome a time delay that takes place between the need for funds and the availability of revenue. Lenders take the financial market risk, interest rate expectations, and the ability of financial institutions to provide funds. Lenders take little risk that involves the question of commercial viability of a project. If there is not an overwhelming comfort with the commercial viability of a company or project, lenders will seek financial support for investors or other organizations who will meet those criteria.

Insurance

Generally, neither class of investor will accept fortuitous risk; this is covered by insurance. Such risks may include damage to a satellite on launch, failure during planned or design lifetime, or liabilities incurred due to damage or injury to others. In today's market, there exists the risk of availability of insurance to cover property loss and injuries to third parties. This latter form of risk is sometimes called *third-party liability*.

Occasionally, insurance availability has created a bottleneck in financing during the construction stage of a system. This phase generally takes about 3 years before generation of revenue. To enter into construction contracts, financing commitments must be in place, that is, 3 years before launch. Insurance underwriters generally will not commit to supply coverage until 3–6 months before launch. This is caused by the uncertainty of appropriate rates caused by a series of launch failures and the corresponding cumulative losses experienced by insurance underwriters. Consequently, with uncertain profitability, there has been uncertain available capacity. Thus, construction will have to commence a significant time period prior to knowing if insurance cover will be available for the launch and related items.

Generally, financing sources will not take insurance availability risks that insurance suppliers are unwilling to take. Consequently, this has resulted in requests for proposals for, or actually contracted "delivered in orbit" satellites, a situation where the satellite supplier assumes the risk of launch insurance availability, and perhaps its pricing. Also, commercial launch service providers have assumed a form of insurance availability risk by committing to provide a relaunch if a launch failure is caused by their rockets. This situation places a large contingent liability on suppliers and consequently limits their ability to grow.

Stages of Financing

Satellites generally take on the order of 3 years to build. Similarly, launchers take approximately the same time to prepare for launch. Unless another organization's satellite is used to generate revenue prior to launch, the investor must wait 3 years before the satellite can make revenue available. The phase, during which no revenue is being produced while the satellite is being built, is called the construction phase, and financing is called construction financing. After the satellite has been launched, financing employed either as a continuation or replacement of the construction financing is called permanent financing. Construction financing is far more difficult than permanent financing because the risk of launch failure and delay must be faced in addition to the normal commercial risk that exists in the permanent phase.

What Are the Factors That Attract Investment?

Clearly, the most important factor is return on investment as measured against alternative opportunities. The intuitive measure of profitability to investors represents how much profit investors can obtain from a satellite system over the time the investor chooses to hold the investment. Equally if not more important is how soon those profits are available. The key factor that makes the concept challenging is the delay between building a satellite and the time until profits are available. For large commercial projects, the time delay factor can overwhelm the variations in cost for achieving adequate investor returns. This is because 1) a commercially competitive project will have high opportunity costs—the earnings forgone from alternative investments; and 2) profits must pay for the satellite and generally much larger investments in ground equipment. The initial time delay issues overwhelm the financial issues surrounding added satellite life available from different launchers or launching locations.

Many of today's most controversial issues concern costs of insurance and transportation. Financially, the delay issue makes them secondary items. It also means that the concept of supplier costs can be quite different as viewed by a buyer and seller. A seller may view price as the measure of cost to a buyer, whereas a buyer may reduce investment and increase profitability by paying "more" for a system but having it available earlier. (This was an important factor in a recent major system purchase decision for a television satellite.)

How Can Investors' Returns Be Magnified?

The large investment in a satellite system causes difficulties in achieving competitive returns when measured against other opportunities. The large size of needed investment means that large revenues are needed to generate profits that allow the investor to reach the competitive returns. For a new application, it may not be possible for a user market to reach a level of revenue fast enough to allow competitive returns to be achieved. This means that the investment must be reduced to a size consistent with the revenue-generating capacity of the market. This is done by using a portion of a satellite or replacing investment with borrowed funds, which reduces the financing cost and the amount of equity that must be covered by adequate profits. Unfortunately, this creates a paradox, because lenders are far more cautious than investors. Therefore, the investors' risk starts to take on the levels of the lenders' risk.

What Are the Effects of the Development of Small Satellites?

Smaller satellites with their faster build and launch times can, in theory, ease some financing and insurance problems as well as commercial risk when they are used as part of small specific-application systems. If, however, they are used as part of a constellation of satellites, with large associated investment, the financial issues will approximate those of large geosynchronous systems.

Sources of Investment

Until recently, there has been a classical ranking of sources of investment based on descending orders of risk. The ranking may change from time to time depending upon circumstances. From highest to lowest risk, and returns, the ranking could be as follows:

1) *Venture Capitalists*—A group of high-risk investors (generally funds from financial institutions or pension funds) who generally make small levels of investment with returns in the "over 40%" rate of return range. They look for high returns because they expect many of their investments to produce losses. Consequently, a few "winners" support many "losers." They tend to invest early and provide "seed money" or later stage financing to cover losses as a company grows until the company turns profitable. Venture capitalists tend to look for profits through sales of equity shares to the public in 3- to 5-year horizons. Risk and rate-of-return criteria depend on opportunities that may exist totally outside space-related areas. For high capital and long-time horizons of space projects, this group has so far not been a major source of funds for space business development.

2) *Corporations with Strategic Interests*—These are generally users of equipment or services that will be provided by satellites. The returns on investment may be due solely to the investment itself, or to opportunities that may be available to employ investors' existing capabilities in conjunction with a new space system. They may also be defensive investments, made to protect an ongoing operation. These investors may have the highest risk tolerance because of limited alternative investment opportunities and the need to strengthen existing business. Investments of this group tend to be large and time horizons for returns long. These factors make this group the primary investor group for commercial space.

3) *Tax-Driven Investments*—These are principally investments in technology that were made possible by tax laws. The theory is that tax partnerships would invest in development projects where a combination of tax benefits from research and developments activities and royalties would be obtained from use of the technology. Changes in the tax laws have made these investment vehicles less popular. When they were available, they were useful as early-stage high-risk investment. The unique problem with this type of investment was that it was based on the durability of a particular technology rather than the ability of a company to adjust to changes in the marketplace. Consequently, a correct business decision to adjust technology could result in a conflict of interest with different classes of investors. (This actually happened in the case of a computer equipment manufacturer.)

4) *Public Equity*—This stage of financing usually comes after the development of a company that has been supported by other forms of financing. It is used to give liquidity or value to investments made by venture capitalists or others. On rare occasions, it can be used as an initial source of funding, such as was done for Comsat, but is usually employed when a more viable company and favorable public securities market conditions exist.

Sources of Debt

Sources of debt have a wide variety of forms including some that are not normally viewed as borrowing. Some of the obvious and not-so-obvious ones are listed below:

1) *Bank, Institutional, and Public Senior Loans*—These funds generally cover time delays between receipts and outflows discussed earlier. The major component of the cost of these funds is generally determined by financial market conditions rather than by the characteristics of the borrowers. These are relatively low-risk funds which have been used as components of major financings where the funds suppliers are not exposed to venture-type risk.

2) *"Junk" Bonds*—These involve high-risk debt and are generally employed with transactions involving the recapitalization, purchase, and sale of existing established companies. The risks accepted are associated with equity risks of established companies, not generally in venture or project situations. Suppliers of such funds have been corporations, pension funds, financial institutions, and some public pooled funds.

3) *Subordinated Debt*—This is sometimes associated with "junk bonds" but generally another form of funding by investors that accept equity risks associated with new ventures or projects to accelerate repayment of projects.

4) *Leases from Financial Institutions*—These are generally a substitute for low-risk borrowing where the effective interest cost might be reduced through use of certain tax benefits . . . not an important form of financing for new ventures. If the availability of the tax benefits makes the difference in the returns to investors such that a project may not proceed, the project is probably marginal anyway. An example of this type of financing is the sale-leaseback used by major communications companies.

5) *Leasing of Partial Facilities or Purchasing Services from Operating Companies*—This is a very important form of reducing risk and investment size of a project. In this type of financing, exotic risks may be taken by the owner of the equipment. An example might be leasing a transponder of a satellite that is already in orbit. (This avoids the launch risk.) Another example would be purchasing time on a satellite transponder to minimize the amount of investment in a new, speculative venture until the venture demonstrates that it will succeed. A third example would be leasing government-owned launch facilities to provide a commercial launch.

6) *Payment Deferral from Suppliers*—This form of financing is valuable when the supplier is willing to accept the risk of other financing being available when deferred payments are due. Under these circumstances, the financing takes some investment risk and reduces the size and sometimes the cost of investment to be raised. Otherwise, the financing is equivalent to low-risk bank loans. An example of valuable supplier financing involves the case when inventory is maintained to lower the build time of the satellite and/or preparation of the launch vehicle. The shorter construction time has the effect of significantly magnifying investors' returns. A recent example is the U.S. government's deferral of launch payments from Spacehab until a successful launch has been accomplished. This deferral has had a significant impact on the financing of Spacehab.

7) *Government Export Financing Programs*—These financing programs are generally loan guarantees or in some cases have been low-interest loans to finance exports. The purpose of such loans is to enable foreign purchase commitments to provide access to low-risk debt. These facilities, provided by the export-import bank, are generally designed to remove financing as a factor in international trade competition. These financing programs are generally not substitutes for high-risk equity or insurance.

8) *Government Progress Payments*—In this case, the government provides progress payments to suppliers. This is analogous to high-risk construction loans. This is most useful when the government purchases satellite and launch hardware, and assumes the risk of launching. When the government has attempted in the past to purchase a service, attempts have been made to substitute bank debt for progress payments. This was not fully successful because of the inherent high risk of the program. This can result in the government becoming the banker, as happened in NASA's TDRSS communication satellite program.

Conditions Necessary for Debt

It is desirable to maximize the amount of debt in financing because of its lower cost when compared to investment. Greater amounts of debt can make the difference between a profitable or marginal activity. The conditions that are necessary for debt are listed as follows:

1) *Certainty of Payment*—This represents the perception to the lenders that there is an expectation of payment that is minimally affected by business risk.

2) *Visible Cash Flow*—Lenders expect to have loans repaid through the normal operations of the business or activity. For a new project, lenders

generally will not lend to projects or companies where repayment is based on some single event that may happen, such as the ability to sell an asset or a future financing. On occasion, lenders might deviate from this criterion if there are assets available to sell that have a ready and predictable market. Space hardware is generally not in this category.

3) *An Adequate "Second Way Out"*—Lenders look to the sale of assets, mentioned above, or a financial guarantee from a substantial party as a secondary method of repayment. The collateral, mentioned above, can sometimes be a method to protect against adversity, but with no clear secondary market value, there will be little motivation by lenders to provide funds (i.e., $75 million of half-finished rocket and half-finished satellite would do little to offer motivation or market value toward encouraging lenders to provide funds).

How Would Total Financing of a Communication Satellite Business Be Arranged?

A possible sequence of events might take place as follows:

1) Initial "seed money" in the form of a few thousand dollars might be used in exploratory studies. The funds would be in the form of investment.

2) If the business is large, the next step would be the assembly of consortium partners who would be the major investors in the business. The partners would be principally involved in system design and regulatory activities—involving a few hundred thousand to several million dollars. Funding would most likely continue to be direct investment in the business.

3) The next step would involve the construction of the satellites and ground equipment procurement. This activity involves several hundred million dollars of commitment. The funding would come either in the form of investor-guaranteed borrowing or direct investment. During this time, a space segment would be built with 2–3 years of progress payments.

4) The next step would involve beginning of operations. Additional investment or investor-supported borrowing would be needed to support ongoing operations. If the business involves entertainment, the size of the investment needed to cover early operating losses during buildup could be larger than the investment needed to build the system. Additional investment in ground stations may be needed to be made by the business' customers. If the investment is not forthcoming in a timely manner, say because users may need time to accept the new service, the company may have to add further investment to stimulate revenue buildup (perhaps another 100 or 200 million dollars).

5) Finally, after all of this outlay, the investors may start in 5 or 6 years, perhaps, to receive some cash dividends from the question. Most likely, shares of the venture would be sold to the public or others in order to accelerate the receipt of additional cash to improve returns to investors. Also some borrowing that is unsupported by the investors might be possible to further increase some of the cash that can be paid to investors.

Obviously all of this outlay and waiting time makes alternative investments very attractive. Some ways that the investment in projects such as these can be reduced is to employ other organizations' satellites to provide limited service to accelerate the buildup of the business' revenues. These revenues are then used as a substitute for the investment. The reduced investment leads to greater profitability and less funds at risk.

What Are the Differences Between Communications Systems and In-Orbit Production Facilities?

Until an industrial space develops, any comparison involves considerable conjecture. Based on some guesses, the main differences will involve principally insurance issues. In a communications system, there is one launch risk per satellite or transponder channel over a long period of time. In industrial facilities, there will be repeated risk as material will be required to be removed and replenished. This magnifies the insurance and places extremely high returns on investment targets to raise funding from institutional sources.

Conclusion

In summary, there should be nothing exotic about financing space projects. What is far more challenging is recognizing the commercial rewards and risk of projects and making them competitive with non-space-related opportunities.

Bibliography

Rycroft, M. (ed.), *The Cambridge Encyclopedia of Space*, Press Syndicate of the University of Cambridge, Cambridge, UK, 1990. (Originally published in French as *Le Grand Atlas de l'Espace*, Encyclopedia Universalis, 1991, Paris.)

U.S. Department of Commerce, *Space Commerce and Industry Assessment*, Washington, DC, May 1988.

Chapter 2. Cost Analysis

Standard Cost Elements for Technology Programs

Carissa Bryce Christensen*
Princeton Synergetics, Inc., Princeton, New Jersey 08540
and
Carl Wagenfuehrer†
NASA Headquarters, Washington, DC 20546

IT is widely agreed that good cost estimates for technology research programs are difficult to derive. The bases of estimates typically used have many flaws; these are analyzed in detail in the following sections, and some innovative solutions are proposed. As an introduction to these analyses, this brief article discusses the appropriate *structure* of a good cost estimate, in this case for a NASA technology program. The cost elements that are identified are specific to a NASA technology research program, although they may, with modification, be applied to other government programs.

This article also discusses an area of the budget process that is often overlooked—the need for program managers to demonstrate that they are able to spend the funds that they are requesting. A high-quality cost estimate may be useless in obtaining project funding if the mechanisms for spending the requested funds within the appropriate time period do not exist, or are not adequately documented in making a budget request.

While this approach focuses on government cost justification, it may be useful to industry researchers to understand the process that NASA uses in assessing budget requests from specific programs. This process is embedded in the recommendations presented here; the recommended approach is designed to provide the type of information that is persuasive to NASA budget analysts.

Copyright © 1992 by Carissa Bryce Christensen and Carl Wagenfuehrer. Published by the American Institute of Aeronautics and Astronautics, Inc. with permission.
*Vice President.
†Program Analyst, Office of Aeronautics and Space Technology.

Developing a Complete and Defensible Cost Estimate

A common approach to presenting and documenting cost estimates may not improve the fundamental validity of the estimate, but can vastly improve the ability of those evaluating the estimate to assess its quality, to compare it to other estimates, and to aggregate information.

Cost Elements

The cost elements that should be included in a defensible cost estimate are listed in Fig. 1. This list reflects three main types of costs: research and development (R&D), research and program management, and construction of facilities. The R&D cost component of a program cost estimate is generally the largest, and requires a significant level of detail. Research and program management (R&PM) costs cover travel and administrative expenses, and construction of facilities (CF) costs cover new facilities.

R&D costs should include both estimates of costs and estimates of required labor hours. Costs should reflect in-house costs, out-of-house costs (generally contracts and grants), and contingency/reserve costs.

In-house costs include:

1) Contractor support (primarily through an on-site support services contract, as distinct from an out-of-house contract for hardware or research)

2) Materials (items that can be purchased off the shelf, through purchase order contracts, as distinct from major procurements)

3) Facility usage/program support (a category covering funds for requirements such as automated data processing and facilities usage (Cost levels are determined using NASA field center-specific algorithms that translate project requirements into costs.)

4) Multiple program support (MPS)/program management support (PMS) (a category applying primarily to flight centers, which assess a head tax on NASA field center programs to cover management, administrative, and overhead costs)

Out-of-house costs include:

1) University grants (research grants to academic institutions)

2) Contracts (includes study contracts, hardware contracts, and, in general, procurements of products or services other than on-site general support which is covered under in-house contractor support).

Contingency/reserve costs reflect the judgment of program managers as to the riskiness of the activity and the uncertainty associated with the cost estimate, and represent some amount of money held in reserve to cover unexpected costs. This is often calculated and expressed as a percentage of the cost estimate.

Cost Estimating Process and Basis of Estimate

The method by which a cost estimate is developed can range from the formal and structured (such as the use of a cost-estimating relationship (CER)) to the intuitive (such as basing costs on general program experience). The formal approach is not always the most appropriate. Technology

STANDARD COST ELEMENTS FOR TECHNOLOGY PROGRAMS 47

COST ESTIMATING

Estimates of Labor Requirements and Costs

Fiscal Year ($ millions): YR 1, YR 2, YR 3, YR 4, YR 5, YR 6, YR 7, YR 8, TO COMPLETE, TOTAL

R&D Labor Requirements:
1. Civil Service Labor (workyears)
2. Contractor Labor (workyears)

R&D Costs:
3. Total In-House Costs
4. Contractor Support
5. Materials
6. Facility Usage/Program Support
7. MPS/PMS

8. Total Out-of-House Costs
9. University Grants
10. Contracts

11. Contingency/Reserve

Total R&D Costs

12. Research and Program Management
13. Construction of Facilities

Cost Estimating Process

14. Basis of Estimate
    Select a, b, or c: _____
    (a) Cost estimating relationships (CERs) or other algorithms or procedures
    (b) The historical performance or track record of the program or division
    (c) Grassroots estimate

15. Major Assumptions and Comments: _____

16. Date Funds Assumed to Be Available: _____

17. Factors Affecting Estimate
    Answer yes or no the following questions:
    New Program? _____ Availability of Civil Service Labor Assured? _____ MPS/PMS Calculated Based on Identified In-House Labor Requirements? _____
    Inflation Factor of 4.5% Used? _____ Availability of Facilities Assured? _____

18. Comments on Factors Affecting Estimate: _____

19. Confidence in Cost Estimate: +/- _____ %
    Select a, b, or c: _____
    (a) High: E.g., similar program implemented recently.
    (b) Medium: E.g., organization has track record of performance in research area.
    (c) Low: E.g., new research area for organization.

Program: _____
Source: _____   Phone No.: _____   Date: _____

Fig. 1 Cost estimating.

research programs involve a high degree of uncertainty and are aimed at achieving innovation. A CER that assumes certain historical relationships may be completely inappropriate to such a program, while an intuitive guesstimate may incorporate important managerial and research experience, and yield a better (i.e., more accurate) result. The key to presenting cost estimates is to justify the approach used in the context of a particular program. Program experience may be the appropriate basis for developing an estimate, but certainly will not (and should not) be accepted as an adequate basis of estimate without additional information about the character of the earlier experience and its relationship to the new program, and some explanation as to why more formal approaches are not appropriate. (There are methodologies that incorporate program experience and the intuitive expertise of technical experts into a formal cost-estimating methodology. A program management tool that incorporates such a cost-estimating methodology, the RADSIM approach, is discussed in an article in this book entitled "Cost Estimating for Technology Programs.")

There are many factors affecting the defensibility of a cost estimate that are not immediately obvious, and that, if addressed, can improve estimates at a low cost (in terms of time spent and information required) to the estimator. For example, the inclusion of a specified annual inflation factor increases comparability across estimates; the inclusion of assurances that required manpower and facilities will be available reduces the concerns that an auditor or analyst might have about schedule delays or unforeseen labor costs; and the specification of the manner in which field center taxes or support costs are calculated increases intra- and intercenter comparability.

The defensibility of cost estimates can be significantly improved through increased documentation. The quality of cost estimates ultimately depends on the cost estimating methods used. Increased documentation will not add validity to an estimate resulting from an inappropriate or poorly applied cost estimating methodology. (Increased documentation will, however, enhance the ability of managers to identify such flaws.)

Demonstrating Ability to Obligate and Cost Funds

Definitions

The *ability to obligate* funds is the ability to enter into contractually binding relationships with industry, universities, or other organizations. Generally, the majority of a program's ability to obligate funds will be its ability to tie these funds to contracts, procurements, and grants. Correspondingly, when a program is unable to obligate a significant percentage of its funds, this is often due to the long lead time required to institute a new contract. Existing contracts also have limitations, in the form of ceilings on annual and total contract expenditures or number of work hours that can be funded through the contract, that must be anticipated in calculating ability to obligate.

The *ability to cost* is the ability to spend obligated funds. Funds that are obligated to a contract are costed only as work under that contract progresses. If work does not proceed under the contract (for example, because the hiring of required personnel takes longer than anticipated, or needed facilities are unavailable), then the obligated funds cannot be costed.

Other types of anticipated expenditures (such as off-the-shelf material purchases and head taxes on employees) also contribute to a program's ability to obligate. Different NASA centers employ different definitions of the point at which funds for noncontract expenditures are considered to be obligated and costed.

NASA Cash Management Policies

NASA budget analysts work to ensure that NASA funds for any particular activity are used soon after the funds are made available, because there is a cost associated with unspent money (the opportunity cost, or lost value associated with alternative activities that could have been funded with the unspent money). This requires that programs be able to obligate and cost their funds in a reasonable time frame. As an example, the target obligation and cost levels used by the Office of Aeronautics and Space Technology are:

1) *Ability to obligate*—Funds should be 95% obligated by the end of the fiscal year in which they are received, and 100% obligated by the end of the 3rd month (i.e., December) of the following fiscal year.

2) *Ability to cost*—Funds should be 70% costed by the end of the fiscal year in which they are received, and 100% costed by the end of the following fiscal year.

It should be noted that these targets hold no matter when funds are received. Funds received at any point during the fiscal year should still be 95% obligated and 70% costed by the end of that fiscal year.

Factors Determining Ability to Obligate and Cost

The ability of a program to use requested funds can be a key factor in the budget approval process. The important aspects of demonstrating ability to obligate and to cost funds are discussed here.

Major Contracts, Procurements, and Grants

As noted above, the largest part of a program's ability to obligate is its commitments through contracts, procurements, and grants. The two key factors that may limit a program's ability to obligate through these methods—and thus the factors that should be addressed in demonstrating an ability to obligate—are contract ceilings and contract lead times.

Contract ceilings on existing contracts may constrain a program's ability to obligate, when requested funds significantly exceed previous or planned budgets. Programs often plan to use existing support services contracts, and these contracts often have annual and overall limitations of the amount of money or the number of hours that the contract is allowed to accommodate.

Contract lead times on new contracts may preclude a program's ability to obligate funds. The process of writing and issuing a solicitation, assessing proposals, and making an award can take over a year for a major hardware procurement. There are some methods of reducing contract lead time, such as preparing a Statement of Work (SOW) or even issuing the Request for Proposals (RFP) prior to the receipt of funds. In demonstrating its ability to obligate, a program should specify any actions it has taken to reduce lead times for new contracts. Lead times vary depending on the contract and contracting organization. As an example, contract lead times in the following ranges are not unrealistic: for a contract value of less than $1 million, about 4 to 6 months; for a contract value of $1 to $10 million, about 6 months; and, for a contract value greater than $10 million [i.e., requiring a Source Evaluation Board (SEB)], about 1 year.

Demonstrating Ability to Obligate and Cost

A program's ability to obligate funds can be demonstrated by identifying the specific contract vehicles through which funds will be spent, with particular attention to showing that these contract vehicles will be in place in time for the funds to be spent, and that contract ceilings are sufficiently high to accommodate anticipated levels of expenditure. If reduced contract lead times are noted, the factors that make these reductions possible should be specified.

A program's ability to cost obviously depends substantially on its ability to obligate; funds that are not obligated cannot be costed. However, the ability to obligate funds does not necessarily imply an ability to cost an equal amount. Factors such as whether personnel are available to perform the contract activities and whether facilities and other resources are available when they are needed affect a program's ability to cost its obligated funds. In asserting an ability to cost funds, these factors should be explicitly addressed where possible. For example, the ability to cost funds will be enhanced if required civil service personnel are already hired and assigned to a project. Often, however, it is difficult to predict whether personnel will be available for a project to begin in several months or years. These other factors are not shown in the Obligation and Cost Planning worksheet.

In addition to contract vehicles, other factors may contribute to the ability to obligate funds. These include anticipated expenditures on materials, facility usage/program support, MPS/PMS, and contingency or reserve funds. These secondary factors are not shown explicitly in the Obligation and Cost Planning worksheet.

A Sample Approach to Consistent Cost Justification

Worksheets that could be used in structuring the presentation and documentation of cost estimates and statements of ability to obligate and cost are shown in the following pages. Each line in each worksheet is defined. These worksheets were developed for NASA's Office of Aeronautics and Space Technology (OAST) as part of a continuing effort to develop more robust cost justification techniques. The authors wish to note, however, that this approach should in no way be regarded as an official statement of policy by NASA or OAST.

The worksheets shown here encompass cost estimating (Fig. 1) and some of the elements of demonstrating ability to obligate and cost (Fig. 2). The

```
                OBLIGATION AND COST PLANNING
                Major Procurements, Contracts and Grants
            1   Name/Description of Contract: _____

            2   Type of Contract
                Select a, b, c, d, or e: _____
                (a) Support Services    (b) Hardware Procurement
                (c) University Grant    (d) Research Contract
                (e) Other _____

            3,4 Start Date _____   End Date _____

            5   Total Contract Value _____

            6   Status
                Select a, b, or c: _____
                (a) Planned        (b) In-Process        (c) Active

            7   If "Planned" or "In-Process" is selected, answer the
                following questions.
                Suppliers/Contractors Identified?    _____
                Expected Contract Leadtime           _____
                Factors Affecting Contract Leadtime:
                                SOW Written?         _____
                                RFP Issued?          _____
                                SEB Required?        _____
                Comments:_____
                _____

            8   If "Active" is selected, provide the following
                information:
                Contract # _____
                If contract is a support service contract, specify
                cap/ceiling: Hours _____   Dollars _____

            NOTE:  Most programs have multiple contracts, procure-
                   ments, and grants. These should each be listed
                   using the format here.
```

Fig. 2   Obligation and cost planning.

most important component of a program's ability to obligate is the availability of contract mechanisms—major procurements, contracts, and grants. The type of information needed to fully describe contract mechanisms available to a program is shown in Fig. 2. Other factors may also contribute to a program's ability to obligate and cost funds, including the transfer of funds to other government agencies, expenditures on materials, costs of program management and administration, and the allocation of contingency or reserve funds. Finally, if required civil service personnel are not available to be assigned to a program at its inception, its ability to obligate and cost funds may be reduced in that year.

Definitions of Standard Cost Elements

See the Estimates of Labor Requirements and Costs section of the worksheet titled "Cost Estimating" (Fig. 1) for the cells defined below.

1) *R&D Labor Requirements—Civil Service Labor*: in-house civil service labor requirements in workyears
2) *R&D Labor Requirements—Contractor Labor*: in-house contractor support requirements in workyears
3) *R&D Costs—In-House*: total of subheadings (contractor support, materials, facility usage/program support, and MPS/PMS)
4) *R&D Costs—In-House, Contractor Support*: in-house contractor support costs; should be based on contractor labor requirements given above
5) *R&D Costs—In-House, Materials*: costs of routine purchases of required materials; does not include hardware procurements, which are listed under R&D Costs—Out-of-House
6) *R&D Costs—In-House, Facility Usage/Program Support*: costs of requirements such as data processing and facilities usage; cost levels generally determined using center-specific algorithms that translate project requirements into costs
7) *R&D Costs—In-House, MPS/PMS*: costs for MPS/PMS; generally applies to flight centers, and costs are determined based on center-specific costing methods reflecting personnel requirements
8) *R&D Costs—Out-of-House*: total of subheadings (university grants, contracts)
9) *R&D Costs—Out-of-House, University Grants*: cost of research grants to universities
10) *R&D Costs—Out-of-House, Contracts*: costs of study contracts, hardware contracts, and so forth awarded to out-of house organizations; does not include support services contracts (which are listed under R&D Costs—In-House, Contractor Support)
11) *Contingency/ Reserve*: amount of money put aside to cover unexpected costs; should reflect the degree of uncertainty and risk associated with the activity
12) *R&PM*: costs of travel and administrative expenses; not part of the R&D budget
13) *CF*: cost of construction of facilities; not part of the R&D budget

See the Cost Estimating Process section of the worksheet titled "Cost Estimating" (Fig. 1) for the cells defined below.

14) *Basis of Estimate*: basis of estimate for R&D costs; may be a) CERs or algorithms or procedures (generally for determining management or support requirements rather than technical requirements); b) the historical performance or track record of the program or division (again, this should be explained); or c) grassroots estimate
15) *Major Assumptions and Comments*: pertinent information justifying the basis of estimate, explaining the basis of estimate in more detail, and noting any important assumptions which, if violated, would significantly affect the validity of the estimate
16) *Date Funds Assumed to Be Available*: date assumed for availability of funds for first year. If meeting this date is critical to either cost

STANDARD COST ELEMENTS FOR TECHNOLOGY PROGRAMS 53

estimates or to projected ability to obligate and to cost, the importance of fund availability should be addressed under Major Assumptions and Comments.
17) *Factors Affecting Estimate*: using the lines provided, answer yes or no to the following questions regarding which factors listed below have had an effect on the cost estimate:
   a) *New Program*: Is the program a new or nearly new program (answer yes), or is it a mostly or completely pre-existing program (answer no)?
   b) *Inflation Factor of 4.5% Used?*: Were outyear estimates calculated using 4.5% inflation (answer yes), or was no inflation factor used (answer no)? (If different inflation factor used, answer no and explain in Comments on Factors Affecting Estimate section). Note that OAST currently recommends a 4.5% inflation factor for technology programs. Due to changing economic conditions, this recommended level could change.
   c) *Availability of Civil Service Labor Assured?*: Will required civil service personnel be available? If no, explain circumstances in Comments on Factors Affecting Estimate section. The importance of the availability of required civil service personnel is that to substitute contractor personnel can increase costs substantially.
   d) *Availability of Facilities Assured?*: Are needed facilities certain to be available at the times and for the level of use necessary? If no, explain circumstances in Comments on Factors Affecting Estimate section. The importance of the availability of required facilities is that if facilities are not available as needed, substantial scheduling delays, and therefore increased costs, can occur.
   e) *MPS/PMS Calculated Based on Identified In-House Labor Requirements?*: applies only to centers that use MPS or PMS.
18) *Comments on Factors Affecting Estimate*: additional information or explanation for any "no" answers to questions in Factors Affecting Estimate section
19) *Confidence in Cost Estimate*: subjective, quantitative estimate of confidence in the cost estimate, $+/- X\%$. An estimate reflecting high variation may require additional explanation in the Comments on Factors Affecting Estimate section. In addition to specifying a percentage, choose among: a) *high*, indicating, for example, that the organization completing the estimate has recently implemented a similar program; b) *medium*, indicating, for example, that the organization has a performance track record in this research area; or c) *low*, indicating, for example, that this is a new research area for organization.

See the worksheet titled "Obligation and Cost Planning" (Fig. 2) for the cells defined below.

1) *Name of Contract*: Use formal name of contract for existing contracts; use descriptive name for in-progress and planned contracts.

2) *Type of Contract*: Identify type of contract, as a) support services (should reflect R&D Costs—In-House, Contractor Support category); b) hardware procurement, c) university grant, d) research contract, or e) other. If other, describe.
3) *Start Date*: contract start date or anticipated start date
4) *End Date*: contract end date (including all option years)
5) *Total Contract Value*: Total contract value for a given fiscal year should generally *not* include options that could be exercised. This question does not arise for most support contracts, which are written with options based on period of performance. That is, an option to spend additional funds cannot be exercised until a given date. (Such contracts generally follow the pattern of 2 base years and 3 option years.) For a support contract written in terms of labor hours, rather than period of performance, an option year can be exercised before that year actually begins. In demonstrating ability to obligate, the exercise of such an option should not be considered to be part of ability to obligate *unless* a compelling reason (such as an unexpected increase in program scope) can be given for using contract hours at a higher rate than was planned when the contract was written.
6) *Status*: Identify contract status, as planned, in-process (this differs from planned in that it indicates that some steps to initiate contract have been taken), or active (that is, currently in progress). Depending on the status selected, additional detail will be requested (see below).
7) *Status—Planned or In-Process*:
   a) *Suppliers/Contractors Identified?*: Are suppliers or contractors that can meet the anticipated need already known?
   b) *Expected Contract Lead Time*: (time prior to award of a contract) estimate, in months, of contract lead time *from date funds are assumed to be available*
   c) *Factors Affecting Contract Lead Time*: yes or no questions asking whether the SOW has been written, whether the RFP has been issued, and whether a SEB is likely to be required. Also provides a space for comments and additional explanation.
8) *Status—Active*:
   a) *Contract Number*: NASA contract number
   b) *Cap (hours)*: maximum number of hours that can be spent in first year under the contract, if such a cap exists; mainly applies to support services contracts.
   c) *Cap ($M)*: maximum amount that can be spent in first year under the contract, if such a cap exists; mainly applies to support services contracts

Conclusion

The approach recommended here is based on a recognition of the types of concerns that NASA budget analysts have about the quality and realism of the cost estimates they evaluate. These concerns include the identification of major assumptions, the specification of secondary analytic assumptions (such as the dates the estimate is intended to cover, the inflation factor used, and the consistent translation of labor estimates to cost estimates), and the status of a number of program factors that typically affect the reliability of estimates. The approach also addresses cost and obligation issues, providing a structure that can aid a program manager in demonstrating that the funds requested can indeed be spent in the appropriate time period.

This article has not discussed the implementation of this approach—not an insignificant concern. Implementation must address questions such as intelligibly conveying the identified requirements to cost estimators, communicating the value of this approach to those responsible for using it, and determining the appropriate level of detail at which the approach should be aimed (program vs research task or even subtask). Continued attention to and analysis of cost justification issues at NASA will, the authors hope, lead to the resolution of these implementation questions and further refinement of this approach.

Cost-Estimating Relationships for Space Programs

Humboldt C. Mandell, Jr.[*]
NASA Johnson Space Center, Houston, Texas 77058

Background

Current Space Industry Cost Estimation Methods

THERE are several methods currently employed by the aerospace industry to estimate costs. As described further under "Perceptions, Precedents, and Politics," all are based on analogies to the past and utilize equations to predict costs from historical data. These relationships are called cost-estimating relationships (CERs). The theoretical development of such relationships as well as examples of models in which they are employed are presented in the Appendix.

Typical methods employ analogies based on historical manpower and productivity data. For example, if it is known that the development or production cost of a spacecraft structure of a given size has always involved 1000 manhours per pound, cost estimates can be made by estimating the weight (volume of the materials multiplied by material density) of a future spacecraft and then simply multiplying the weight by the historical manhour-per-pound ratio and the current labor rate for the types of labor employed.

A second analogous method is based on the activities that constitute a project (this type of method is called an activity-based cost model, or ABC model). This method examines the job to be done based on the individual activities, usually derived from a logic diagram of the new project; once activities are identified, historical analogs are sought for each. For example,

---

Copyright © 1992 by the American Institute of Aeronautics and Astronautics, Inc. No copyright is asserted in the United States under Title 17, U.S. Code. The U.S. Government has a royalty-free license to exercise all rights under the copyright claimed herein for Governmental purposes. All other rights are reserved by the copyright owner.
[*]Acting Deputy Manager, Lunar & Mars Exploration Program Office.

if the new job involves the creation of crew training manuals, data can be employed from a past program that produced crew training manuals of approximately the same complexity. Once all activities are identified and analogies found for each, a composite cost estimate is made by simply adding the costs of all activities.

A third analogous method, that which will be the primary subject of this chapter, uses analogous relationships between physical and performance parameters to estimate future costs. For example, if experience has shown that booster engines in the 30,000-lb thrust range cost $100,000 per pound to develop, and similar engines in the 5000-lb thrust range cost $150,000 per pound to develop, a relationship between development cost and engine thrust can be constructed simply by connecting the two points of data.

As more data points are added to the historical data base, experience has shown that most of these CERs assume log-linear forms (i.e., linear relationships in the transformed logarithmic space defined by the performance parameter and the cost). The usual curve form is Cost = A × (Parameter) B, where A and B are constants. Typically, the power to which the parameter is raised, for space programs, has been found to be roughly 0.5 (the square root of the parameter value) for development costs, and roughly 0.75 (the three-fourths power) for production costs of hardware.

CERs are combined into cost models using the work breakdown structure of the new project. For elements of the work breakdown structure for which no physical parameter exists (such things as management and integration costs), the usual practice is to utilize either an ABC model or, more often, a percentage of the sum of all hardware costs (see "Cost Models", for a description of a typical parametric cost model).

Context of Cost Estimation

The greatest barriers to American space programs today are probably more programmatic than technological. Over the past two decades, as space programs became fewer, funding less plentiful, and existing programs longer and longer, the tendency was sometimes to blame the economy and the national budget for the reduction of American activity in space. This rationale is not totally accurate.

Besides the economy and the national budget, large influences on space activity levels are exerted by the costs of existing ventures, as well as the perceptions of the costs of future ventures, costs that are more driven by cultural influences than by technological ones.

Cost overruns have been a perennial problem of doing business in space. Often, overruns are attributed to the inability to predict the costs of future ventures. That also is only partially correct.

Therefore, before discussing further the techniques of cost estimation, a short discussion is presented of the context of proper use of estimation methods. Very often, the inability to predict costs accurately is not the result of cost estimation technology, but is, rather, the failure on the part of program advocates to fully understand the context of the estimation process.

While a complete discussion of that context is beyond the scope of this chapter, or even of this book, it must be mentioned here that the ability to predict accurately the costs of space programs goes far beyond the ability to construct and use CERs. A few of the more important contextual issues are described.

Research by NASA has shown that the most powerful influence on development costs is the culture of the developing organization, more powerful, even, than the technical and technological drivers on a particular venture. But, until recently, the cultural effects have not been quantifiable.

Within the space industry, cultural changes have often occurred very slowly. Hence, the use of costing methods based on historical costs have worked relatively well. But, in today's environment, where the total activity of the space industry tends to be limited by the budgetary availability, study is underway to identify ways to produce more activity for the same amount of cost. Or stated another way, ways are being sought to lower the costs of any given space project without sacrificing capability, content, schedule, or mission success, recognizing that funding constraints will always be present.

It will be demonstrated that so pervasive is the cost influence of the development culture on cost that program cost reduction activities are being focused on the development cultures, as opposed to technologies. This gives rise to the situation where the utility of historical analogies would be completely useless, *unless* the cultural influences on the analogies were to become a part of the statistical data base.

Within the space community today, therefore, there is a major effort devoted to understanding the relationships among development culture and the technological and structural factors of historical programs, to make the data useful in predicting the future.

In performing this research, it has been found that ways of doing business exist, in fact are well known and widely practiced, which can reduce the costs of spacecraft development by 50% or more. Such business practices are being used more and more by other American industries. This subject is described further under "Influence of Organizational Culture on Program Cost."

Cost as a Function of Funding Availability

Particularly in large programs, the costs of space ventures can be more a matter of funding availability than of any characteristic, physical or programmatic, of the venture to be performed. No program can cost more than the funding available to it. Therefore, the first step in estimating the cost of any space venture is to identify, at least in a probabilistic sense, a time-phased profile of available funding.

The major design driver in the Space Shuttle program is generally acknowledged to have been the peak annual funding requirement (of approximately 1.2 billion 1971 dollars). Designs for the Shuttle had evolved to the point of entering into the design and development phase of the program with a two-stage fully reusable configuration. Early cost estimates showed that this configuration could not be developed within the expected budget (doing business as NASA then did business).

This unexpected finding caused a rapid search for a configuration that would fit the Office of Management and Budget (OMB)-mandated funding profile. The search ended, after dozens of iterative design studies, with the current partially reusable configuration. Likewise, in the case of most major new space ventures, funding constraints will almost certainly determine the pace, if not the content, of the initiative.

The answer to the question, "What *can* a space program cost?" is then a complex one, involving at least the following variables:
1) Statement of the mission (requirements of the job)
2) Technologies available to the designer (inheritance)
3) Culture of the developing organization
4) Amount of money made available and the rate of availability
5) Schedule

Cost-estimating relationships, as described above, and as usually employed in the space industry, generally only deal with the first two of these variables.

Perceptions, Precedents, and Politics

Other factors must also be considered in the cost estimation process, including precedents, perceptions, and politics. In spacecraft development, cost can become the price. Price is determined by market conditions. If the congress and the administration place a high value on a space venture, its eventual cost will undoubtedly be higher than if the inverse were true.

Price is also determined by precedent. If it is generally known that a space venture in the past cost a certain amount, the perception will very likely have been created that all future ventures of its type will bear similar costs. This can be a totally false perception, obviously, as parameters such as inheritance also have a heavy influence.

And in the political arena, perception often *becomes* reality. Today, a perception exists that human exploration of the planet Mars is an undertaking so costly that no single nation can bear its expense. Despite the fact that this has been disproven analytically, the perception still exists.

The cost estimator must take great care that his or her estimating process is not overly influenced by perceptions and expectations of cost. In today's environment, being overly influenced by the perceived cost of a venture could be the very thing that prevents the venture from ever being undertaken. In a very real sense, the cost estimator has a life or death role in the evolution of a major program.

Cost Estimation

Now that the contextual variables have been discussed, a brief discussion of analytical estimation theories, as they pertain to space program cost estimation, will be presented.

Estimation Principles

Cost estimation is prediction. Prediction can be done in a number of ways. Physical laws and models are helpful to engineers in predicting masses, sizes, and performance of spacecraft. These, in turn, can sometimes, but not always, be useful in predicting costs.

As mentioned previously, much estimation is done by analogy. The more precise the analogy, the more accurate the prediction. Thus, to predict the cost of an automobile battery on one day, one only has to know what a similar battery cost the previous day to make an accurate prediction. Estimating uncertainties increase as the type of battery changes, the car model varies, or the time elapsed increases, giving rise to probable cost estimation variances.

Another effective tool of the estimator is statistics, although one observes much misuse of statistics in space program cost analysis. The primary reasons for misuse are twofold. First of all, in the spacecraft development industry, very few analogous data points are usually available. Analysts often fit curves to two or three data points, which is statistically very risky (see Figs. 1 and 2 for pictorial representations of the risks and distribution of risks associated with data samples of various sizes).

The second misuse stems from the omission of organizational culture as a variable (i.e., the assumption that cultures of the past will continue into the future, and that historical data will therefore be predictive of the future). In a situation where cultural change is required, culture must be introduced into the regression analyses as a variable. This is discussed further under "Influence of Organizational Culture on Program Cost."

Cost Models

A cost model is a logical accumulation of estimating methods. This accumulation is based on a work breakdown structure tailored to the project or program being costed. In a typical model, costs are estimated at the subsystem or system level of a program utilizing a set of CERs, often regression equations based on cost and technical histories from analogous

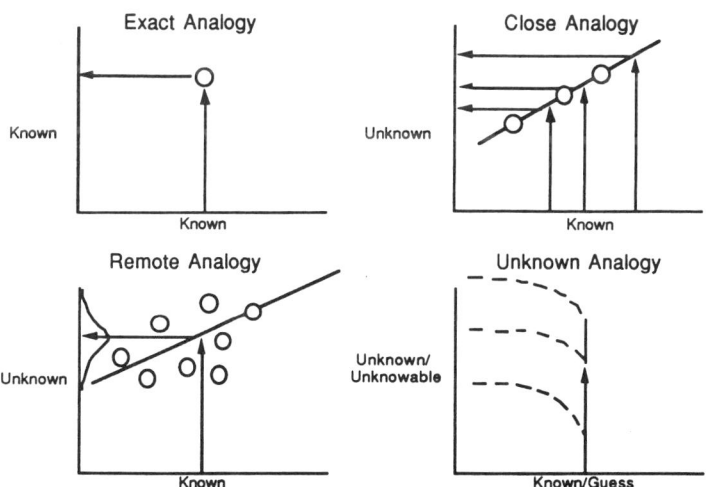

Fig. 1 **Role of analogy: applicability of statistics.**

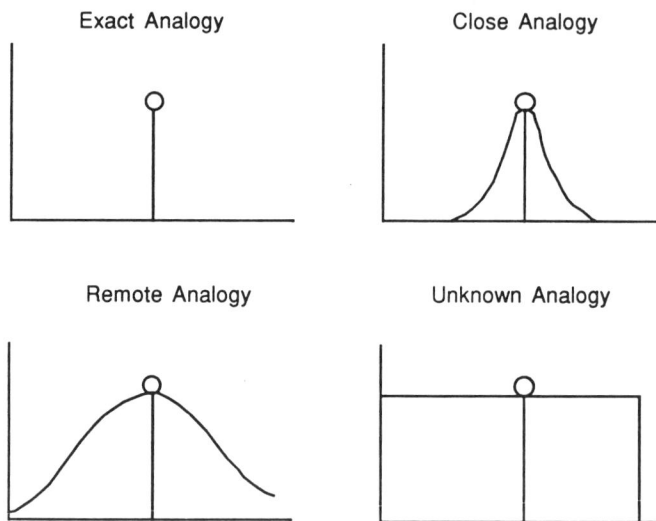

Fig. 2 Role of analogy: distribution of uncertainty.

systems or subsystems. The dependent variable (cost or manpower) is often predicted from performance- or size-related parameters, such as weight or thrust.

Systems or subsystems costs are combined algebraically to higher work breakdown levels, and integration costs are added as each level is combined to the next. Management and overhead costs are generally added as percentages of the sum of system and integration costs or estimated from ABC models. This is illustrated in Fig. 3.

Cost models are only as good as the logic used in their assembly. Often, models are created for a specific purpose (e.g., a launch vehicle cost model, or a manned spacecraft cost model). In concept, these models, based on historical analogs, are assumed to have predictive power for future systems. But there are at least four major sources of error in the estimation process.

Sources of Error

Anyone interested in the discussion of statistical errors, beyond the simple presentation of Figs. 1 and 2, should consult any of the many good texts on statistics. This discussion will be limited to errors of methods, rather than to errors of the mathematics of cost models.

First, most parametric cost models are based on logarithmic transformations of the data set (see the Appendix), which tend to obscure divergences in the historical analogs. Particular care must be taken to ensure that the transformation process results in predicting or forecasting expected costs, since only expected values are additive.

Second, historical data are often not homogeneous for a number of reasons, most of which are associated with the lack of rigor in data collection and the absence of standards, such as accounting standards, which

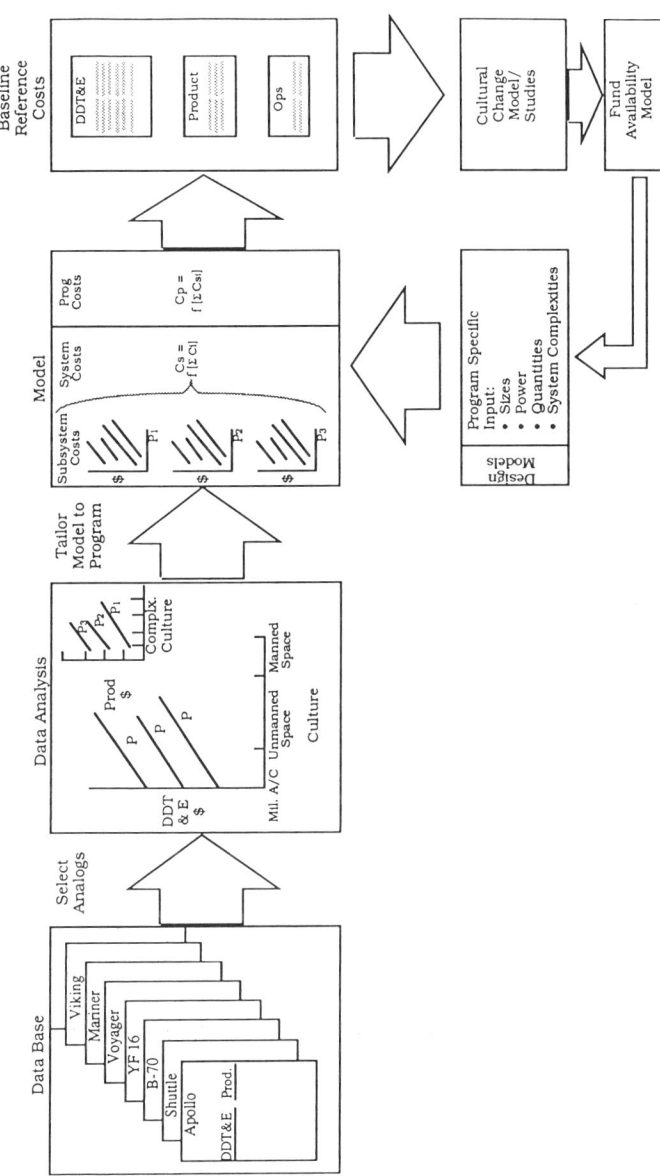

Fig. 3 Cost model description. P_i, parameter i; c_i, cost of subsystem i.

would make McDonnell Douglas data comparable to Rockwell data, and to Northrop data, and so forth.

Third, historical data are seldom purged of cost growth. The theory is that all programs will experience cost growth proportional to that in the data base of the model; this certainly is a weak assumption. In fact, in today's space program environment, the scarcity of resources will demand that cost growth be controlled much more closely than has been the norm in past spacecraft development programs. The means for controlling cost growth are discussed later.

The final source of errors, that associated with the culture of the developing organization, is so powerful that it will be dealt with separately.

Influence of Organizational Culture on Program Cost

Space programs have historically borne high development costs as compared to similar programs in other industries. At first, this was assumed to be a natural consequence of the high risks, both technological and engineering, of space flight.

As space budgets began to decline at the end of the Apollo lunar landing program, the understanding of the causes of high cost became a priority. The first thing discovered was that space technologies, while highly advanced, were not unique. For example, at approximately the same time NASA was developing the rather conventional aluminum and steel structures of the Apollo lunar vehicles, Lockheed was developing the all-titanium SR-71 aircraft (and doing it in 22 months).

Components of spacecraft environmental control and life support systems were discovered to be outgrowths of aircraft technology in many cases. Where new components were developed, very often the technologies were not highly advanced. Of course there were a great many exceptions, but when added together, the exceptions did not dominate the development cost picture.

Gone, then, was the most simple explanation for high costs. At this point, a study was performed by RCA Price Systems Division (for NASA) to determine the real causes. RCA had the advantage of having developed analogous devices for a wide variety of customers, from commercial aircraft companies, to the spacecraft industry, to the consumer market.

It was hypothesized that perhaps the existing NASA requirements for parts traceability (NASA has a standard requirement that every component of every spacecraft flown must be traceable to the basic materials from which it was produced) was one reason; another guess was that high safety, quality assurance, and reliability requirements added to the costs; and another guess was made that the heavy documentation requirements placed on all contractors contributed to NASA costs being higher than those in comparable industries, if indeed there was a comparable industry.

The results did find correlations with all of the hypothesized reasons for higher cost. However, a large residual error remained in the regression equations. Only after a variable was introduced to identify the developing agent did high correlations emerge.

The final report[1] makes the following statement: "Organizational work habits do not readily change completely as a result of imposing a different specification level. Residual costs were attributable only to the organizational manner of doing business [culture]."

These results have been recently verified independently in the Advanced Mission Cost Model research effort which is ongoing at the Johnson Space Center.[2]

The latter research has produced quantified correlations between development culture and program cost (Fig. 4). This figure is presented not to suggest that spacecraft programs can be produced with the same management cultures as aircraft programs (although it should be asked, "Why not?"), but to illustrate the large "leverage" available by slight variations in the culture dimension.

Based on an extensive interview process,[3] directions have been identified for evolution of space program management which can be expected to produce much lower costs. These involve giving more responsibility to the private sector, incentivizing product performance, and utilizing the power of the competition of the marketplace to minimize costs.

CERs and Organizational Culture

Research has revealed the pervasive nature of development culture on the costs of products, not only in the spacecraft industry, but in many sectors of the economy. But simply knowing a cause and effect, when it comes to dealing with human cultures, is not sufficient. The changing of a culture is a task as demanding as designing a new spacecraft, if not more so.

At the beginning of the Space Station Freedom Program, NASA was aware of the need for cultural change and what needed to be done to

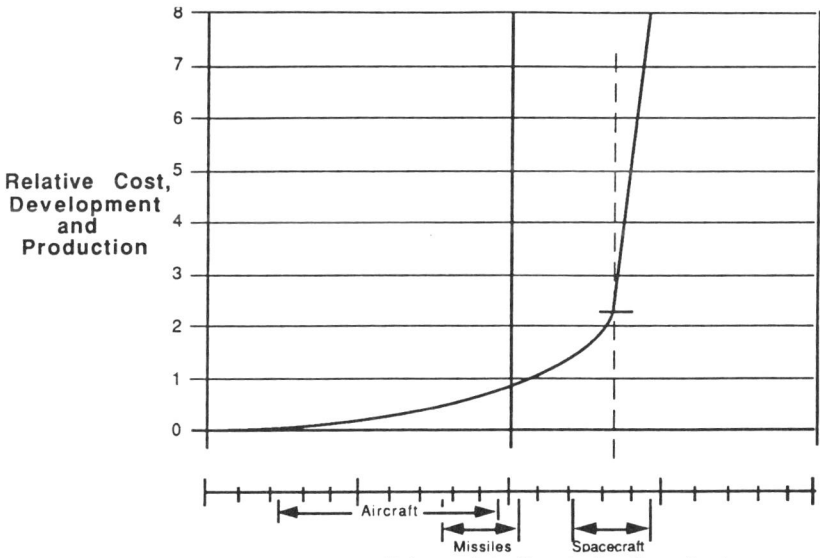

Fig. 4 Effect of development organization type on program development and production cost.[2]

produce low costs. Members of the initial task forces established to begin program development were aware of the lessons learned from previous programs and subscribed to the need for cultural change. However, other influences were at work.

First of all was the inertia of the culture itself. Behavioral science has dealt with this problem for decades, without producing prescriptive remedies. It is known that cultures change only in response to their environments, and that threatening environments produce more rapid change rates than unthreatened environments. Rapid change, in the absence of major threat, will not happen.

Xerox Corporation received the Malcolm Baldridge award for the successful cultural changes in that corporation. David Kearnes, ex-Xerox executive, tells the story of how, in the process of "benchmarking" a competing Japanese copier, it was discovered that Xerox could not even manufacture the copier for the selling price of the competitor. This very real threat provided all the stimulus required to force that successful change experience.

But in the Space Station time frame, there was no such perceived threat to NASA. Like many governmental organizations, NASA is a monopsony, i.e., a sole customer able to dictate the behavior patterns of its suppliers, which all but rules out unilateral cultural change on the part of the private sector suppliers. And since monopsonies, by definition, have no competition, threats are harder to recognize.

Instead, the influences that were present at the beginning of the Space Station Freedom program were more institutional, with each NASA installation recognizing and responding to the threat that it perceived, i.e., the declining budget for that installation (center). The natural response to that threat became one of each installation struggling to maintain the maximum share of the greatly reduced NASA budget. The result was to become the competition of the various field installations for parts of the relatively small program.

As a consequence, the program was unnaturally divided into four "work packages," and program costs grew. Budgets negotiated with the congress and the OMB have never been fully commensurate with the projected runout costs of the program. This program has now been forced by the congress to undergo major "restructuring" designed to deal with the shortage of funds.

Another lesson learned in the interview process was never to begin a program until the requirements for funds are matched by the available resources. In the space program culture, the uncertainties involved in initiating a program have sometimes obscured the importance of matching program cost requirements with available resources. Space program planning must therefore be commensurate with *realistic* predictions of program budgets.

A final lesson to be learned from studying space program cultures is that cultures themselves resist change. The individual who violates a norm is identified by the group quickly as being away from the norm, or "ab-normal"; the individual either conforms to the norm eventually or is purged by the group, whether or not the norm is rational. And norms can be destructive to the enterprise of the group.

Cultures are changed by changing their norms, i.e., setting new standards for the behavior of the group. It is therefore essential that any attempt to lower space program development costs by cultural change be accompanied by willing acceptance by the group of each change. The alternative would be the failure of the change and the eventual purging of those involved in making the change.

Meaning to Users of CERs

What does the understanding of the influence of culture have to do with the user of CERs?—almost everything. The examples cited above can be viewed as failures of the planners of the programs to appreciate fully the cultural effects on the program. Some of these are:
1) The cultural response to lower budgets
2) The false assumption of cultural changes to lower program development costs (changes not to be realized)
3) The erroneous use of CERs based in one culture to estimate costs in another.

CERs are generally derived from data taken from a very slowly evolving culture. Either the effects of the slow evolution are too small to be measured or are often obscured by greater influences during the regression process.

Therefore, the use of CERs based on past aerospace program experiences can be safe in many instances. But what of instances where cultural changes are either needed or required to enable a given program to be performed within a given budget? Single-culture CERs can preclude dealing with the very variable that should be changed in order to sell or conduct a successful program.

Two Paradigms for Parametric Cost Estimation

Today's state of the art in cost estimation techniques for space programs has progressed down two lines of development. The first of these is what have been termed the *Rand paradigm*, statistical cost estimation methods that are generally single-culture and single-system in nature.

During the 1950s and 1960s, the RAND Corporation of Santa Monica, California, created the first economics department dedicated to the prediction of costs of future aerospace and military vehicles.[4] David Novick staffed the department with people well versed in the use of statistical methods, and proceeded to build an extensive data bank from aerospace programs dating at least to World War II.[5]

Methods developed by this organization were expectably from the paradigm of the statistician. Methods developed often dealt with a limited range of technologies and vehicles (e.g., aircraft engines). And because of the slowly changing nature of the culture of the industry, little attention was paid to the time-related effects of cultural change. RAND did, however, begin to notice and to document time-related trends of technology.

Typically, CERs developed in this paradigm contained one or two predictive independent "parameters" (giving rise to the name of "parametric" methods), typically weight or mass, engine thrust, payload (or some combination of these parameters), and a measure of complexity to predict the dependent parameter, usually cost or manpower required to develop or build a given system. Dependent parameters were related to independent parameters through regression equations, often employing logarithmic transformations of the data. (For a more complete explanation of the methodology, see Ref. 3 and the Appendix.)

As aerospace products began to acquire systems that had few or no technological precedents, and for which there were little data to develop single-culture CERs, a quest for better methods began. This quest was led by Frank Freiman, then of RCA Corporation. Freiman developed the hypothesis that similar devices that were developed and operated in varying cultures could provide more universally useful data. Thus, avionics developed for the commercial aircraft industry could provide data useful for electronics components developed for spacecraft, or for ships, or for submarines.

Despite the differences in application of the various families of components, Freiman found that, by the introduction of a set of "cultural" variables, data from one culture could indeed be utilized to make accurate cost estimates in another, thus enormously expanding the data base available to make statistical cost predictions. This breakthrough and others enabled Freiman and RCA to develop and market an entire new product line of cost estimation methods.[6] Methods of this type are referred to as *PRICE paradigm* methods.

Analysts throughout the world continue to develop RAND paradigm methods and to use them with great success. And RCA has a world wide customer base utilizing their successful family of models.

Choosing the Correct Method

Which of these two sets of methods is, then, best for a given application? There is not a clear answer to the question. It has often been found that use of "secondhand" RAND paradigm types of estimation methods have failed to produce good results. But the failures have often been attributable more to the lack of skill of the estimator than to the methods. In general, single-culture, single-system estimation techniques are useful in the environment where the next generation of development can be reasonably expected to duplicate the culture of the prior one.

But what of an undertaking where a cultural change is necessary to make the endeavor even possible? (The current NASA space exploration initiative falls into this category.) In this case, technical parameters become less dominant than cultural parameters. It has been found, by RAND and others, that development costs scale (roughly) as the square foot of independent parameters such as mass (exponential equations with negative second derivatives). Culture, on the other hand, has a positive exponential influence (positive second derivatives) on cost, as is illustrated in Fig. 4. Therefore, in situations where even small cultural changes are anticipated, a PRICE paradigm method should be employed to quantify the effects of the cultural change and to assist in design of the required changes to the culture.

Keys to Making the Estimate Come True

Program cost estimates are probabilistic variables. There is not a deterministic model that will predict costs with absolute certainty. Thus, cost models only predict costs *subject to the exact reproduction of the assumptions made* in the estimation process.

If assumptions are made as the numbers of development test articles, mass properties, technological requirements, and development culture variables, all of these assumptions must be reproduced in the execution of the program for the estimates to have any chance of being realized. If NASA were to predicate the costs of American planetary exploration on the introduction of a certain amount of cultural change (e.g., more delegation of development risk to the private sector), it must be willing to satisfy the assumptions made.

Conclusions

The usual processes utilized by the aerospace industry will not be sufficient to estimate the costs of future space programs. For example, the Space Exploration Initiative is too massive in scale to submit to traditional methods. The usual cost prediction parameters of size, technological difficulty, and scale have been superseded by interactions between the mission and the political environment, which will determine the resources available (which will become the costs), the timing of events, and the development culture employed.

Once these parameters and their cost influences have been defined, then the more traditional estimation techniques can be called into play. In economic terms, the question of how much large space programs will cost is a supply-side issue. Supply will be determined by the benefits promised, as well as by the demands of competing ventures on the national scene. Costs, particularly of large space programs, are then not only a function of the parameters of the hardware to be built, but are a function of how much the nation is willing to spend on the venture.

Cost-estimating relationships, in the right hands, are powerful tools. However, improperly used, they can be the tools of destruction of a noble venture.

This chapter provides an overview, not only of the derivation and use of CERs, but of their limitations. Cost models are a vital tool of the program planner.

But cost models must be tailored to each situation. General-purpose models are available, but depend on the exact reproduction of the assumptions of the model to result in a successful outcome.

In the modern space industry, where cultural change will be the norm if the nation is to remain competitive with foreign competition, CERs founded in a single culture can have the effect of becoming self-fulfilling, i.e., in the perpetuation of a noncompetitive culture. It is strongly recommended that cost estimators become interactive with their cultural environment to provide the tools essential to guide the nation to the improvement of development cultures. This can become a major influence in maintaining the United States space industry in its current position of world leadership.

References

[1]RCA Price Systems Division, "Equipment Specification Cost Effect Study" (Phase 2, final rept.), RCA, Cherry Hill, NJ, Nov. 30, 1971.

[2]Cyr, K. J. "The Role of Cost Analysis in Manned Spacecraft Development, SAE International, (SAE Paper 901863), Warrendale, PA, Oct. 1990.

[3]Mandell, H. C., *Assessment of Space Shuttle Program Cost Estimating Methods.* The Univ. of Colorado, Denver, 1983.

[4]Novick, D. "Cost Estimating, Cost Analysis, and Systems Analysis: A Historical Perspective," *National Estimator*, Vol. 1, No. 2, p. 8.

[5]Slavinski, S. C., *The Rand Cost Analysis Department Data Bank*, The Rand Corp., Santa Monica, CA: Sept. 1964.

[6]Kaufmann, G. A., *The PRICE Life Cycle Cost Model.* RCA Corp., Cherry Hill, NJ, March 1978.

Appendix: CERs and Models

Joel S. Greenberg
Princeton Synergetics, Inc., Princeton, New Jersey 08540

Development of CERs

Mathematical Development*

The basic concept underlying the development of CERs is that cost is a function of one or more physical attributes of the item being costed and that this relationship is similar to or the same as that of other similar items. For example, the cost of a new spacecraft's attitude control subsystem is related to its mass in the same way that previously developed spacecraft attitude control subsystem costs are related to their mass. This is illustrated conceptually in Fig. A1. (The methods described here are those referred to in the text as RAND paradigm, i.e., methods based on relationships between cost and physical attributes. Later, PRICE paradigm methods introduced development culture as an explicit variable in the estimation process. The principles of CER development are, however, generically the same otherwise, and the discussions of regression analysis processes are fully applicable to both paradigms.)

Referring to Fig. A1, the points indicated by + have the values of X_K and Y_K where K represents the Kth item (e.g., a communications/telemetry, tracking, and command subsystem) having a physical attribute measure (e.g., mass) of X_K and associated cost of Y_K. It is assumed that an equation of the general form $Y = A + BX$ or $Y = AX^B$ can reasonably represent the functional relationship between cost and the identified phys-

---

*This section of the Appendix is based primarily on Ref. 1.

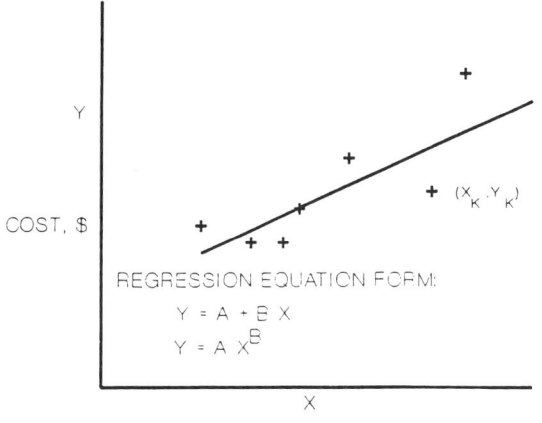

Fig. A1 Cost-estimating relationship concept.

ical attribute. A regression analysis is performed to establish the equation coefficients (e.g., A and B) in terms of the specific historical data points. To illustrate, the ordinary least squares regression "best" fits a straight line $Y = A + BX$ to a set of ordered pairs (X_K, Y_K) of data points in two-dimensional space. Procedures based on the same philosophical and mathematical principles extend the ordinary least squares regression to the case of curved lines, such as logarithmic, and to the multidimensional case. Because this Appendix presents a statement of principles, rather than a guide to mathematical computations, only the straight-line case is discussed herein.

The N data points are labeled $(X_1, Y_1), (X_2, Y_2), \ldots, (X_N, Y_N)$, where Y_K is the actual cost (e.g., nonrecurring or unit recurring) associated with an existing program that has mass X_K. Whereas the line $Y = A + BX$ is the general predictor of cost of the program in question, the specified cost estimate would have been $A + BX_K$ rather than the actual cost Y_K. The equation $Y = A + BX$ is, therefore, called a CER.

The error in the estimate of the cost of any program in question is the difference $D_K = Y_K - (A + BX_K) = Y_K - A - BX_K$ between the actual cost Y_K and the cost estimate $A + BX_K$. The principle of least squares asserts that the constants A and B, which determine the CER, should be such that the sum of squared errors

$$\sum_{K=1}^{N} D_K^2 = \sum_{K=1}^{N} (Y_K - A - BX_K)^2$$

is as small as possible. By considering this as a two-variable minimization problem, taking partial derivatives with respect to A and B, setting the partial derivatives equal to 0, and solving the resulting simultaneous equations for the two unknowns A and B, ordinary least squares determines explicit expressions for the slope and intercept of the CER:

$$B = \left[N \sum_{K=1}^{N} X_K Y_K - \left(\sum_{K=1}^{N} X_K\right)\left(\sum_{K=1}^{N} Y_K\right)\right] / \left[N \sum_{K=1}^{N} X_K^2 - \left(\sum_{K=1}^{N} X_K\right)^2\right]$$

$$A = \left[\left(\sum_{K=1}^{N} Y_K\right)\left(\sum_{K=1}^{N} X_K^2\right) - \left(\sum_{K=1}^{N} X_K\right)\left(\sum_{K=1}^{N} X_K Y_K\right)\right] / \left[N \sum_{K=1}^{N} X_K^2 - \left(\sum_{K=1}^{N} X_K\right)^2\right]$$

The first step in the development of CERs is to identify potential cost drivers. This is done by generating a hypothesis relating cost to those underlying parameters believed to be cost drivers. The hypothesis requires a sound understanding of the engineering principles that drive cost and should identify parameters, as well as the equation form that relates those parameters to cost. The strength of any CER emerging from this process derives as much from the soundness of the engineering hypothesis as from the favorable statistics testing the regression fit.

The second step in developing CERs is proper model specification. CER specification is a process that involves selecting from many possible equations the one equation that best represents the estimating relationship. Normally, only equations that can be expressed as linear models, either in terms of the original variables or transformations of the original variables, are considered. Forms requiring nonlinear regression analysis are normally excluded.

The linear relationship is discussed above. A frequent alternative is the exponential relationship of the form $Y = AX^B$. In order to estimate the parameters of this equation, it is necessary to transform both Y and X by expressing them in logarithmic form. The equation thus becomes

$$ln(Y) = ln(A) + B\, ln(X)$$

where ln represents the natural logarithm.

An adjustment factor is necessary for those CER functions that are based on natural logarithmic transformations. This adjustment is necessary because the exponential curves are modeled as linear curves in log-log space and then transformed back to Cartesian space. The mean and median are one and the same for the linear CER in log-log space. However, when the curve is transformed back into Cartesian space the mean and median differ, with the CER predicting the median instead of the mean. A multiplicative correction factor is used to adjust each CER so that the mean or expected cost is predicted. This is necessary when adding costs, because mathematically only means are additive.

Several typical CERs are illustrated in Fig. A2 with other examples described in following sections. Among cost models, space cost models are rather unique for at least the following reasons[2]:

1) Space systems are more prototype than production; it is difficult to distinguish between development and fabrication costs.
2) It is difficult to distinguish between nonrecurring and recurring costs.

COST-ESTIMATING RELATIONSHIPS FOR SPACE PROGRAMS

INTEGRATION & ASSEMBLY COST:

$$Y = 0.0058 * (X1^{1.54}) * (3.60^{X2}) * [1.36]$$

WHERE X1 = SPACE VEHICLE DRY WEIGHT (LBS)
X2 = 1 FOR COMMUNICATION MISSION
X2 = 0 FOR NON-COMMUNICATION MISSION

COMM./TELEMETRY, TRACKING & COMMAND

$$Y = 2798.12 + (0.066 * X1)$$

WHERE X1 = SUBSYSTEM WEIGHT * POWER REQUIRED (LBS*WATTS)

Fig. A2 Several typical CERs for recurring cost.[1]

3) Small quantities mean fewer data points for statistical regression.
4) Schedule is often a cost driver.

Data Requirements

The data bases used for establishing CERs are normally comprised of cost, technical, and programmatic data from a wide range of time periods and contractors. However, the data cannot be used for parametric analysis without being normalized (i.e., comparisons must be apples to apples). In particular, the following specific questions must be answered[2]:
1) Are the costs comparable in currency, year, and elements of cost?
2) Are the products grouped into homogeneous categories?
3) Are the programs/products discriminated as to culture (specification level)?
4) Are the programs discriminated as to state of the art?
5) Are consistent sizing and technical parameters used?
6) Are consistent nonrecurring/recurring cost splits observed?

The objective is to produce a set of cost data that is consistent for purposes of comparison. This implies that the costs need to be based on the same standard set of elements; for example, the costs should include direct labor, overhead, material, subcontracts, burdens, general and administrative (G&A) costs, and allocated prime costs, but not fee. All cost data should include the associated year so that they can be translated (using escalation tables) from then-year to constant-year base dollars. (This is not required when all data are in man-years rather than dollars, as is discussed in Ref. 4.) For foreign currencies, the conversion tables are necessary to generate U.S. dollar figures at convenient economic conditions.

In parametric analysis, the sizing variables are significant; for example, it is common to utilize cost per pound of hardware or cost per line of software source code. Both the size and cost dimensions must be consistent. For example, Johnson Space Center (JSC) cost model uses only the dry weight of systems, which removes the extraneous influence of propellant and fluids costs, which have a low bulk price.[3] Solid rocket motor sizes are, however, reported as fully loaded weights because solid propellants do have significant costs.

The variables found to be important indicators of cost will effect the manner and type of collected data. For example, the JSC model develops a variable electronic composition factor because model building has shown this to be a cost driver.[3] This factor is defined as weight of subsystems containing electronics divided by total dry weight.

In parametric analysis it is critically important to analyze homogeneous products. This homogeneity must exist across product groupings and within individual products. An approach is to classify missions by application and objectives. Homogeneity within products may be maintained through the use of a work breakdown structure (WBS) which controls the end-item dimension of cost. A typical WBS is shown in Fig. A3.

Past practice in parametric analysis has been to model separately the nonrecurring and the recurring costs. A preferred method for nonrecurring/recurring cost separation is detailed analysis at lowest levels of cost data collection, preferably by the prime contractor. This provides the best insight into design, manufacturing, and support cost transitions.

Lacking detailed assessments, two alternative methods may be used for assigning nonrecurring and recurring costs. The first is referred to as time phasing, wherein the stream of costs is referenced to some agreed-on schedule milestone and all costs before this milestone are considered nonrecurring and those afterward are considered recurring. The second performs a pro rata of total acquisition costs using an equivalent units calculation. In this method every prototype article is assigned an equivalent quantity of flight units, and an equivalent is also estimated for design and development activity (taking into account difficulty, inheritance, etc.). The ratio of DDT&E to production units then determines the ratio of nonrecurring to recurring costs.

In addition to the above, some models attempt to normalize the state of development (that is, differences in the development effort that ranges from simple follow-on buys to complex, multipath developments with each development history invoking different nonrecurring costs) as well as normalizing the program culture (that is, program practices or specification levels).

This discussion illustrates the level of complexity of establishing appropriate data for use in CER development. Extreme care must be taken to avoid inconsistent data that would lead to significant cost estimation errors.

A number of cost models are discussed briefly in the following section. It is worth noting here that none of these models include data that relate to performance attribute changes during the different phases of program development, yet the developed CERs (based on programs that have been completed) are frequently utilized to estimate the cost of future programs during the early stages of their development. Thus, CERs developed for apples are utilized to estimate the cost of oranges.

Typical Cost-Estimating Models

Many cost estimation models have been developed and are in use. A number of these are summarized in the following paragraphs. These models have been selected to illustrate the differences in their general characteristics as well as their areas of applications.

```
01  Space Segment
    01  Spacecraft Platform
        01  Bus Hardware
            01  Structures/Mechanisms/Thermal
                01  Structures
                02  Mechanical
                03  Thermal
            02  Electrical Power
                01  Power Source
                02  Power Conditioning
                03  Power Distribution & Control
            03  Attitude Control
                01  Attitude Determination
                02  Attitude Control
            04  Propulsion
                01  Main Propulsion (Less Engines)
                02  Secondary Propulsion
                03  Reaction Control Propulsion
                04  Common
                05  Instrumentation
            05  Communications & Data Handling
                01  Tracking
                02  Telemetry
                03  Command
                04  Data Management
                05  Instrumentation
            06  ECLSS/Crew Accommodations (If Manned)
            07  Booster Adapter
        02  Non-Bus Hardware
        10  Non-Bus Software
    02  Common Services
        01  Management & Support
            01  Program Management
            02  Financial/Schedule Control
            03  Configuration Management
            04  Data Management
            05  Manufacturing Plans
        02  Systems Engineering
            01  Requirements Definition & Allocation
            02  System Analysis
            03  Reliability, Maintainability, Quality
            04  Other "Llities"
            05  Test & Ops Planning
        03  Integration & Test
            01  Integration Management Plans
            02  Mission Support
            03  On-Orbit Servicing Operations
        04  GSE
        05  Support
            01  Other Direct Cost
        06  Facilities
    03  Operations Support
        01  Launch Support
        02  Mission Support
        03  On-Orbit Servicing Operations
    04  Segment-Level Integration & Test
        01  Integration Hardware
            01  Fairing
            02  ASE
51  Payload
```

Fig. A3 Typical work breakdown structure.[3]

Unmanned Space Vehicle Cost Model

The United States Air Force Space Division's Unmanned Space Vehicle Cost Model (USCM6)[1] is a parametric estimating tool that provides CERs for estimating unmanned, Earth-oribiting space vehicle costs. Each CER was developed through rigorous statistical analysis of hypothesized cost drivers. This was done by generating hypotheses relating cost to those underlying parameters that were thought to be cost drivers. Each tested hypothesis was based on a sound understanding of the engineering principles that might drive cost. Selected CERs had to demonstrate favorable statistics and, from a behavior standpoint, be consistent with engineering expectations. Although a CER obtained solely by exhaustive testing of potential cost drivers might have exhibited higher statistical explanatory power, if it did not contain true causal variables from an engineering standpoint, it was not selected.

USCM6 contains CERs at the subsystem and the component level for both nonrecurring and recurring costs. The scope of the model is indicated by the following:

1) The model addresses only Earth-orbiting space vehicles, with two exceptions—the structure and thermal CER data bases include data points for large structures such as the Apollo Adapter, Apollo Module, Lunar Module Descent, Shuttle, Spacelab, ERBS, STOTA, UARS, COBE, GRO, and Space Telescope. These data were collected during an SD/ACC-sponsored research effort which examined potential estimating methodologies for large space structures.[5]

2) The model is an approximation of the real world based on mathematical relationships derived from analyses of historical cost data. Implicit in the construction of any model derived from historical cost data is the assumption that historical costs will properly reflect current and future costs.

3) All costs included in the model are end-of-program actual costs or estimates of mature programs (with at least one launch).

4) The model's emphasis is on spacecraft (platform/bus) hardware costs. Additionally, it addresses communications payloads, but does not include other types of mission hardware (e.g., observational sensors).

5) Launch vehicles, stage vehicles, and their associated ground equipment are not included within the scope of this model.

6) CERs are based upon burdened costs (direct plus indirect) with G&A costs included. In other words, this model consists of total cost through G&A application CERs.

7) A 95% cumulative average learning curve was used to derive data base first unit costs.

8) Except for program level CERs, the recurring CERs represent first unit costs and require appropriate consideration for production learning. All program-level recurring CERs estimate total recurring costs.

9) In using this model, cost estimates from the space vehicle subsystem/component CERs, the integration and assembly CER, the program level CERs, the aerospace ground equipment (AGE) CERs, and launch and orbital operations (LOOS) CERs, are summed to determine a space vehicle

program cost subtotal. This cost estimate is expressed in terms of U.S. government fiscal year 1986 dollars, and must have appropriate inflation costs added if costs are to be expressed as future expenditures.

10) Target and incentive fee costs are not included in the model. These costs should be added to the model output based on the contractual scheme anticipated, or in effect, for the system being estimated.

11) The model yields a "starting point estimate" which represents the "average" cost for a program with average problems, average technology, average schedule, average engineering changes, and so forth.

12) The model does not include costs for technology development and preliminary design studies. Furthermore, the USCM6 costs must be adjusted if there is dual source competition during the development or production phase of the program.

The USCM6 work breakdown structure is illustrated in Fig. A4 and the scope of the data base is summarized in Fig. A5. Subsystem and component CERs are provided in Ref. 1.

TRANSCOST Model

The TRANSCOST Model[4] was initially developed in 1980 by Messerschmitt-Bolkow-Blohm GmbH and has been updated several times. The model is based on a comprehensive launch vehicle cost data base established over more than 30 years. TRANSCOST is limited to propulsion—related space systems and all types of engines. It develops launch vehicle development, production, and operations cost estimates. All costs are in man-years of effort and not direct units of currency (i.e., dollars).

The TRANSCOST Model is organized in three submodels: 1) development cost or nonrecurring cost; 2) vehicle recurring cost (flight units fabrication, assembly and verification costs); and 3) flight operations cost (direct and indirect operations cost). These submodels can be used separately or combined, depending on the application. All three result in life-cycle cost.

The TRANSCOST Model has been conceived for and is considered to be especially useful as a design tool for the economic optimization of future launch vehicles. Cost estimation relationships exist for liquid-propellant engines, propulsion modules, solid rocket motors/boosters, expendable ballistic vehicles/stages, winged orbital vehicles, and advanced high-speed aircraft (winged first-stage vehicles). Typical developed CERs are summarized in Figs. A6–A9. The operating costs indicated in Figs. A8 and A9 take into account refurbishment/maintenance cost, prelaunch ground operations cost, launch and mission operations, propellant cost, transportation/recovery cost, vehicle cost amortization, and indirect operations cost.

Advanced Missions Cost Model

The Advanced Missions Cost Model (AMCM)[3] is being developed by the Exploration Programs Office of NASA's JSC. The JSC activity is aimed at updating the assumptions and art of costing major initiatives that have a characteristic of being at the concept stage,

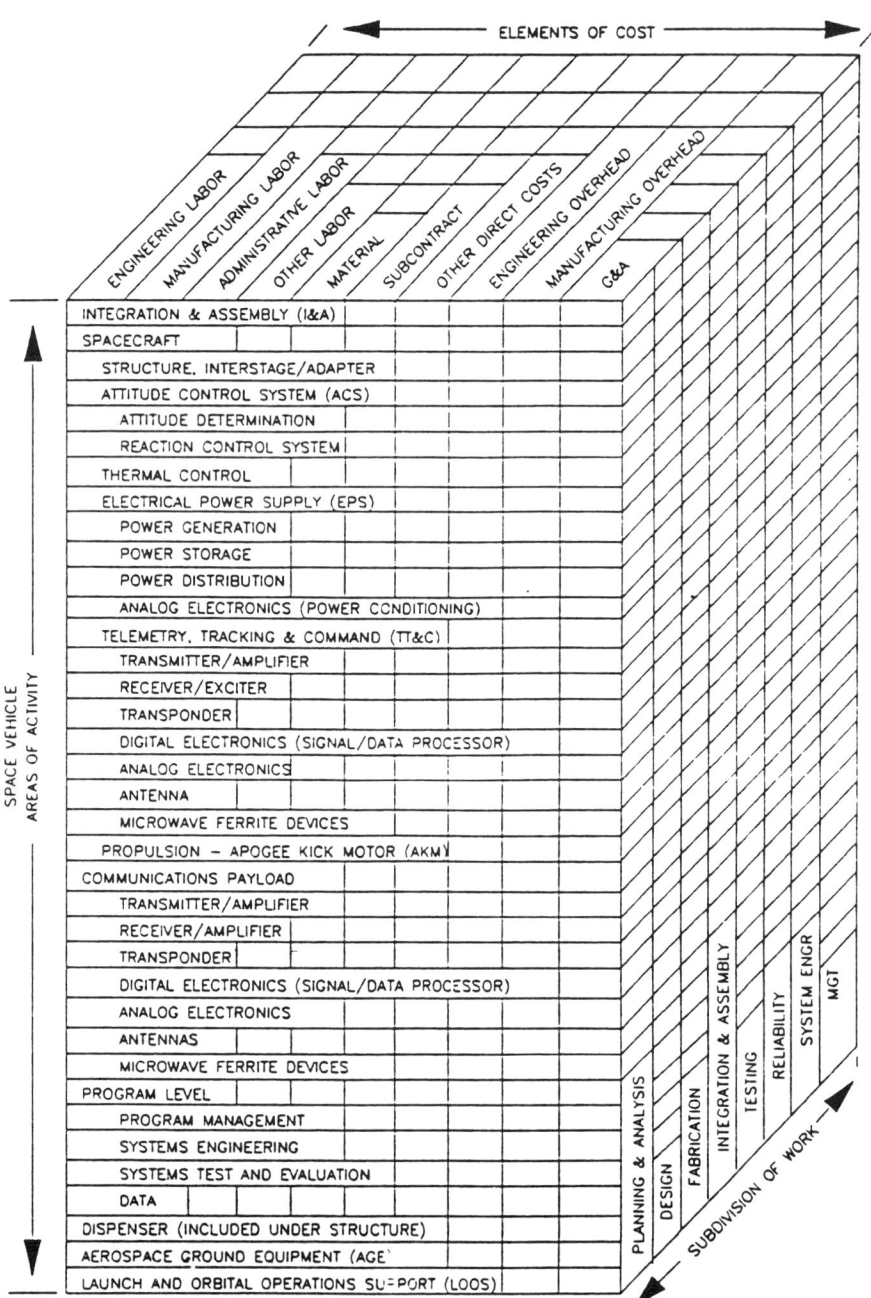

Fig. A4 Work breakdown structure—USCM6.[1]

COST-ESTIMATING RELATIONSHIPS FOR SPACE PROGRAMS

	'69 FIRST EDITION	'71 PHASE I UPDATE	'72 PHASE II UPDATE	'73 SECOND EDITION	'75 THIRD EDITION	'77 INTERIM REPORT	'78 FOURTH EDITION	'81 FIFTH EDITION	'88 SIXTH EDITION
(A)	VELA VASP IDCSP IDCSP/A TACSAT	VELA VASP IDCSP IDCSP/A TACSAT DSP I	VELA VASP IDCSP IDCSP/A TACSAT DSP I DSCS II	VELA VASP IDCSP IDCSP/A TACSAT DSP I DSCS II	VLLA VASP IDCSP IDCSP/A TACSAT DSP I DSCS II DSP II DMSP (5D-1) P72-2 S3	VELA VASP IDCSP IDCSP/A TACSAT DSP I DSCS II DSP II DMSP (5D-1) P72-2 S3 NATO III	VELA VASP IDCSP IDCSP/A TACSAT DSP I DSCS II DSP II P72-2 S3 NATO III	VELA VASP IDCSP IDCSP/A TACSAT DSP I DSCS II DSP II DMSP (5D-1) P72-2 S3 NATO III P78-1 P78-2 GPS-1 FLTSAT 1-5	IDCSP S3 NATO III P78-2 DMSP (5D-1) GPS 9-11 FLTSAT 6-8 DSCS III A/B
(B)	OGO SYNCOM LUNAR ORBITER ATS A (M/G) ATS B,C (S/S) ATS D,E (S/S)	OGO SYNCOM LUNAR ORBITER ATS A (M/G) ATS B,C (S/S) ATS D,E (S/S) OAO TIROS M	OGO SYNCOM LUNAR ORBITER ATS A (M/G) ATS B,C (S/S) ATS D,E (S/S) OAO TIROS M	OGO SYNCOM LUNAR ORBITER ATS A (M/G) ATS B,C (S/S) ATS D,E (S/G) OAO TIROS M NIMBUS A,B,C NIMBUS E,F ERTS A,B	SYNCOM LUNAR ORBITER ATS A (M/G) ATS B,C (S/S) ATS D,E (S/S) TIROS M NIMBUS E,F ERTS A,B SMS OSO I ATS F AE	SYNCOM LUNAR ORBITER ATS A (M/G) ATS B,C (S/S) ATS D,E (S/G) TIROS M NIMBUS E,F ERTS A,B SMS OSO I ATS F AE	SYNCOM LUNAR ORBITER ATS A (M/G) ATS B,C (S/S) ATS D,E (S/S) TIROS M NIMBUS E,F ERTS A,B SMS OSO I ATS F AE HEAO	SYNCOM LUNAR ORBITER ATS A (M/G) ATS B,C (S/S) ATS D,E (S/G) TIROS M NIMBUS E,F ERTS A,B SMS OSO I ATS F AE HEAO	ATS-AE ATS-F OSO-I AE HEAO TDRSS
(C)	INTELSAT III	INTELSAT III	INTELSAT III	INTELSAT III	INTELSAT III INTELSAT IV	INTELSAT III INTELSAT IV	INTELSAT III INTELSAT IV MARISAT	INTELSAT III INTELSAT IV MARISAT	INTELSAT IV INTELSAT V-A MARISAT

A = MILITARY B = NASA C = COMMERCIAL

Fig. A5 Data base history—USCM6.[1]

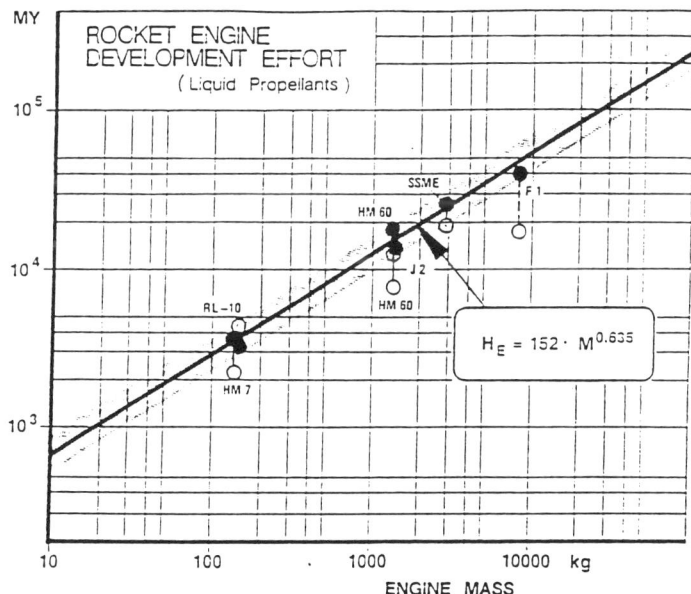

Fig. A6 CER for rocket engine development effort.[4]

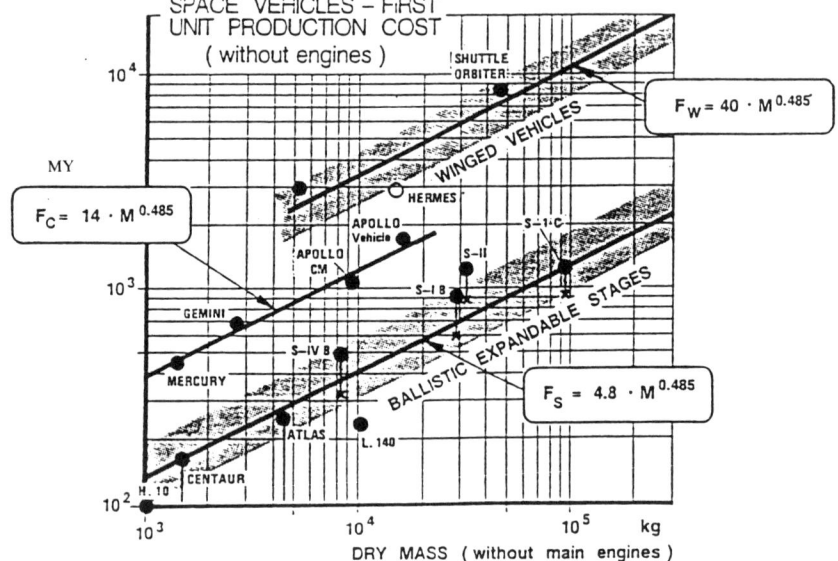

Fig. A7 CER for space vehicle first unit production cost.[4]

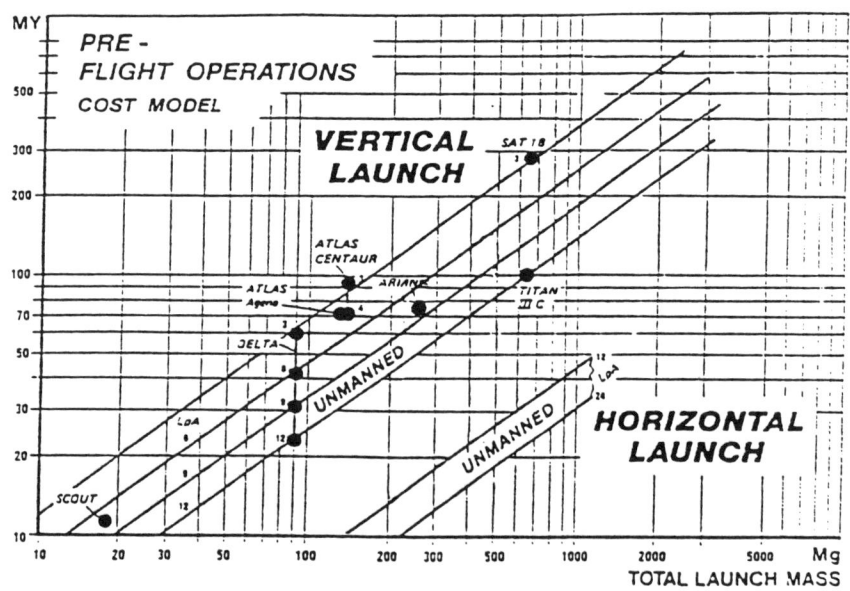

Fig. A8 CER for preflight operations.[4]

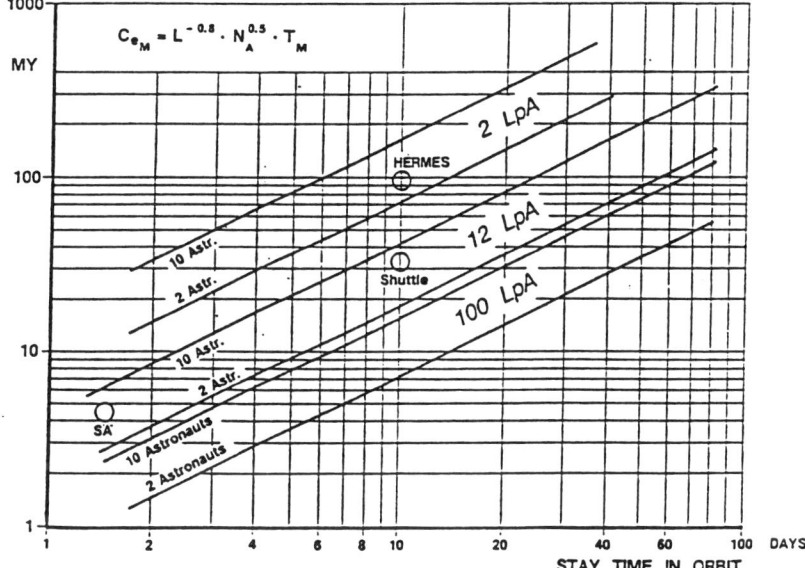

Fig. A9 CER for manned missions operations cost.[4]

with implementation far into the future, where experience and techniques are very different. "Since the new initiatives are not expected to start until late in this century, the cost models must predict much further into the future than is normal. Some of the initiatives may also require technology that has not been developed yet; hence, there are no historical analogies on which to base estimates. Finally, the programmatic factors, rather than hardware characteristics will probably be the major cost drivers." (And yet, the AMCM assumes "doability" and does not explicitly allow for uncertainty in schedules, resulting costs, and level of achieved performance.)

The AMCM expands the number of data points in the data base by developing a "culture" variable that is effective in combining different technologies in the same data base. "For future space developments, Culture may be the most significant variable the cost analyst has to select. Weight and quantities will usually be given, but the particular hardware may not fall into any of the historical subcategories. It may be possible to estimate Culture for future programs using deterministic methods, such as a function of the ratio between work and quantity." The effect of culture on cost is illustrated in Fig. A10.

The inclusion of a time variable causes the effects of time to be removed from the other variables in the model. Hence, the model can be used for long-range planning if the future effect of time can be predicted.

In addition to the culture variable, it is possible to specify a complexity variable that allows for the possibility that systems (within a given category) may vary considerably in terms of performance, capacity, level of technology, complexity of design, and other factors.

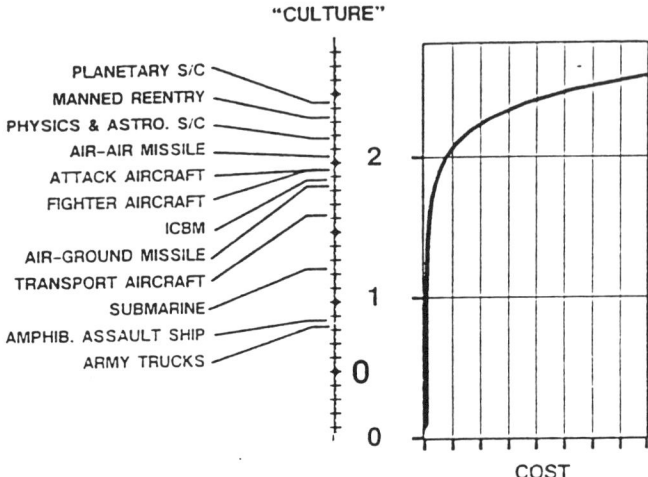

Fig. A10 Effect of culture on cost.[3]

AMCM, as opposed to the static concept of CERs, was designed and is operated under the concept of a generalized system. Specifically, the CER is a static concept in that each CER represents a unique and exclusive product in terms of weight or other parameters. The generalized concept of AMCM incorporates the observations of a multitude of CERs and determines a pattern of cost dimensional movement across the entire spectrum of products relative to that dimension. For example, the movement in cost opposed to weight for satellite structures, the movement in the cost of cables relative to weight, and the movement in the cost of propulsion relative to weight are not observed as separate and distinct models, but rather as items of different complexity in a continuous complexity plane, and the differences in the cost over weight slopes are reference points on a surface of potential slopes for other items in the complexity plane. This concept allows the interpolation of cost as a function of weight coefficients for items not in the historical calibration library. The concept is a product of the generalized system developed by Dr. Frank Frieman at RCA in the 1960s. During implementation, this concept generates a complexity factor in lieu of a linear relationship by the removal of the independent variable as a relative factor due to normalization.

The process of normalization is the extrapolation of an input parameter to a neutral point in the cost hyperplane as determined by the model designer. For example, in the comparison of the historical costs of two different satellites, the cost analyst modifies the historical costs to a common economic point in time (i.e., 1985 dollars) before relative cost complexity comparisons are performed. The analyst would also normalize the historical cost as a function of the build quantity to preclude inaccurate assessments of relative cost, due to the economies of quantity scale.

For example, a build quantity of 10 satellites may be reduced to a build quantity of one satellite and a build quantity of three satellites is also artificially reduced to the build quantity of one. In both cases, the modifications to actual history are artificial and for comparative purposes only. In the sophisticated parametric system, all relevant common cost dimensions are neutralized in this manner to an artificial reference point. The same procedures as outlined for economic escalation and build quantity are utilized in the normalization of weight, state of the art, schedule, and other variables to determine the relative cost of the subject item at a single point in the cost hyperplane. This point, commonly referred to as *cost complexity factor* is then used by the model as a seed for the generation of similar technologies where the cost parameters lie at different distances on each dimension of the cost hyperplane.

The process of calibration is a user-instigated process where the independent variable, cost, is known and the complexity factor is the dependent variable and is determined by the model by the iterative method as several of the algorithm coefficients are a function of the calibration of the dependent variable, complexity factor. This process is illustrated in Fig. A11.

The forecasting process in the model operations is the reverse concept of the calibration process in that the complexity factor derived from historical calibration is now the independent variable, and cost is the dependent variable. This process is illustrated in Fig. A12.

Figure A13 illustrates the algorithms developed for AMCM as a result of extensive iteration design and testing. The left column of the figure indicates user inputs. The second column provides the processing algorithm. The third column provides the primary output variables.

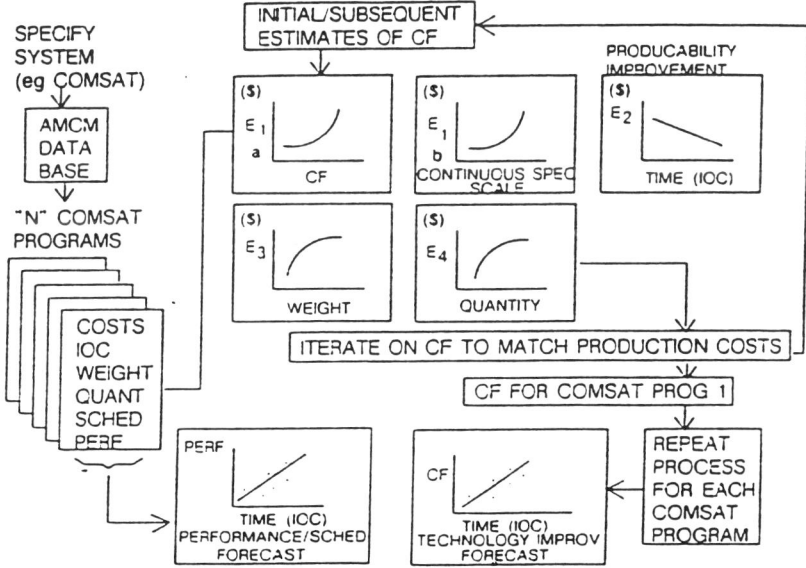

Fig. A11 Evaluating complexity.[3]

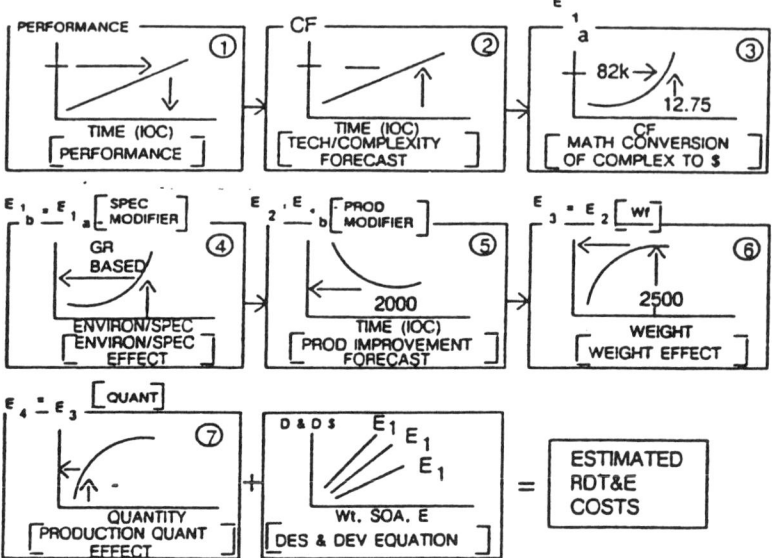

Fig. A12 Estimation process.[3]

INPUT	SPACE SHUTTLE ORBITER	OUTPUT
cf = 8.263 s = 1.80	$E_1 = 69.323\, e^{.73(cf)\,.3436\,.2172(s)(cf)\,.6333}$ (cf) e	264.497
IOC = 1981	$E_2 = E_1\, e^{.0036(1987-IOC)E_1^{.2}}$	343.853
f = .2 w = 154.950	$E_3 = E_2\, w^{.3[1-e^{-(.15\ln(E_2)-.37)^3}]\,+\,.3(f)\,=\,.3}$	835.9C$, 543
Q = 2	$b_1 = [.9085Q^{.0065} - .1755 - .0361Q\,\ln(E_2/1000)]$ $E_4 = \dfrac{E_3\, Q^{[\ln(b_1)/\ln(2)\,+\,1]}}{1,000,000}$	.7343 Production $ 1,229 m
SOA = 3	RDT&E = $.251\, E_1^{-.74}\, w\, E_1^{.63}\, [.3(SOA) + .1]$	RDT&E $ 12,004 m

Note:
cf = Complexity Factor   s = Specifications
IOC = Initial Operational Capability   f = Electronic Fraction
w = Weight   Q = Quantity
SOA = State-of-the-Art $ = Constant 1987

Fig. A13 AMCM algorithms.[3]

PRICE Parametric Cost Models

The GE PRICE Parametric Cost Model[6,7] is undoubtedly the most commonly used set of CERs in use by the aerospace industry. The PRICE family of models addresses estimating, scheduling, and planning problems from hardware and software acquisitions to the support of systems throughout their operational lifetime.

The application of mathematical modeling to simulate traditional estimating practices is the foundation of the PRICE family of models. This family consists of the following parametric estimating models:

1) PRICE H—Estimates development costs, production costs, and schedules of hardware products and systems

2) PRICE HL—Estimates maintenance and support costs of hardware products and systems

3) PRICE S—Costs and schedules are estimated for design, implementation, and test and integration for all types of computer software; software maintenance, enhancement, and growth during operation; software size (lines of code) estimating; DOD-STD-2167 compatible

4) PRICE M—Estimates development and production costs and schedules of microcircuit components (chips) and electronic assemblies (boards)

5) PRICE PM—A network scheduling model for projects that involve hardware development and production, and software design, implementation, and test

PRICE H is used primarily to estimate development and/or production costs and schedules for items at the subassembly or higher level. Larger systems are also estimated with the aid of the supporting system model that aggregates many levels of assemblies into a system total.

PRICE H relies upon the size characteristic as an input to establish goals for cost and schedule purposes. Specific inputs allow for the identification of an item's weight in two domains—the active electronic weight and the mechanical/structural (nonactive electronic) weight. Accurate inputs for weight are critical to the accuracy of the PRICE output. Although not critical to cost, density and volume inputs may be provided to establish consistency checks for the trained PRICE analyst.

PRICE H uses three qualitative, or empirical, inputs to distinguish among classes of hardware that are of similar size: structural/mechanical, electronics, and engineering complexity. These parameters are called complexity variables, two of which are manufacturing complexity, and are measures of cost per unit weight of the structural/mechanical and the electronic portion of an item. The other complexity, the engineering complexity, pertains to the scope of the hardware development task (if pertinent) and the skills of the development team. Together with the weights, the manufacturing complexities are the most critical inputs to the model. The engineering complexity is critical to the cost estimate for development programs.

Manufacturing complexities are selected by one of two methods. They may be taken from reference tables for structural/mechanical or electronic assemblies, or obtained from the results of model calibration.

Model calibration is preferred by many for several reasons: 1) The critical inputs are based solely on programs of interest to the using organization. 2) It allows for hardware identification at the level most desirable to the organization (e.g., module vs unit vs subsystem, etc.). 3) It provides the greatest ease in modeling the organization's cost accounting. PRICE H devotes a special operating mode to calibrating manufacturing complexities to known costs.

Processing the data through PRICE H yields complexities that match the input cost. The identical exercise run on similar products provides the PRICE analyst with sets of manufacturing complexities that will then form the foundation for applying PRICE H to analogous hardware systems.

Other input data include:

1) Schedule milestones for both the development and production phases of a program. This includes start of effort, first article completion, and end of effort.

2) Both the amount of design repetition and design inventory considerations are addressed.

3) For development programs, the number of prototypes, and for production, the number of units are required as inputs. A production learning curve may also be specified.

4) The economic basis for the PRICE analysis and yearly rates of escalation/de-escalation purposes may be input or deferred to model selection.

Additional inputs allow for calibration to the individual cost elements, for modeling multiple manufacturing facilities, for customizing production engineering changes, and for building systems of any number of integration levels.

Estimating systems with PRICE H involves the preparation of inputs for each identifiable hardware item of a system, the construction of an aggregate system with a utility stacking model, and the accumulation of costs into subsystem, system, and major system totals with PRICE H. Any number of integration levels is possible.

Special inputs identify each hardware item's impact upon system integration costs. These inputs specify the quantity of units per system and the level of mechanical and electronic integration. Additionally, the PRICE H model has the capability through a special mode to estimate the costs for integration (installation) of computer software within the hardware system.

Additional operating modes account for the acquisition of hardware by way of purchase from a vendor, furnished by the customer, and/or modification of existing hardware. The cost integration and system testing is estimated after all system components have been identified.

References

[1]Hillebrandt, P., et al., "Space Division Unmanned Space Vehicle Cost Model SD TR-88-97," 6th ed., Nov. 1988.

[2]Apgar, H., "Developing the Space Hardware Cost Model," Paper IAA-CESO-04(90), IAA Symposium on Space Systems Cost Methodologies and Applications, May 1990.

[3]Cyr, K., "Cost Understanding," Vol. III, Space Exploration Annual Rept., SAA-C-R-001, Johnson Space Center, Oct. 1988.

[4]Koelle, D. E., "The TRANSCOST Model for Launch Vehicle Development, Production, and Operations Cost Estimation (1990 Update)," IAA Symposium on Space Systems Cost Methodologies and Applications, May 1990.

[5]Strope, D. H., "Large Structures Costing Estimating Methods Development," Tecolote Research CR-0251, Dec. 1987.

[6]Kaufman, G. A., "PRICE Life Cycle Cost Model," RCA/PRICE Systems, March 1978.

[7]RCA PRICE Systems, *An Executive Guide to PRICE*, Moorestown, NJ, 1987.

Cost-Estimating Relationships: A DCAA Perspective

Michael Thibault*
Defense Contract Audit Agency, Alexandria, Virginia 22304

Introduction

THE U.S. Defense Contract Audit Agency (DCAA) was established in 1965. DCAA's mission is to perform all necessary contract audits for the Department of Defense and, when requested, to perform contract audit services on a reimbursable basis for other government agencies. In essence, DCAA provides accounting and financial advisory services for procurement and contract administration activities. Contract audit activities include providing professional advice on accounting and financial matters to assist in the negotiation, award, administration, repricing, and settlement of contracts. DCAA audits are conducted in accordance with generally accepted auditing standards established by the General Accounting Office.

This chapter discusses the government's expectations when defense contractors use parametric cost-estimating relationships for estimating government contract costs. The chapter also emphasizes that companies must meet all the adequacy criteria set out in the *Federal Acquisition Regulations* (FAR) and applicable supplements to obtain approval for their estimating systems. Companies must apply the same criteria to their parametric cost-estimating relationships to ensure they are acceptable for use in estimating systems.

DCAA believes now, as it has always believed, that parametric estimating techniques using cost-estimating relationships are acceptable in the appropriate circumstances for proposing costs on government contracts. DCAA is ready and willing to work with industry in the evolution of parametrics. Operation Desert Shield and Operation Desert Storm dramatically demonstrated that our government must be capable of responding

---

This paper is declared a work of the U.S. Government and is not subject to copyright protection in the United States.
*Assistant Director, Policy and Plans.

quickly to changing procurement requirements. Parametric systems can help us do just that. Future estimating systems must be responsive, accurate, and cost effective.

Background

DCAA was issuing official guidance on parametric systems as early as 1978. Parametrics was broadly defined as a technique that employs one or more cost-estimating relationships to estimate costs associated with developing, manufacturing, or modifying an end item. In the 1980s, DCAA auditors reported an increase in the number of contractors using parametric cost estimating. DCAA developed and issued audit guidance to assist the field auditors in this new area. Studies conducted by DCAA; the Office of the Secretary of Defense Cost Analysis Improvement Group; Headquarters, Air Force Contract Management Division; Headquarters, Aeronautical Systems Division; and the Space Systems Cost Analysis Group provided the basis for DCAA audit guidance issued in 1982.

Parametric Criteria

This guidance was also the subject of an article written for the Spring 1982 issue of *Journal of Parametrics* published by the International Society of Parametric Analysts. Charles O. Starret, Jr., then-Director of the Defense Contract Audit Agency, wrote the article entitled "Parametric Cost Estimating—An Audit Perspective." The guidance contained in that article is essentially the same as the guidance given to DCAA auditors today. It reiterates DCAA's long-held view that parametrics is an acceptable estimating technique. The 1982 article included the criteria a contractor should apply before submitting a contract price proposal using parametrics. The criteria are still on point today, and they are:
 1) Logical relationship
 2) Significant statistical relationships
 3) Verifiable data
 4) Reasonably accurate predictions
 5) Proper system monitoring

Logical Relationship

Contractors are expected to demonstrate that cost-to-noncost-estimating relationships are logical. "Logical relationship" is often difficult to determine in a finite sense, yet is very important. DCAA's primary concern in this area is that a contractor consider all reasonably logical estimating alternatives and not use only the first apparent set of variables. Contractor analysis may disclose multiple alternatives that appear logical. Statistical testing should be used to help identify the best alternative.

Significant Statistical Relationships

Contractors are also expected to demonstrate that a significant statistical relationship exists among the variables used in a parametric cost-estimating relationship. There are several statistical methods such as regression analysis that can be used to validate a cost-estimating relationship; however, no single uniform test can be specified. Statistical testing may vary depending on an overall risk assessment and the unique nature of a contractor's parametric data base and the related estimating system. Proposal documentation should describe the statistical analysis performed, including the contractor's explanation of why the cost-estimating relationship is statistically valid.

Verifiable Data

There must be a system in place for verifying data used for parametric cost-estimating relationships. In many instances, the auditor will not have previously evaluated the accuracy of noncost data used in parametric estimates. For monitoring and documenting noncost variables, contractors may have to modify existing information systems or develop new ones. Information that is adequate for day-to-day management needs may not be reliable enough for contract pricing. Data used in parametric estimates must be accurately and consistently available over a period of time, and easily traced to or reconciled with source documentation.

Reasonably Accurate Predictions

The contractor's demonstration that the parametric cost-estimating relationships predict costs with a reasonable degree of accuracy is also important. The key is that if the contractor's analysis of historical estimating and cost performance data shows that the parametric estimating system is as accurate as a discrete estimating system, then the government has increased assurance of receiving a fair and reasonable price.

As with any estimating relationship derived from prior history, it is essential for the contractor to document that the work being estimated using parametric cost-estimating relationships is comparable to the prior work from which the parametric data base was developed.

Proper System Monitoring

The contractor should also ensure that cost-to-noncost parametric rates and factors will be monitored periodically in the same manner as is expected for cost-to-cost rates and factors. Because of improved technology, production changes, or better pricing alternatives, cost-estimating relationships can and do change. The contractor should be prepared to revalidate any parametric cost-estimating relationship whenever system monitoring discloses that the relationship has changed.

Audit Planning and Requirements

The old expression, "The more things change, the more things stay the same," seems to apply. Government procurement procedures may change to accommodate changes in the services and products it buys, but the basic procurement goal of getting products and services for fair and reasonable prices remains the same. And so it goes with auditing. Whether DCAA audits proposed costs for space stations or for missiles, DCAA's basic aim of ensuring that the proposed costs are allowable, allocable, and reasonable and therefore acceptable for government reimbursement is still the same. How DCAA accomplishes its audit objective varies with the sophistication of contractor accounting and estimating systems.

Auditors begin the audit by ensuring they have the requisite familiarity with DCAA guidance on estimating systems and techniques. This guidance is contained in our *Contract Audit Manual* (CAM), which is available to the general public. The auditor then proceeds to do the following:

1) Ensure they are familiar with the company's estimating policies and procedures.

2) Identify the estimating methods used to develop the proposal.

3) Determine that the supporting cost and pricing data for the individual proposal was derived in accordance with the contractor's estimating system and is in compliance with applicable regulations.

The auditor plans the audit scope using what is known about the contractor. For example, the audit scope will vary depending upon the estimating methods the contractor uses. In addition, the auditor will consider the following types of questions: What is the dollar amount and type of contract contemplated? Has the contractor established strong internal controls and sound accounting and estimating systems? What kinds of testing does the contractor do to ensure compliance with these systems? What does our prior audit experience tell us about the contractor's internal controls or estimating practices?

Audit planning requires the auditor to answer all of these questions and to make a determination regarding the government's risk. Judgment is then exercised in deciding the degree of risk that the estimate could be materially overstated. This assessment of risk will be used to decide what audit procedures to employ.

The auditor identifies the method of estimating the contractor uses to determine the kind of support that should be available. A contractor could be using any or all of the following methods:

1) *Detailed*—also known as the bottoms-up approach. This method divides proposals into their smallest component tasks and are normally supported by detailed bills of material.

2) *Comparative*—develops proposed costs using like items produced in the past as a baseline. Allowances are made for product dissimilarities and changes in such things as complexity, scale, design, and materials.

3) *Judgmental*—subjective method of estimating costs using estimates of prior experience, judgment, memory, informal notes, and other data. It is typically used during the research and development phase when drawings have not yet been developed.

Parametric estimating techniques may be used in conjunction with any of these methods. Whatever the method selected by the contractor, it must comply with applicable laws and regulations.

The laws and regulations most often encountered in dealing with parametrics are:

1) CAS 401, "Consistency in Estimating, Accumulating, and Reporting Costs"

2) Truth in Negotiations Act

3) DFARS 215.811, "Estimating Systems"

Cost Accounting Standards (CAS) provides guidance in accounting for contract costs at larger contractors. CAS 401 requires that a contractor's estimating practices be consistent with those governing the accumulation and reporting of costs during contract performance. Some contractors see parametrics as being inconsistent with CAS 401. Contractors must ensure both cost and noncost information used in estimating is separately accumulated and reported as required by CAS 401.

The purpose of the Truth in Negotiations Act, 10 U.S.C. 2306(f), is to provide the government with all facts available to the contractor at the time it certified the cost or pricing data was accurate, complete, and current. Parametric estimates must meet the same basic disclosure requirements under the act as discrete estimates. Although the principles are no different, proposals supported in whole or in part with parametric estimating will have different types of cost or pricing data than traditional discrete cost estimates.

Fundamental to the definition of cost or pricing data are "all facts . . . which prudent buyers and sellers would reasonably expect to have a significant effect on price negotiations" (FAR 15.801). Reasonable parallels may be drawn between the data examples provided in FAR for discrete estimating approaches and the type of data pertinent to parametric estimating approaches. The contractor is also expected to provide all factual data for the parametric cost estimates. This data must be accurate, complete, and current.

Many contractors use parametric cost estimating for supplementary support or validation of estimates developed using other methods. This requires judgment in selecting which data will be used in developing the total cost estimate relied upon for the price proposal. In distinguishing between fact and judgment, FAR states the certificate of cost or pricing data "does not constitute a representation as to the accuracy of the contractor's judgment on the estimate of future costs or projections. It does apply to the data upon which the contractor's judgment or estimate was based" (FAR 15.804-4b). Thus, if a contractor develops a proposal using both parametric data and discrete estimates, it would be prudent to disclose all pertinent facts to avoid later questions about completeness of the submission.

Auditors are also required to evaluate estimating systems of major Department of Defense (DoD) contractors. This includes ensuring that, if parametric estimating procedures are part of the estimating system, they are properly disclosed. Of key concern to the auditor in evaluating the estimating systems' disclosure of parametric procedures are the following:

1) Do the procedures clearly establish guidelines for when parametric techniques would be appropriate?

2) Are there guidelines to ensure the consistent application of estimating techniques?

3) Is there proper identification of sources of data and the estimating methods and the rationale used in developing cost estimates?

4) Do the procedures ensure that relevant personnel have sufficient training, experience, and guidance to perform estimating tasks in accordance with the contractor's established procedures?

5) Is there internal review of and accountability for the adequacy of the parametric models, including the comparison of projected results to actual results and an analysis of any differences?

DCAA believes parametric estimating approaches are acceptable when they are properly implemented. Auditors encounter it most often as a technique used in conjunction with other estimating methods. For example, parametrics are often used for estimating costs of scrap and other such factors. The majority of the proposals audited are not developed solely based on parametric estimating techniques. The use of parametric estimating may be appropriate in such circumstances as when the program is at the engineering concept stage, or when no bill of materials exists and the program definition is unclear.

Observations and Suggestions

The contractor can consider some observations made by DCAA auditors as to the pitfalls contractors fall victim to when employing parametric techniques. The first is when a contractor fails to do a cost-benefit analysis before implementing an elaborate parametric estimating model. Key questions for any contractor considering implementing a complex parametric model are:

1) How often can we reasonably expect to use it?
2) How much time can we expect to save?
3) What are the costs of maintaining the model?
4) Will the model produce the necessary precision?

Contractors should be satisfied that implementation and monitoring costs do not outweigh the benefit of reduced estimating costs. Moreover, it is critical that the environment is appropriate for the use of parametrics. It would not be prudent to rely exclusively on parametric techniques to estimate costs when directly applicable historical cost data are available. Such is the case of follow-on production for the same or similar hardware. Contractors manufacturing mature weapon systems already have a record of the actual costs. The increased precision achieved by using this history in projecting costs is critical when you consider these weapon systems proposals are very significant, sometimes in the billions of dollars. The exclusive use of parametrics is generally not appropriate for economic forecasting of such elements as labor and indirect cost rates. These require separate forecasting considerations as to time and place of contract performance.

Another problem encountered is contractors failing to properly disclose their parametric estimating practices. Auditors have experienced instances where the first time a parametric model is disclosed is during the evaluation of the proposal. This is often too late. As mentioned earlier, larger DoD contractors have an obligation to disclose in writing their estimating procedures. Making proper, timely disclosure will minimize problems and expedite the negotiation process.

Contractors can take the lead in helping to streamline the oversight process. In a few words, they should *practice self-governance*! DCAA has been a leading proponent of the Contractor Risk Assessment Guide (CRAG) program. This self-governance program was initiated to encourage contractors to establish and maintain good systems of internal control in five key areas, including estimating systems. It requires contractors to provide their own oversight—to detect system weaknesses and take corrective action as necessary. This initiative recognizes that prudent contractors already have the means in place to ensure their operations are efficient and cost effective.

DCAA has another initiative that contractors should consider as a part of streamlining the oversight process. It is called "coordinated audit planning." DCAA defines coordinated audit planning as a voluntary process wherein the DCAA auditor and the contractor's internal and external auditors consider each other's work in determining the nature, timing, and extent of his or her own auditing procedures. Coordinated audit planning considers the extent to which reliance can be placed upon work performed by the other auditor to minimize duplication of audit effort. In addition, this process strengthens the evaluation of internal control systems.

Understanding estimating system controls, assessing risk, and transaction testing are common objectives of DCAA, and the internal and external auditors associated with contractors. This often results in duplicative audit procedures. In the coordinated audit planning conducted to date, DCAA found that the instances of duplication are significant. DCAA auditors are more than willing to rely on the work done by contractor internal and external auditors, providing DCAA has the opportunity to evaluate and test their work.

Contractors are expected to establish and maintain reliable estimating systems. Departmental procurement officials, Members of Congress, and the average American citizen hold Defense contractors to high and exacting standards. The funds involved can be enormous.

The expectations are equally high for the auditor whose job it is to protect the taxpayer's interest. The DCAA auditor must comply with the American Institute of Certified Public Accountants' *Statements on Auditing Standards* and the GAO's *Government Auditing Standards*. These standards require that the auditor be independent in fact and in appearance. They also require the auditor exercise a healthy degree of professional skepticism. These requirements, however, should not render the DCAA auditor and the contractor enemies. Such polarized and adversarial relationships are dysfunctional and not in either party's best interest.

Both parties are taking significant actions to improve relationships. These changes are producing a culture change that is very positive: positive because contractors are beginning to more fully accept their responsibilities; positive because auditors are beginning to more effectively communicate audit plans and objectives.

Defense procurement is taking on a new, streamlined look in the 1990s. Government and industry are both concerned with quicker, less costly means of procuring goods and services. This is one reason why parametric methods continue to stir up so much interest. Parametric techniques, properly applied, can assist contractors and the government alike in streamlining the acquisition process. In addition, the ability of the government to place greater reliance on contractor oversight of contractor systems will also result in meeting procurement needs more timely.

Summary

In today's and tomorrow's procurement environment, a great challenge facing us all is the development of a cooperative work climate conducive to quickly acquiring quality products and services at fair and reasonable prices. New estimating techniques such as parametrics can cut estimating costs. Adequate estimating systems, fully supported and self-governed by industry, can cut audit costs. Quick estimating and audit turnaround times can cut procurement costs. All of this, however, requires communication and teamwork. Whether our buying effort is for innovative space equipment or for recurring maintenance, we must all work to meet the challenge.

Cost Estimating for Technology Programs

George A. Hazelrigg*
National Science Foundation, Washington, DC 20550
The views expressed here are strictly those of the author and do not necessarily reflect the views or policies of either the federal government or the National Science Foundation.

Need for Cost Estimating

EARLY in the planning phase of all major endeavors, it is important to estimate cost. This is true whether the endeavor is the construction of a building, the development of a commercial venture, or a space mission. No project is ever begun without some estimate of its cost and also its value. Cost estimation of space programs has long been a topic of concern, and a great deal of effort has gone into producing cost estimates and improving the technology of cost estimating. Indeed, cost estimation is of such importance that cost estimators have their own professional society, their own journals, and their own conferences devoted to the technology and application of cost-estimating methods.

Over the past 20 years, a cost-estimating methodology of particular application to the aerospace industry has emerged. This is the methodology of cost-estimating relationships (CERs), which are discussed in detail in an earlier chapter. The idea behind the CER methodology is that aerospace systems are comprised of physical subsystems that have measurable properties such as weight, size, power consumption, and so on. It is felt that engineers can generally estimate the physical properties of proposed subsystems with some degree of accuracy. The cost estimator then relates these parameters to subsystem cost through the CER. The CER in turn is derived from historical data for similar subsystems, adjusted for variations in performance. Total system cost is taken to be the sum of the costs of

---

This paper is declared a work of the U.S. Government and is not subject to copyright protection in the United States.
*Deputy Division Director, Electrical and Communications Systems.

the subsystems, with added costs for activities such as systems integration, assembly, and test.

The CER approach to cost estimating has become widely accepted over the past several years, perhaps more because of the traceability of the method than its accuracy. The method is relatively easy to apply and produces results that can be traced to historical data. Yet it leaves much to be desired. It does not work well for systems that rely heavily on new technologies, or that require significant research and development as a part of the overall program. One problem with the CER approach is also its major benefit: it isolates cost-estimating activities from engineering activities, thereby enabling cost estimators to perform their job with little knowledge of the engineering details of the system. The cost of an engineering system is the result of a very complex decision-making process, and CERs never attempt to understand or model this process. They merely model the aggregate results of the process, and cannot predict well the variations in cost with variations in design strategy.

Technology programs present special cost-estimating problems that are not well addressed by the CER methodology. To begin, the products of technology programs are knowledge and information. These are not physical things with easily measured parameters that one could use in a CER. Furthermore, new knowledge is just that: it is new. It is not a function of previous knowledge, and it cannot be measured in units of previous knowledge. And, even if it were possible to measure "increments of knowledge" and the cost of obtaining them, it would be inappropriate to use these data as the historical basis for forecasting the cost of "buying" the next increment of knowledge. Consequently, one cannot even hope to develop a reliable and well-founded approach for the direct use of CERs to estimate the cost of technology programs.

It is important to give credit to the CER method where appropriate. But it is equally important to recognize the assumptions upon which the method is based, and the limitations that these assumptions impose:

1) In gathering data for the estimation of CERs, only data from successfully completed programs are included. This is reasonable because the cost to completion of a failed program cannot be measured—it simply does not exist. Because of this, however, *an inherent assumption in the CER method is that it must be an absolute certainty that the program can be completed successfully*. This means that CERs can be applied legitimately only to estimate the cost of systems that use established technology, that is, within the set of things that have already been built. Any attempt to extend CERs beyond this mundane use violates a major condition under which the method was developed.

2) In gathering data for the estimation of CERs, actual costs and physical system parameters are taken, not estimates of these parameters made early in the program. Thus, it is assumed that physical parameters can be estimated with precision and certainty, even as early in a program as the time when the first cost estimates are needed. This again reinforces the notion that CERs apply only to systems that are comprised entirely of tried and proven technologies.

3) The accuracy of the cost estimates given by a CER is sometimes measured in terms of the goodness-of-the-curve fit that represents the CER. There is no sound mathematical basis for this use of a CER.

It follows that the application of CERs within the assumptions that enable their derivation is highly restrictive. In practice, many applications of CERs lie outside these restrictive assumptions. Inappropriate use of CERs is a major reason why their results are sometimes so inaccurate.

As noted above, technology programs do not possess properties that would make them directly amenable to cost analysis by means of CERs. Their product has no physical parameters by which to measure them, and they provide a result that is always new. Thus, another method of cost analysis must be used for the estimation of technology program costs.

Attributes of Technology Programs

Technology programs possess many attributes that set them aside from programs whose output is a physical system:

1) The product is technology, that is, it is knowledge and information, which lead to the ability to make good engineering decisions and to carry them through to their desired conclusion. Technology cannot be measured in terms of physical parameters.

2) Unlike a program for building a physical system, which seeks a single design for implememtation, a technology program may pursue its goals through several parallel approaches. Each approach may itself have parallel and overlapping activities.

3) Where there exist parallel and overlapping activities, there may be correlation in the probabilities of success or failure between these activities.

4) There might be a variety of ways of achieving "success" in any particular technology program. Each successful outcome would provide a technology capable of satisfying the goals of the program.

5) There might be a significant probability of failure.

6) There can be considerable uncertainty in the schedule by which the program reaches success or failure. Along with the uncertainty in schedule, there can be a significant uncertainty in the cost of a technology program.

7) A technology program can have several phases, across which programmatic decision making can be very important. For example, in the early phases of a technology program, the objective might be to prove the feasibility of certain approaches. Then, based on proven feasibility and other factors, a decision might be made to eliminate certain approaches from further development, carrying forward to completion only one technology capable of satisfying the requirements of the program.

Any research program can be measured in terms of three parameters: technical accomplishments, cost, and schedule. And, at least in theory, it is possible to hold two of these three constant, allowing the third to vary. Clearly it is possible to hold cost (or cost and schedule) constant if technical accomplishments are allowed to vary. Less clearly, technical accomplishments and cost could be held constant and schedule could be allowed to vary. The latter may not always be possible, but the point is that, if the schedule can be varied, some variation in technical accomplishments is possible. A research program cost methodology must take into account the management approach, what will be held constant and what will be allowed to vary.

Concepts of Cost Estimating

To begin to understand the fundamentals of cost estimating, a most fundamental concept must be recognized: *the cost of a system is not a physical parameter of that system.* Rather, cost is a concept that relates to human values. All human activities take time, and individuals place values on that time. But, the time of different individuals, presumably with different skills, may be valued differently, and thus there does not necessarily exist a one-to-one correspondence between time and value. Economists generally recognize three basic categories of cost: labor, capital, and resources. But all three reduce, essentially, to time—labor directly and the other two indirectly. Capital is the time of others invested at an earlier stage in a project or even before the project begins, and resources reduce to the time of others invested more or less concurrently for the recovery and processing of resources plus an economic "rent" that relates to the increased time that it will take future workers to find, recover, and process incremental amounts of the resources as supply is depleted. Thus, cost estimating can be reduced to a process of estimating the time that will be invested in a project and the placing of a value on that time.

Yet, because cost is not a physical parameter of a system, the estimation of cost is neither simple nor unique. To be sure, as the value structure of each individual is unique, so, in general, will be the cost that individual assigns to a project, even in the absence of uncertainty. There are fundamental issues of costing that are unresolvable. For example, how does one assign capital costs to specific projects when the capital is used over several projects? How does one amortize capital over time? These are issues with which economists have wrestled, and many theories have emerged. All theories make use of parameters that involve the measurement of human preferences, such as discount rates, and none is expressly correct in the sense that the others are wrong.

Uncertainty further complicates the estimation of cost. The presence of uncertainty implies that errors in cost estimates are likely. Thus, a cost estimate should be viewed as a probability distribution, not as a single number, and analyses should deal mathematically with cost estimates accordingly. This fact has several implications, for example, with respect to the addition of subsystem cost estimates to obtain a total system cost estimate. Care must be taken in the addition of costs to recognize that one is adding probability distributions, not numbers, and the addition must proceed within the context of probability theory.

Cost estimating, be it for hardware programs or research, is a form of forecasting. The cost estimator is attempting to predict the cost of accomplishing some objective. Accordingly, all of the principles and pitfalls of forecasting apply. And it has been said that *predicting things is very difficult, especially the future.*

Some things can be predicted only in terms of probabilities. For example, without detailed and precise knowledge of all relevant conditions, the outcome of a coin toss cannot be predicted. It can be said only that it will be heads with probability 0.5 and tails with probability 0.5. On the other hand, a prediction can be made with reasonable accuracy that approximately half of 1000 coin tosses will result in heads. As the number of tosses increases, confidence increases in the prediction that half of the tosses will result in heads. But at the same time, the likely deviation from precisely half will grow in absolute terms. (Remember that with a single coin toss, the estimate of heads can be wrong by a maximum of one toss. The potential error increases as the number of flips increases.) Thus, as the number of events increases, the ability to predict percentage improves, but the ability to predict exact numbers lessens.

Some physical processes behave according to rather well-established laws. It is possible to use these laws to reduce the uncertainty in many predictions. For example, the law $F = ma$ can be used to determine the position, with a high degree of accuracy, of a ballistic object as a function of time. Yet, it is important to recognize that uncertainty always remains in this and almost any other prediction.

With these concepts in mind, three categories of predictions are identified:

1) *Problem Solving*—In problem solving, known laws are used, such as the laws of nature or scaling laws based on experience, to predict costable parameters of an activity. This is what CERs do. Engineering analyses are performed to estimate physical parameters of a subsystem, and these are used as the input to scaling laws that convert these parameters into costs.

2) *Law of Large Numbers*—Predictions based on the law of large numbers recognize that, when probabilistic activities are repeated numerous times, statistical parameters such as averages can be predicted with reasonable accuracy. Law of large numbers predictions have been applied routinely to the prediction of such things as population and cost learning. The population of the United States in the year 2000 can be predicted applying the law of large numbers, but this prediction does not give any insight into specifically who will be alive and who will not be alive in the United States in the year 2000. Similarly, cost learning occurs because of the application of many cost-cutting approaches, some of which succeed and some of which fail. The prediction is an estimate of the average expected learning; it does not predict which approaches will succeed and which will fail.

3) *Probabilistic*—Probabilistic forecasting is the only approach to forecasting that may be available for single events or events that occur only a few times, such as the coin toss discussed above. In this case, the outcome of an event is not predicted, but is described by a range of possible outcomes each with an assigned probability of occurrence.

The problem-solving approach applies to events that appear to be unique, but that can be related through a unifying concept. CERs are such a unifying concept. They note that the cost of similar systems relate to each other through the CER, and using the CER they predict the cost of a system that has yet to be built. Thus, CERs can be applied to one-of-a-kind systems, provided that the technology is tried and proven, as noted above. The law of large numbers applies to predictions of groups of similar events that in themselves are random, but that have nonrandom parameters, such as means. This is the case of the coin toss illustrated above. The probabilistic approach applies to one-of-a-kind predictions where specific experience is lacking, and unifying concepts do not exist. Research and technology programs are examples of activities that fall into this category. Probabilistic forecasts do not forecast a deterministic future, rather they provide the range of possible futures and place probabilities on the occurrence of each.

In any case, a complete forecast provides not merely a single number or set of numbers, such as the expected population of the United States in the year 2000. It must include an estimate of its accuracy as well.

Simulation Approach to Costing Technology Programs

An approach to the costing of research programs that incorporates these considerations has been developed and tested for the case of magnetic confinement fusion research under support from the National Science Foundation and the Department of Energy.[1] The approach is called RADSIM, Research and Development Simulation Model. The RADSIM methodology is much like the program evaluation review technique (PERT) methods. However, unlike PERT, RADSIM accounts for uncertainty in schedules, degree of technological accomplishment, and correlation between related activities, and allows for major decisions to continue or discontinue research and development (R&D) activities at each decision date. The RADSIM methodology uses the Monte Carlo technique to simulate an R&D program. Each R&D activity and each decision is simulated forward in time from the present through either completion of the program with mission success, termination of the program because of technical failure, or termination because of time or budget constraints. Each program simulation represents one possible outcome of the R&D program. By conducting several (perhaps 1000) such simulations, a statistical sampling is obtained of the possible outcomes, including the probability of success and program cost.

The RADSIM formulation begins by defining the general structure of an R&D program as shown in Fig. 1. A total program may be comprised of research activities on a number of alternative approaches. Each approach has associated with it a number of generic problem areas (A, B, C, . . .) within which significant technology advancement is required in order to achieve program success. Within each alternative approach, technological advancements are structured into a sequence of R&D phases (I, II, III, . . .), each referred to as a project, ending in the development of a useful technology. The overall research program can thus be viewed as a set of alternative approaches, each of which consists of a set of projects that together could satisfy the objectives of the mission.

Finally, the activities may progress through R&D *phases* such as basic research, applied research, subsystem development, and prototype and test from concept to technology readiness. Research program cost esti-

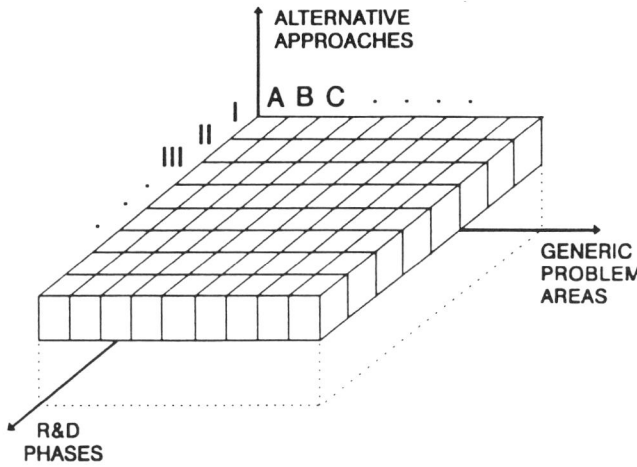

Fig. 1 Dimensions of an R&D program.

mating must account for these three dimensions of an R&D program, and it must take into account the uncertainties, overlaps, and correlations between disparate activities and across the alternative approaches. Each project is comprised of a set of subprojects, of which each is a discrete activity with a definable set of potential outcomes. Subprojects are illustrated as the small blocks in Fig. 1. Projects are illustrated as a row of blocks in a particular horizontal plane of this figure, including all subprojects in a given R&D phase. An alternative approach is represented as all blocks in a given horizontal plane, and the program consists of all blocks in all planes of the figure.

There are five elements of the RADSIM methodology. These elements relate to modeling: 1) the research process, 2) technological breakthroughs, 3) relationships among subprojects, 4) the performance of alternative approaches, and 5) the decision or program management process.

Modeling the Research Process

The RADSIM methodology models the research process at the level of the R&D subproject. The manager, program planner, or cost estimator using RADSIM first identifies each R&D activity that might take place in the overall R&D program, that is, each plausible subproject that could be conducted relating to R&D in a particular generic problem area, R&D phase, and approach. The cost estimator can accomplish this by listing the alternative approaches that appear viable, and then listing generic problem areas associated with each approach. Using these lists, the program planner then associates each problem area with each approach using an incidence matrix as shown in Fig. 2.

Based on the incidence matrix, the estimator then, in each generic problem area for each alternative approach and for each R&D phase, sets minimum technology requirements that represent a successful level of tech-

GENERIC PROBLEM AREAS	ALTERNATIVE APPROACH – NUCLEAR PROPULSION EXAMPLE			
	SOLID CORE	GAS CORE	ELECTRIC PROPULSION	DUAL MODE
CRYOGENIC TANKS	✓	✓		✓
ELECTRIC PROPELLANT TANKS			✓	✓
NOZZLE	✓	✓		✓
ELECTRIC THRUSTERS			✓	✓
THERMAL ENERGY CONVERSION CYCLE			✓	✓
SHIELDING	✓	✓	✓	✓

Fig. 2 An incidence matrix.

nology development. It is this set of requirements against which both subprojects costs and probability of success are gauged. Within each subproject, the technology requirements are given as a lexicographic set; failure to achieve the value stated for any element of the requirements is interpreted as failure of the subproject.

Having identified the minimum technology requirements, the estimator must then assess the probability of success and budgetary requirements as a function of time, given a specific set of R&D activities. In general, the planning and evaluation process proceeds for the set of technology requirements to a specifically planned activity designed to achieve the technology requirement. R&D activities can be planned at a variety of different activity levels, as well as with different approaches. Each approach or activity level will have a unique relationship between success and time, and it will have a unique set of associated costs. Thus, for example, a very vigorous program of technology development may result in success in a very short period of time, whereas a less ambitious level of activity and cost might have both a lower probability of success and longer duration.

Figure 3 shows a format that the estimator could use to input subproject probability of success and cost data. Shown is the probability that a particular subproject activity has been successfully completed as a function of time from the start of activity on the subproject. The cost estimator could develop these curves in any of several ways. One relatively easy approach is to obtain from appropriate technical specialists answers to the four questions:

1) What is the probability of ultimately achieving the stated technology requirement?

2) What is the minimum subproject activity time required to achieve the stated technology requirements?

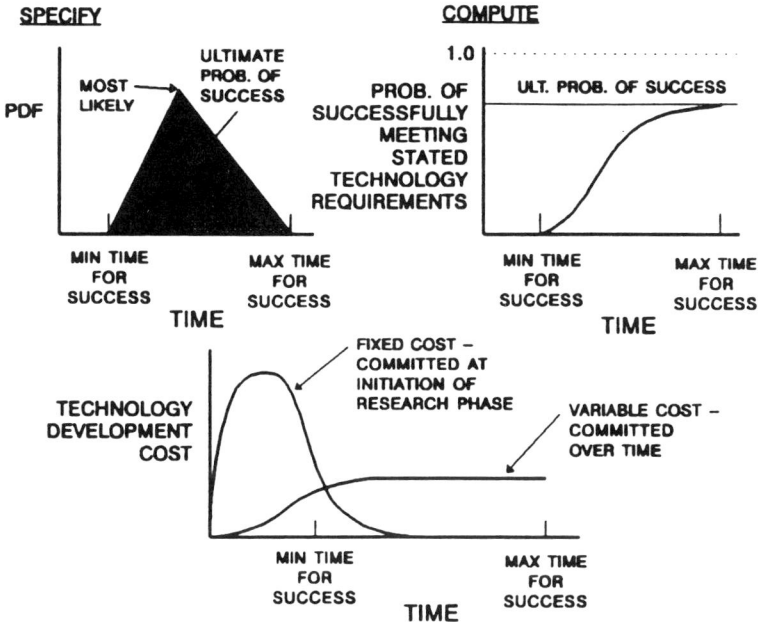

Fig. 3 Subproject input data.

3) What is the most likely subproject activity time required to achieve the stated technology requirements?

4) What is the earliest subproject activity time at which the subproject would be deemed a failure if the stated technology requirements have not yet been met?

Based on the four numbers obtained from these questions, the estimator can determine the probability distribution of the successful completion time as a triangular function, with the four numbers representing, respectively, the area under the triangle, the left corner of the triangle, the peak of the triangle, and the right corner of the triangle. The integral of this triangular probability density function is the probability distribution required by RADSIM.

The estimator must also specify subproject costs. These can be estimated after the subproject activities are clearly defined. Subproject activities, in general, call for the procurement and installation of capital equipment and subsequent support and operation of the equipment. Accordingly, as shown in Fig. 3, subproject costs are conveniently divided into two parts, a fixed part and a variable part, with the fixed costs being committed upon initiation of the subproject and the variable costs comprising an ongoing component of cost that terminates upon completion of the subproject.

The subproject costs outlined here are the marginal or incremental costs of the subproject, above and beyond the basic costs of the R&D phase. The basic cost of performing each R&D phase must also be accounted for. R&D phase costs include major capital investments in large test facilities, and the basic cost of operating these facilities over the duration of the R&D phase. These costs may be input in a manner similar to the subproject costs.

The *conduct* of research requires a set of activities such as construction of facilities, the fabrication of equipment, purchase of instrumentation and other capital items, and labor. All of these things can be estimated with the aid of CERs. Thus, CERs can play a major role in the estimation of costs that form the inputs to RADSIM.

Modeling Technological Breakthroughs

If the nature of R&D was such that research subprojects were planned only to accomplish the stated technology requirements for each R&D phase, the process of program evaluation would be considerably simpler. Often, however, subproject R&D activities are planned to provide an opportunity for technological breakthrough while, at the same time, assuring that the minimum technology requirements are met. Expressed differently, R&D is planned as a stepwise process in which each of the steps is sufficiently small that success is reasonably well assured, and yet the activities undertaken are such that there is some nonzero probability of technology advancement well beyond the minimum required level for success in each R&D phase. When such breakthroughs occur, it might not be necessary to undertake certain future R&D subprojects that had originally been designed to provide for the technology development accomplished by the breakthrough. The process of skipping future R&D activities as the result of breakthroughs is referred to here as leapfrogging. The RADSIM approach accounts for leapfrogging through the use of a parameter referred to as the technology advancement factor. This factor specifies the extent to which technologies in each generic problem area are developed beyond the minimum stated technology requirement.

A technology advancement factor of unity (1.0) indicates that the stated technology requirements are precisely met and not exceeded. A technology advancement factor of 2 indicates that the minimum stated technology requirements were exceeded by an amount such that the minimum technology requirements of the succedent R&D subproject in this generic problem area have been accomplished, and therefore, it is not necessary to perform the subsequent subproject. A technology advancement factor of 3 indicates that the technology requirement of two subsequent subprojects in this generic problem area have been met.

Technology advancement is not a deterministic process. The RADSIM approach requires the program planner to specify a probability distribution on the technology advancement factor. A typical such distribution is shown in Fig. 4. As this figure shows, some subprojects may be planned in such a way that the minimum technology requirements are likely to be met but not exceeded.

Noninteger values of the technology advancement factor are also permitted. A technology advancement factor of 1.5, for example, indicates that, while the technology has not totally leapfrogged the next subproject within a particular generic problem area, the probability that the next

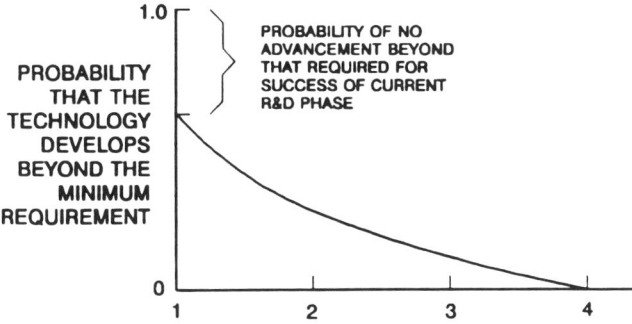

Fig. 4 The technology advancement factor.

subproject will succeed is increased. RADSIM calculates an updated estimate $P^*(t)$ of the probability of success for the next subproject in each phase and each approach having a noninteger technology advancement factor, using the formulas

$$P^*(t_f) = P(t_f) + r[1 - P(t_f)]$$

and

$$P^*(t) = P(t)\left[\frac{P^*(t_f)}{P(t_f)}\right]$$

with $P(t)$ being the estimator's original estimate of the probability that the subproject would succeed by time t, t_f the final possible time for success, and r the fractional part of the technology advancement factor. Figure 5

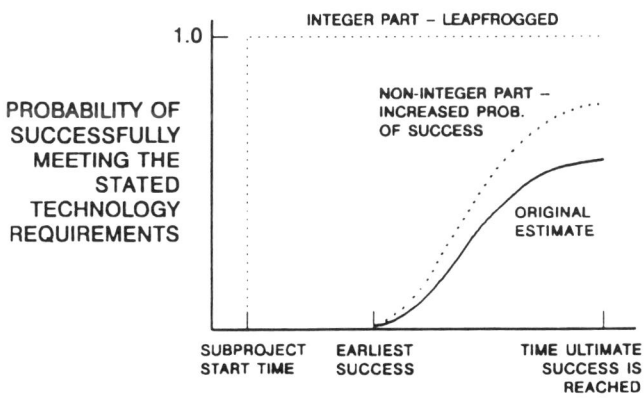

Fig. 5 The effect of leapfrogging on the probability of success in a subsequent subproject.

shows the result of applying the technology advancement factor to the subsequent subproject within a particular generic problem area.

The technology advancement factor described here is arbitrary; other descriptions could be used. This model was chosen because of its relative simplicity and its ability to capture the essence of the phenomena that it models. It is based on the approximation that, if the goals of a particular research activity have not been accomplished by earlier research, the time required to accomplish the goals will be about the same as originally planned. The probability of ultimate failure is, however, decreased by a factor of one minus the fractional part of the technology advancement factor. RADSIM can allow for specification of alternative relationships for technologies that are not well approximated in this way.

RADSIM computes a specific technology advancement factor, based on the estimated probability distribution, each time the simulation finds that a subproject is successfully completed. This technology advancement factor is then used to modify future subprojects according to the rules set forth above.

Modeling Interrelationships Among Subprojects

Another aspect of R&D that the RADSIM approach accounts for is the correlation between activities that occur in projects associated with alternative approaches. For example, if a particular material is under development as part of two different approaches, and if it is successfully developed under the research of one approach, then it is also successfully developed for the other approach as well. RADSIM accounts for these correlations by allowing subprojects for different approaches to be identified as equivalent subprojects. Where subprojects are so identified, RADSIM equates successful completion of one subproject with successful completion of the equivalent subproject or subprojects. This holds not only for ongoing subprojects but also, through the technology advancement factor, for leapfrogging future subprojects.

There are two additional interrelationships among subprojects that must be accounted for. These are referred to as prerequisite subprojects and excluded subprojects. Prerequisite subprojects are subprojects that must be completed before another subproject may begin. The RADSIM approach allows several subprojects to be identified as prerequisites of another subproject. All prerequisite subprojects must be successfully completed before the subproject to which they are prerequisites may begin. Excluded subprojects allow the formulation of parallel research activities, but prevent precise duplication of research activities in the parallel projects. For example, suppose a research strategy to develop a solid core nuclear thermal propulsion system provides for concurrent development of two concepts. Both may use the same nozzle, and therefore each program taken separately must have a nozzle development program. But if both configurations are developed concurrently, it is reasonable to have only one nozzle development program. By specifying nozzle research subprojects as exclusory, it is possible to enable RADSIM to configure the reseach program so as to have only one subprogram on nozzle research common to the research on both reactor configurations.

The subproject property of equivalence is bidirectional. That is, if subproject A is equivalent to subproject B, then subproject B is equivalent to subproject A. Bidirectionality does not apply, however, in the case of prerequisite or excluded subprojects. Clearly, the prerequisite property is not bidirectional. The exclusory property may sometimes be bidirectional. In those cases where it is, the property can be made bidirectional by specifying each of two subprojects as exclusory of the other.

Modeling Performance of Alternative Approaches

A key aspect of the RADSIM approach that sets it apart from many other costing methodologies is that it normatively models the decision or selection process relative to the program manager's preferences. To do this requires: 1) a method for identifying options, 2) a method for ranking all options presented by the model at every decision point, and 3) a decision-making strategy. Normative decision modeling is needed to account for the variability in technical success, cost, and schedule that is always present in R&D programs. Options at each decision point are identified through the specified structure of the program as shown in Fig. 1 and the specified program logic (prerequisite subprojects, excluded subprojects, etc.). This subsection discusses the generic problem of ranking options.

The ranking of options must take into account their significant differences in each relevant attribute. New space technologies typically must pass three tests to win acceptance: they must be technologically successful, they must have an acceptable cost, and their implementation must have an overall beneficial effect on the overall mission. One means of ranking options is in terms of their expectations for success at these three tests. To do this for all possible options at all possible decision points, as is necessary for an automated simulation, requires some form of mathematical model. It is appropriate for this model to reflect the preferences of the manager. Although any such model is likely to be neither totally accurate nor relatively simple, so long as the differences between options are substantial, even approximate models may well provide similar rankings. If the differences between options are small, on the other hand, errors in ranking may not be highly consequential.

RADSIM provides for two different measures of performance, one taking into account only the likelihood of a technological success and the second combining cost and implementation success with technological success. Other measures of success could be established if necessary. In most R&D programs, the first measure of success would be used for assessing decisions to continue or terminate R&D phases whose purpose is to demonstrate technological feasibility of a new idea or concept. Measures of cost and implementation become important after technological feasibility has been demonstrated. Each of these measures is used in two different ways: 1) to indicate when an alternative approach should be terminated (if the measure drops below a specified threshold, the approach will be deemed to have failed and will thus be terminated), and 2) to set priorities for alternatives as candidates for new starts.

For the first measure of performance, the likelihood of technological success of a specific alternative approach, RADSIM uses the estimated probability of successful completion of a specified research phase, which in turn is dependent on successful completion of specific subprojects. For an R&D project yet to begin, the probability that one of its subprojects will succeed or fail is the a priori probability of success of that subproject within the final time. Subprojects that have begun in the simulation, but have not yet succeeded or been terminated, require different analysis, as illustrated in Fig. 6. For example, consider a subproject having a probability of 0.1 of not succeeding in its allotted time of 5 years. If the subproject has not succeeded as the 5th year approaches, one's expectations of ultimate failure will increase.

According to Bayes's law, the probability $P_t(t_f)$ that ultimately a subproject will not succeed, as estimated at time t, given that success has not occurred prior to time t, is

$$P_t(t_f) = \frac{P_0(t_f)}{P_0(t)}$$

where $P_0(t)$ is the probability that the subproject will not succeed by time t as estimated at the beginning of the subproject, and t_f is the final possible time for success. [Note that $P_t(t)$ may depend on the status of other subprojects, and thus may change during the course of the subproject.] Thus, as the subproject proceeds without successful completion, the probability of that subproject ultimately not succeeding increased toward 1 (unless $P_0(t_f) = 0$). At some point in time, the probability of not succeeding may become so close to 1 that the manager will conclude that it is not desirable to continue the subproject. If the subproject were a crucial link in achieving success for a particular approach, then it would no longer be desirable to continue to pursue that approach or any of the R&D activities associated with it.

RADSIM's second ranking measure, which combines the ultimate usefulness of the technology with technological success, is based on the benefit

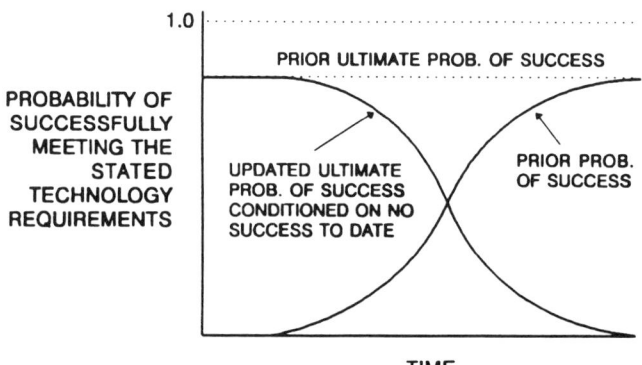

Fig. 6 Probability of subproject success vs time in a subproject.

of successfully completing an R&D program. This benefit may depend on the date when technology development is completed and on the particular technological alternative that is developed. These dependencies may be expressed, for example, in terms of a total value that varies with time, a measure of the extent to which the new technology is adopted and used, and some measure of benefit to each mission in which the technology is used. The benefit to each mission can be expressed in terms of cost savings to the mission, improved mission reliability or probability of success, enhanced mission capability, or some combination of the above.

The RADSIM approach uses the expectation of a value/cost ratio as its second measure of performance, based on the benefits of successful completion estimated above. This measure is an extension and adaptation of the benefit/cost ratio to multiphase programs and is based on the notion that research is a process of buying information for future decisions. It makes use of an independent valuation of each alternative approach in a program as of a given date, computed by the following procedure (Fig. 7): 1) the benefit B of overall success, discounted to present value, is estimated; 2) the costs $C_1, C_2, \ldots$, for phases 1, 2, ... of the R&D process are calculated as the value discounted to the present value of costs associated with that R&D phase; 3) the current estimates of the probabilities $P_1, P_2, \ldots$ of go decisions for each R&D phase are calculated; and 4) the expected net value V of the alternative approach at that time is calculated. At the beginning of a three-phase program, for example, V is equal to

$$V = P_1 P_2 P_3 B - C_1 - P_1 C_2 - P_1 P_2 C_3$$

To obtain the desired measure of performance, V must be compared to a cost. A standard benefit/cost ratio would compare a project's benefit to its entire cost. However, V is the expected benefit of pursuing an approach net of the expected cost of all pursued future R&D phases, taking into account that future phases might not be pursued. Since the only commitment under consideration is to the next R&D phase, the nonsunk cost of that next phase is thus the cost to which V is compared, and the economic measure of performance used by RADSIM for the nth phase is V/C_n.

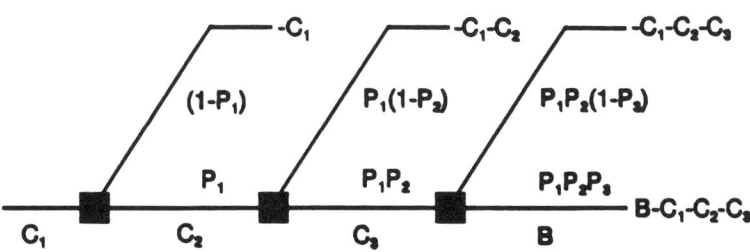

Fig. 7 A project evaluation decision tree.

Modeling the Decision Process

Decisions are assumed to take place on a periodic basis (such as yearly) with respect to every ongoing and potential R&D effort on every alternative approach. In particular, every ongoing research activity is checked against the value measures outlined above and is terminated if the value of continuing the research drops below the specified threshold. Additionally, after each year that an R&D phase is successful or meets a particular milestone, the opportunity arises to begin new R&D phases, and the decision either to begin or not to begin a new phase must be made. The RADSIM approach permits the manager or program planner a great deal of flexibility in considering this decision.

Figure 8 shows some of the process involved in making such a decision with respect to an R&D program that includes two alternative approaches. Alternative 1 is in phase I with research proceeding in four generic problem areas. The shaded bars indicate the time over which the R&D continues in each generic problem area prior to successful achievement of the required technology developments. When the subprojects in each generic problem area of the project succeed, phase I is completed and the decision to begin phase II can be made. In the example shown, phase II begins immediately upon completion of phase I. In the meantime, research is continuing on phase II for alternative 2. As shown, phase II for alternative 2 is completed before phase II for alternative 1. Upon completion of phase II for alternative 2, the decision could be made to proceed with phase III directly or, as shown, to wait until phase II for alternative 1 is complete,

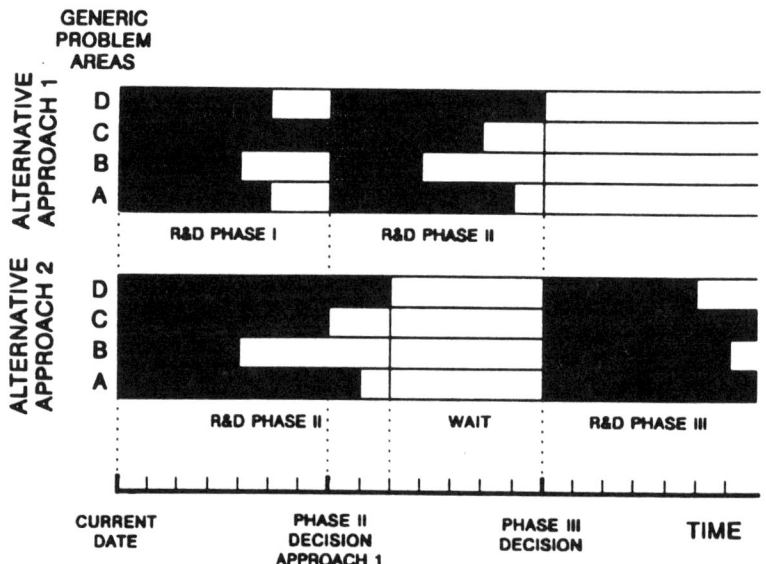

Fig. 8 Simulation status and decisions.

at which time the decision is made to continue with only one of the two alternatives. Thus, the decision to undertake a new R&D phase can occur upon completion of the previous R&D phase, upon completion of certain milestones in the previous R&D phase, or upon successful completion or failure of specific phases of other approaches.

At the time that it becomes possible to begin a new R&D phase, the decision to initiate that phase is made consistent with a set of specified constraints. RADSIM allows establishment of any number of constraints of the class that mandate that the number of subprojects in any specified subset of the full set of possible subprojects at a specified state of completion must be no greater than a specified number. This class of constraints permits a broad range of practical program management constraints to be imposed on the selection of new R&D phase starts. For example, the RADSIM approach permits such constraints as:

1) Allow R&D on only one alternative after phase III.
2) Allow R&D on only one alternative in each of two classes after phase II.
3) Do not allow R&D on more than one alternative in each R&D phase beyond phase II.
4) Do not allow any new R&D phase starts after one alternative reaches a specified phase.

These decision rules are merely examples chosen to show the range of possibilities available within the framework of RADSIM. Decision rules appropriate to the circumstances of a variety of particular research programs can be constructed.

New R&D starts are selected from the set of candidate new starts according to the following logic: 1) the candidate new starts are divided into two groups depending on whether or not technological feasibility has been successfully demonstrated; 2) the group representing those approaches that have not yet demonstrated technological feasibility is ranked using the probability of ultimate technological success measure and the group representing approaches beyond this phase is ranked on the expected value/ cost ratio measure; and 3) new starts are chosen from among the candidates according to a list of priorities and within the specified set of constraints. Priority would generally be given to the potential new start that is farthest advanced in the overall R&D program. Thus, candidates for the final R&D phase are selected first, those for the next-to-last phase are selected next, and so on. Once decisions have been made on new R&D phase starts, the appropriate initiation of relevant subprojects begins. Subprojects may be started upon initiation of the R&D phase of which they are a part or at some specified time after the initiation of the R&D phase, or they may be keyed to the accomplishment of specified prerequisites.

Typical Management Results

RADSIM is a simulation approach to the evaluation and costing of large-scale research and development programs. As such, it obtains data on events at the level of detail of the input data. All events occurring in the simulations are stored for retrieval and analysis. Thus, RADSIM can produce a vast array of data on virtually any aspect of the program. It must be kept in mind that neither RADSIM nor any other simulation can generate scientific data, nor can RADSIM predict specific events. But RADSIM can display the consequences of the input data, that is, of the estimates of costs and probabilities provided at a highly disaggregate level. Estimates of specific RADSIM outputs are the following:

1) Curves of the probability of successfully completing the R&D objectives of the program as a function of time for various management strategies.

2) Curves of cost vs time for various management strategies. Cost is often a random variable. Thus, probability distributions of cost may be obtained for each time period (typically, year) in the simulation. There are a variety of ways of displaying such distributions. One way is to provide cost probability distributions for each year. Another is to plot expected value of cost as a function of time together with 10 and 90% confidence cost curves. Other formats are also possible.

3) RADSIM provides the probabilities of each technological approach succeeding or failing. If a particular approach has a very high failure rate, the program manager might want to see if program cost can be reduced without significant effect on the probability of success by removing that approach from the program. The manager could also investigate the effects of increasing the level of activity on those approaches that are most successful.

4) Sometimes R&D programs hang up on needed technologies that are not ready in time. This can be discovered in the RADSIM output, and variations in the program strategy can be examined in which added emphasis is given to such technologies to assure their readiness in a timely manner.

5) Sometimes R&D programs fail because, in the manager's quest to conserve resources, program options are eliminated too soon. RADSIM enables these problems to be uncovered and the merits of broader strategies to be examined.

RADSIM can produce many other outputs, including counts and probability distributions on virtually all programmatic variables. The above list is only a sample of the capability of the RADSIM approach.

Summary of RADSIM Approach

RADSIM is an approach to the costing and evaluation of large-scale research and development programs that takes into account the uncertainties inherent in the research process. In the RADSIM approach, the overall costing and evaluation problem is approached by disaggregating the entire program into a set of research subprojects, each of which can be evaluated by experts in the particular field of the subproject. The underlying philosophy is that the manager or program planner seeks to make use of the expertise that resides within the individual disciplines of which the program is comprised. The RADSIM software then facilitates the aggregation of the data obtained at the subproject level to provide cost and evaluation information for decision making at the aggregate level of the entire program. Thus RADSIM is a method for making use of the full range of expertise available across the scientists and engineers who will staff the research program and framing that expertise in a format that can be useful for both cost estimating and program planning.

Reference

[1] Hazelrigg, G. A., and Huband, F. L., "RADSIM—A Methodology for Large-Scale R&D Program Assessment," *IEEE Transactions on Engineering Management*, Vol. EM-32, No. 3, Aug. 1985, pp. 106–115.

Space System Life Cycle Cost and Availability Analysis

Joel S. Greenberg*
Princeton Synergetics, Inc., Princeton, New Jersey 08540

Introduction

A PREVIOUS chapter ("Financial/Investment Analysis") discusses financial analysis of a profit-oriented business where a typical objective was the maximization of net present value of cash flow or return on investment. It is pointed out that, because of uncertainty and resulting risk, it is most desirable to describe the net present value and return on investment as probability distributions. The selection from among a number of alternatives should thus consider differences in both the expected value and risk dimensions. Fortunately, in most cases, the probability distributions of important financial performance measures tend to be normally distributed; thus the necessary comparisons are between different sets of expected values and standard deviations. The decision maker's risk avoidance preferences may be exercised to choose the "best" alternative from among the set of alternatives lying on or near the frontier of best alternatives.

This chapter discusses a similar problem but one in which there is neither revenue nor profit; for example, the government needs to maintain a system of global positioning or weather satellites. The basic problem is similar to the previously discussed investment decisions with the exception that there are no revenues nor taxes. Thus, cash flow has only a negative component, i.e., cash outflow or costs, and the objective is thus the minimization of life cycle costs within performance constraints. Frequently the performance constraints can be simply expressed in terms of desired availability of sensors or transponders where availability represents the chance, or fraction

---

Copyright © 1992 by Joel S. Greenberg. Published by the American Institute of Aeronautics and Astronautics, Inc. with permission.
*President.

of the time, that at least N out of M sensors or transponders are operational when required. The objective is to select the minimum life cycle cost approach that meets the availability constraints.

Life cycle cost is defined as the present value (i.e., discounted to the present) of all current and future costs associated with the mission or system being considered. This encompasses all nonrecurring and recurring costs and their specific timing. For a space mission, nonrecurring costs include those costs incurred in satellite research, development, testing, and engineering (RDT&E); specific modifications that must be made to launch vehicles in order to provide compatibility; and the purchase of capital assets such as buildings and other special facilities. Again, for a space mission, recurring costs include the cost of all satellite acquisitions (including initial, replacement, and spares), launch costs, engineering and research and development (R&D) costs incurred in continuing support of the mission, mission control and operations costs, and related administrative and support costs. When there are impacts of the mission in other nonmission areas, the resulting cost impacts must also be considered. For example, if additional costs will be incurred to maintain compatibility of other system/missions, then these costs need to be considered. If, on the other hand, other system or mission costs will be eliminated or foregone, then these cost reduction streams must also be considered (as a contra-expense or the equivalent of a revenue stream in the mission cash flow). These other costs may include both nonrecurring and recurring costs.

Life cycle cost analysis is usually an important part of the process associated with the justification of investments in new technology, missions, and systems. It is used frequently to demonstrate the magnitude of the economic improvement that will result from the investment, taking into account the often-faced problem of trading off increased near-term nonrecurring cost with longer term reductions in annual operating costs. Life cycle cost analysis frequently plays an important role in the evaluation and comparison of alternative courses of actions aimed at selecting the minimum cost approach for achieving a set of objectives. It is this latter area that is the concern of this article.

In the space business, annual costs, and therefore the present value of life cycle costs, must be considered as probability distributions. The reasons for this are many, but are primarily due to uncertainties associated with both nonrecurring and unit recurring cost (in no small part resulting from the use of new technology); less-than-perfect reliability of launch vehicles,[1,2] failures of which can significantly affect costs and schedules; schedule uncertainty resulting from both failures and necessary rescheduling to recover from failures and to respond to uncontrollable delays (i.e., weather and stand-down time required to correct previous failures); and random and wearout failures of operational satellites which require replacement and therefore affect the number of cost-incurring events and their timing.

As will be seen, the analysis of life cycle costs is further complicated by a myriad of operational alternatives involving sparing and maintenance/repair strategies. In addition to satellite design considerations relating to sparing and redundancy, there are operational considerations such as the use of on-orbit active or dormant spare satellites, ground spares, and the consideration of on-orbit servicing or repair at a transportation node. Since sensor/transponder availability is a function of overall satellite configuration (including redundancy), sparing and maintenance strategies, and transportation system reliability and availability, it is necessary to consider the many complex interrelationships that exist when configuring a satellite and estimating mission life cycle cost.

In the following paragraphs, a number of factors are discussed that require consideration when estimating life cycle costs. These include random and wearout failures, cost spreading, learning effects, operational analysis, and the explicit consideration of uncertainty and resulting risk. All are concerned with establishing the number of cost-incurring events and their timing and the magnitude of the associated cost. Each of these topics is discussed briefly. These are then followed by a discussion of a life cycle cost and availability model that builds on these concepts and adds a formal and consistent structure to life cycle cost analysis. Finally, the general flexibility and applicability of such a model are described together with typical results.

General Considerations

Operational Analysis

Space system operational analysis is the quantification of the probability distributions of the number of recurring cost-associated events required to establish and maintain a space mission. These events may include the number of satellites purchased, the number of launches, the number of satellite retrievals or refurbishments, and so on. Each of these events has associated with it a cost that, at least in the planning phase, includes some degree of uncertainty. The random nature of the number of events can be due to hardware failures, failure to accomplish certain prescribed events (e.g., a rendezvous and docking maneuver), or variability in certain wearout phenomena such as running out of station-keeping propellant. Thus, in addition to analyzing the number of events, operational analysis must also be concerned with their time of occurrence which will ultimately affect system availability.

Operational analysis explicitly considers the possibility of failures both in the transportation system and the satellites, the chance of these failures occurring, and the consequences if the failures do indeed occur. More specifically the analysis considers: 1) the number, sequence, and complexity of operations to be performed; 2) the recovery modes, that is, given that a failure has occurred, the possible resulting sequence of events; 3) the probability of successfully performing each of the required operations, both in the success and failure recovery sequences; and 4) scheduling (or lack thereof) of events. The results of an operational analysis in toto are referred to as the *operational risk*.

The transportation system operational analysis includes all of those events associated with the orbital placement, refurbishment, and retrieval of satellites, including the satellite deployment and initial operation. These last two events are included as transportation system events, because if failures occur during these operations, the recovery modes sometimes involve components of the transportation system.

The first step in performing an operational analysis of a given mission (e.g., a satellite placement, a satellite placement and retrieval, etc.) is to establish the success-oriented mission profile. The success-oriented mission profile consists of the nominal mission timeline or sequence of events where each event may be successfully accomplished or unsuccessful. If the event is accomplished successfully, the mission proceeds on to the next event. If the event is unsuccessful, a failure recovery mode must be adopted. Sometimes the failure recovery mode can return the mission to the normal timeline and sometimes major modifications are necessary in the mission. Some failures, such as the loss of a redundant subsystem, cause mainly inconvenience and possibly a minor cost item; others, such as the loss of a transfer stage and satellite, result in major cost items. It is generally possible and helpful to sketch the resultant success-oriented mission profile in cartoon form as shown by the example of Fig. 1. (Note that the reusable system shown can be made into an expandable system by setting recovery probabilities to zero.)

The second step of an operational analysis requires an explicit definition of what constitutes completion of a mission. In a service-type space program, if a satellite is required to provide the service and the mission is to place that satellite in orbit, then the satellite must be successfully placed and operational for the mission to be complete. If a particular flight fails to accomplish this, it must be repeated. On the other hand, if a satellite is to be retrieved from space and the retrieval attempt fails, one may elect not to make a second retrieval attempt, but instead to augment the ground-

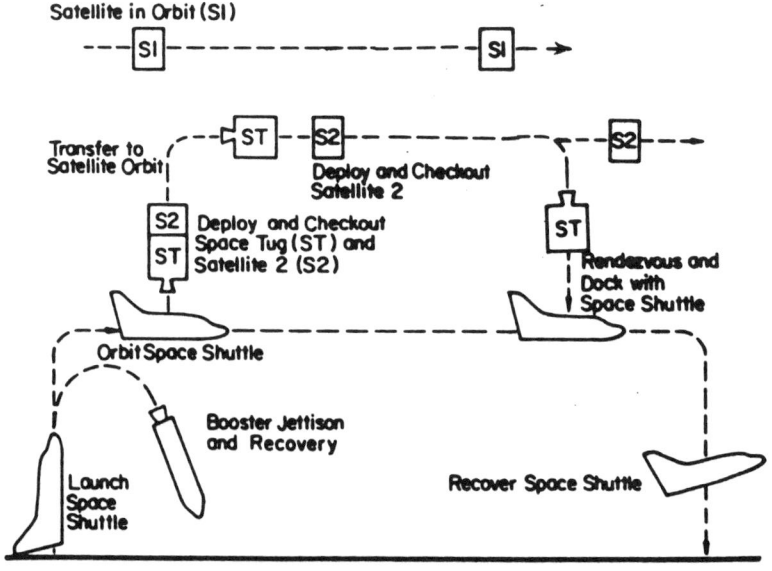

Fig. 1 A generalized satellite placement mission profile. (From Ref. 2.)

based inventory via the purchase of a new satellite. Here it becomes apparent that one of a variety of failure recovery modes must be chosen in order to proceed. Ultimately, each of the alternatives should be investigated and the choice made on an economic basis. [Reference 3 describes a model that incorporates maintenance/repair decisions that depend upon the type and timing of failures (i.e., a random failure near the end of wearout life will have different consequences than the same random failure occurring at the beginning of life).]

The third step involves establishing the mission scenario. The mission scenario is defined as the timeline sequence of all possible events (within the desired level of detail) that can occur from start to completion of a mission. The success-oriented path through the mission scenario is the mission profile; however, the mission scenario includes all of the pertinent failure recovery paths as well. The mission scenario can be thought of as a series of nodes connected by branches. Each node is a decision point representing a group of events. Emanating from each node are branches for the success and failure recovery paths. The probability of departing the node on one branch or another depends upon the probability of success (or failure) of the events represented by the node. The failure recovery paths must ultimately provide a route to mission completion as defined in step two above. In any event, mission completion requires a proper restoration of the inventory to its premission level. For example, if the mission is to place a satellite and retrieve a satellite, and if there is a failure to retrieve the satellite, then the purchase of a new satellite is necessary to restore the inventory to its proper level.

The mission scenario can also be shown as a logic flow diagram as in Fig. 2 for the example mission shown in Fig. 1. In Fig. 2, the nodes are represented by the diamond-shaped boxes (representing probabilities of success, or reliability) and the branches as lines with major cost-associated events given in the rectangular boxes.

When the mission scenario is established and the corresponding probability data are available, the operational analysis can be programmed for computer analysis. One method of computer analysis is by Monte-Carlo techniques wherein many random walks (typically 1000) are made through the mission scenario and events counted. Decisions are taken at each diamond-shaped box depending upon the value of a computer-generated random number relative to the associated probability of success. (If the computer-generated number is equal to or less than the probability of success, the success path is chosen, otherwise the failure path is chosen.) Because of the extremely low probability associated with certain of the paths, the finite Monte-Carlo analysis can yield only approximate results.

The space transportation system operational analysis results in both the determination of the number of events (and their timing) that will be used in the life cycle cost analysis (for each of the Monte-Carlo runs) and histograms or probability distributions (across all of the Monte-Carlo runs) of the number of occurrences of each major cost-associated event required to perform a given mission. (It is important to note that costs must be determined prior to performing the next Monte Carlo run, otherwise incorrect costs will be established because of the loss of event correlation effects.)

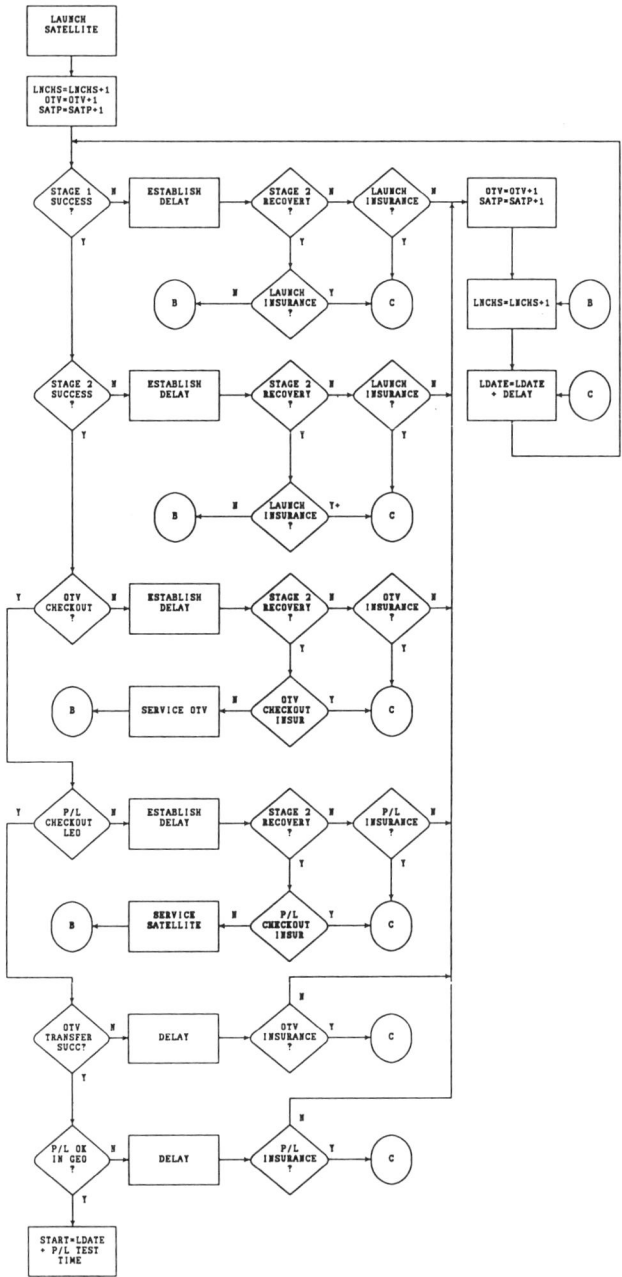

Fig. 2 Satellite placement scenario.

Satellite Operational Analysis

A service mission typically consists of two phases, namely the establishment of a desired level of service and the maintenance of that level of service. The establishment of the service is concerned with successfully placing a number of satellites in orbit over a period of time, as determined by service needs. If it is assumed that both the need and the capability of the satellites are known, it follows that the number of satellites required to establish the service is deterministic. The maintenance of the service is concerned with maintaining the desired number of operational satellites in service. As a result of less-than-perfect reliability and variability associated with wearout phenomena, such as attitude control gas depletion, satellite failures will occur in a random manner with the result that the traffic associated with the maintenance of the service will be known only in a probabilistic sense. The probabilistic nature of the traffic due to satellite failures and replacements adds an important degree of uncertainty and risk to the service mission. It is, therefore, necessary to consider the impact of satellite reliability on the overall mission.

Satellite failures occur as the result of four causes: 1) improper design or manufacture generally due to an incomplete understanding of all related physical principles; 2) imperfect quality control during manufacture; 3) uncertainties in the environment during storage, transportation, and operation; and 4) design and technology limitations generally associated with lifetime. These failure causes generally manifest themselves in three different types of failures: early, chance, and wearout. In this article, early satellite failures are considered in the space transportation system operational analysis. Their consequences are dependent upon the space transportation system and the operational modes.

The general satellite subsystem failure model, considering both random and wearout failures, is given by

$$R(n) = \frac{\exp(-\lambda n)}{\sigma(2\pi)^{1/2}} \int_n^\infty \exp\left[\frac{-(a - M)^2}{2\sigma^2}\right] da$$

This expresses the combined effects of the random and wearout phenomena as the probability of surviving through the nth time period (years) and λ is the failure rate (failures per year) or reciprocal of the mean time between failures (MTBF). M is the mean or expected wearout life (years) and σ is the standard deviation of wearout life (years). The general shape of the reliability function is illustrated in Fig. 3. The probability of a satellite subsystem failing in time period n, $F(n)$, is thus

$$F(n) = R(n - 1) - R(n)$$

The reliability model, taking into account the arrangement of subsystems (i.e., their series/parallel relationships), is used to establish the number and timing of satellite failures, based on the number in orbit and operational, and thus need to be replaced. This, together with the space transportation system operational analysis, establishes the probability distributions of all pertinent cost-associated events as a function of time.

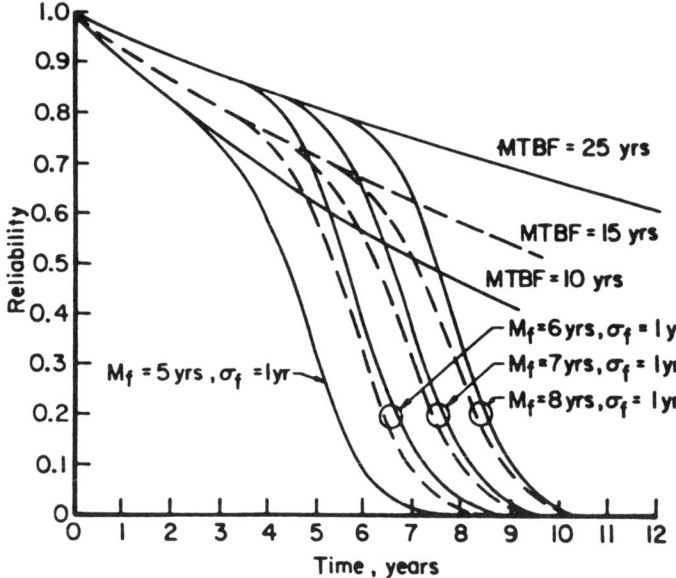

Fig. 3 Satellite subsystem reliability functions. (From Ref. 2.)

It should be noted that it is necessary to define a satellite failure, that is to specify the combination of sensor/transponder failures that will result in a cost-incurring event (e.g., a satellite replacement or servicing). Since satellites may be comprised of multiple sensors or transponders, it is necessary to specify critical failures or combinations of failures either in terms of specific failures or the loss of value where value is related to the relative importance of the information obtained from the sensors/transponders.[3] Thus, when the total satellite value is reduced below a specified threshold, a cost-incurring event is required.

Cost Analysis

Costs can generally be categorized as either nonrecurring, that is, one-time costs such as RDT&E, or recurring, namely those costs that are activity related. During the planning phase of a program, both nonrecurring and recurring costs are nondeterministic quantities. Nonrecurring cost uncertainties arise from all aspects of a program with which there may be any amount of uncertainty: the outcome of a test, the purchase of various equipments, manpower requirements, and so forth. Recurring cost uncertainties arise from the difficulties of predicting the cost of producing an item before it has been designed (this is particularly true when dealing with products that utilize new technology). A life cycle cost analysis must be concerned with the uncertainties in both nonrecurring and recurring costs.

The problem addressed is how to quantify uncertainty. This requires that informed estimates be made of the ranges of uncertainty of key cost variables and their probability distributions within the range. The estimates of uncertainty might be made, for example, at the cost-estimating relationship (CER) level or they might be made at the unit cost (transportation, satellite, etc.) level. The uncertainty assessments can be made by individuals with the assistance of an experienced analyst or they can be made by an experienced group of individuals using Delphi-type techniques.[4,5] The estimates are very subjective in nature and quantitatively express the attitudes regarding the uncertainties. The estimates reflect past experience with similar efforts, problems that have been encountered in the past, and insights into problems that might develop.

A methodology for establishing the shape of the uncertainty profiles, illustrated in Fig. 4, has been employed in numerous analyses.[6] The first step is to establish the range of uncertainty based on knowledgeable persons assessing what can go right and what can go wrong. The range is then divided into five equal intervals and a relative ranking of the likelihood of the cost variable falling into each of the intervals is performed. The general shape (skewed left, skewed right, central) of the uncertainty profile is thus established. The next step is to establish relative values of the chance of falling into each of the intervals. For example, in the illustration, the chance of falling into the first interval is estimated to be half as likely as falling into the second interval. This is repeated for each interval relative to the previously considered interval. The last step is to solve the illustrated equation for the quantitative values by substituting the data from the previous step.

In order to simplify this procedure, a large number of typical uncertatinty profiles may be stored in the computer. The evaluator may thus simply specify the range of uncertatinty (minimum and maximum values) and the name of the uncertatinty profile which reasonably represents his or her feelings. If none of the stored profiles is suitable, then the previously described procedure may be followed and the appropriate uncertatinty profile data provided as part of the input data.

In some cases, such as satellite unit recurring costs, it may be assumed that future procurement costs will be highly correlated with first unit costs. Historically, the cost of the nth satellite of a procurement will be less than that of the $n - 1$ satellite. As additional satellites are manufactured, costs decrease as a result of a "learning" process. Typically, the learning process is such that every time the quantity produced doubles, the cost is reduced by a specified percentage, i.e., one minus the learning rate. Therefore

$$\frac{n\text{th unit cost}}{\text{first unit cost}} = N^{[\log_{10}L - 2.0]/0.301}$$

where N is the cumulative number of units produced and L is the learning rate (%). Thus, if a 90% learning curve is followed, the unit cost of the second unit is 90% of the first unit cost; the cost of the fourth unit is 81% of the first unit cost; and so forth.

Normally the cost of a satellite or its launch does not all occur in a single year. These costs are spread over several years, typically 2 to 4. Thus, the computation of annual cost and cash flow must allow for this "cost spread-

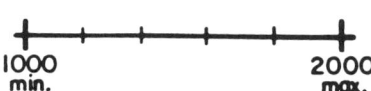
a) Specify Range of Uncertainty

b) Perform Ranking (Qualitative)

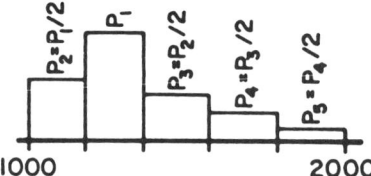
c) Establish Relative Values

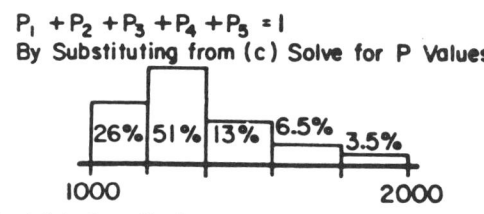
d) Establish Quantitative Values

Fig. 4 Methodology for establishing shape of cost uncertainty profile. (From Ref. 2.)

ing." Typically the time of an event is established and the cost spread backward in time according to an anticipated spending profile.

Economic Analysis

The economic analysis of space programs involves cash flow patterns that occur over several (n) years; thus, it is desirable to present the results of the economic analysis in terms of the present value of costs. The present value, which explicitly takes into account the magnitude and the timing of the cash flow patterns, is defined as the summation of future annual costs discounted to the present and is given by

$$PV = \sum_{i=1}^{n} \frac{C_i}{[1 + r/100]^i}$$

where PV is the present value of cost, C_i is the cost in the ith year, and r is the discount rate (%) or the cost of capital. When working with constant dollars, the real cost of capital should be used.

The costs entering into the above equation, however, are not deterministic quantities. Variations in the yearly costs C_i occur because of the uncertatinties in predicting future item (or per event) costs due to the uncertatinty both in the number of events necessary to perform the desired program and the time of occurrence of these events. Thus the present value of costs must also be characterized by a probability distribution.

Since the objective is to achieve a specified level of availability of sensors or transponders, comparisons of alternatives must be made at the desired level of availability. Since each present value of life cycle cost is characterized by both an expected value and a standard deviation (i.e., the risk dimension), a tradeoff between expected cost and risk must be made to select the best alternative. This is illustrated in Fig. 5 where all of the points represent alternatives that will result in the required level of availability. Here alternatives 1 and 2 have the same level of risk (i.e., $\sigma_1 = \sigma_2$), but the expected PV of cost of alternative 2 is greater than that of alternative 1. Therefore, alternative 1 is preferable to alternative 2. In a similar manner, it can be argued that alternative 3 is preferable to alternative 4. Also in a similar manner, alternative 3 is preferable to alternative 1 since both have the same expected PV, but alternative 1 is riskier. This process can be continued with all alternatives being considered. In the limit, it can be seen that a frontier of "best" alternatives can be established. Each of the points or alternatives represented by the frontier are different in the respect that the risk and expected PV are different. The class of best alternatives has thus been obtained and the "best" alternative can be selected based on the decision maker's risk judgment. That is, the decision maker must decide what the tradeoff is between a reduction in expected PV of cost and an accompanying increase in risk.

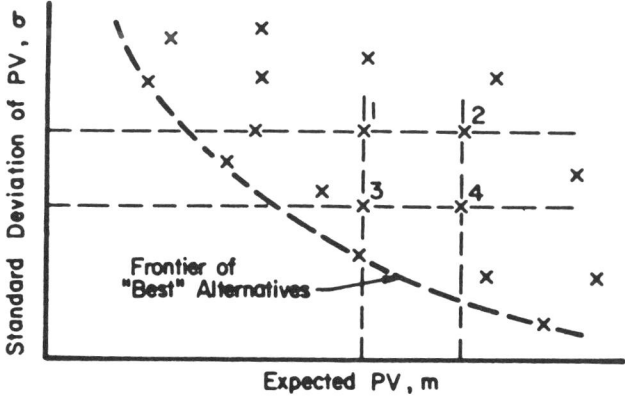

Fig. 5 General problem of decision making under uncertainty. (From Ref. 2.)

Satellite Life Cycle Cost and Availability Modeling

Utilizing the general considerations discussed in the previous paragraphs, a generalized space mission life cycle cost and availability model was developed[7-9] and is discussed in the following paragraphs. Typical results are then described.

Since mission requirements are frequently specified in terms of required sensor availability, the objective then is to select that system which minimizes the present value of life cycle cost (and risk) and satisfies mission requirements, including sensor availability. Sensor availability is a function of overall satellite configuration (including redundancy), sparing and maintenance strategies, and transportation system reliability and availability. It is therefore necessary to consider the many complex interrelationships that exist in order to select the system and associated operational approach that minimizes life cycle cost and achieves availability objectives. The following paragraphs describe methods and procedures that are required and have been developed for considering this complete system framework. This is accomplished by describing a developed simulation model (SATCAV) and presenting typical results that consider alternative maintenance and sparing strategies.

The Satellite Cost Availability Model (SATCAV) is a dynamic stochastic life cycle cost and availability model that simulates the launch and on-orbit operations associated with the initiation and continuing operation of a generalized space mission comprising multiple satellites with multiple sensors. The model operates on an IBM PC microcomputer and utilizes a LOTUS 123 user-friendly menu-driven input/output system to create a data file that is accessed by a FORTRAN Monte Carlo program that performs the computations. The model develops cost, event, availability, and cause of failure statistics reports. It also develops a typical event timeline report.

SATCAV utilizes the previously described techniques. It simulates satellite launch operations using expendable or recoverable launch vehicles and upper stages and takes into account the consequences of a set of defined failures in terms of cost incurring events and time delays. It simulates the random and wearout failures of a multisensor satellite, determining when specific failures occur and when maintenance actions are required to respond to critical failures. The model encompasses alternative maintenance scenarios that include both ground and on-orbit active or dormant spares. Both launch on failure and launch in anticipation of wearout failure alternatives are available. Different transportation scenarios may be selected for placement and maintenance flights. Maintenance also includes on-orbit servicing, and repair at a transportation node. The SATCAV Model provides the means for assessing alternative transportation scenarios, sparing strategies, maintenance/repair concepts, and satellite configurations, in terms of life cycle costs and availability statistics.

Launch operations of a multistage vehicle are simulated by considering the reliability associated with the performance of major operations (including payload checkout and testing) and the consequences of a priori specified types of failures in terms of cost-incurring events and time delays. Delays that may result from different types of failures are taken into account in the simulations. Both expendable and recoverable launch vehicles and upper stages may be considered. Random and wearout characteristics of a multisensor satellite are simulated, determining when specific failures occur and when maintenance actions are required to respond to critical failures.

The consideration of subjective uncertainties, transportation system reliability, and satellite subsystem random and wearout failures results in the establishment of the probability distribution of annual events and the probability distributions of annual costs and present value of life cycle costs. Sensor availability statistics are developed as the probability density function of the number of sensors available (for each sensor type) annually as well as over the mission duration. The statistics take into account sensor and subsystem random and wearout failure characteristics, sparing strategy, transportation scenarios, delays associated with different types of failures, and duration of on-orbit testing. Cause-of-failure statistics are also developed that indicate the probability that different subsystem failures were the cause of the sensor failures.

Important attributes of a space system life cycle cost and availability model include:

1) Reasonably long time horizon (up to 25 years)
2) Consideration of multiple operational satellites
3) Consideration of multiple sparing strategies that include:
    a) Multiple ground and on-orbit active or dormant spares
    b) Launch on failure of critical sensors
    c) Launch on random failure but in anticipation of wearout failure of critical sensors.
4) Identification of critical sensors (individually and in combination)
5) Specification of desired launch dates for initial operational satellites and on-orbit spares (actual launch dates take into account failures and associated delays)
6) Specification of completion date of ground spares
7) Specification of the probability that a dormant satellite will operate successfully when activated and of the time delay to move the dormant satellite into replacement position
8) Consideration of satellite cost learning effects
9) Choice of transportation scenario for initial placement and maintenance flights from the following:
    a) Direct placement using Earth-based assets
    b) Placement and return using Earth-based assets
    c) On-orbit repair using Earth-based assets
    d) Replacement/return/repair using Earth-based assets
    e) Direct placement using space-based assets
    f) Placement and return using space-based assets
    g) On-orbit repair using space-based assets
    h) Replacement/return/repair using space-based assets
    i) Replacement/return/repair on and from space-based assets
10) Ability to describe transportation scenarios in terms of the probability of successfully performing each major operation in the launch/

recovery sequence taking into account the consequences of failures in terms of cost-incurring events and time delays
11) A priori defined uncertainty variables (i.e., subjectively specified as ranges of uncertainty and the form of uncertainty) include:
   a) Delays (function of failure type)
   b) Transportation cost from Earth to low Earth orbit (LEO)
   c) Transportation cost from LEO to Earth
   d) Transportation cost from LEO to payload (P/L) orbit
   e) Transportation cost from P/L orbit to LEO
   f) Satellite unit recurring cost
   g) Satellite repair kit cost
   h) Transfer vehicle repair cost
   i) Multiple capital expenditures
   j) Satellite control operations cost
   k) Engineering expenses
12) Satellite configuration consisting of multiple sensors, each with multiple support subsystems and multiple satellite bus subsystems
13) Ability to describe each sensor and subsystem in terms of its random and wearout failure characteristics
14) Cost spreading for nonrecurring capital items and satellite, satellite repair kit, and launch recurring cost
15) Ability to develop cost, event, availability, and cause of failure statistics, and satellite expected life and associated standard deviation

With the above factors included in the life cycle cost and availability model, it is possible to analyze space mission life cycle costs and availability and to evaluate the effects of alternative launch and procurement schedules, achievement of different subsystem reliabilities (random and wearout), alternative sparing strategies, alternative transportation scenarios (for placement, replacement, service and repair), utilizing expendable or reusable launch vehicles, launch vehicle reliability and cost, and launch delays on developed life cycle cost and availability statistics. As a result, their impacts on life cycle costs can be established. The result is the determination of annual cost (expected value and standard deviation), present value of annual cost and risk, and sensor availability statistics, which serves as the basis for comparing and evaluating alternatives in order to achieve minimum cost approaches within acceptable levels of risk for achieving mission objectives.

SATCAV utilizes Monte Carlo simulation techniques. Monte Carlo implies the performance of an experiment or simulation many times, such as rolling two dice (either actually or through the use of simulation model) repeatedly to determine the chance of seven or more occurring. In the SATCAV Model the experiment consists of simulating a space mission for up to 25 years. This mission is simulated typically for about a thousand different situations where each of the specific situations is developed by random sampling of a set of probability density functions or selected subjective uncertainty profiles, and specific event timing is established from specified random and wearout failure characteristics. The specific values obtained from the sampling of the selected subjective uncertainty profiles and the reliability characteristics then establish the parameters of the space mission that is simulated. Results from all of the simulations are saved and histograms are developed (or summarized by expected values and standard deviation) of pertinent cost and availability measures. These histograms represent the result of combining the selected areas of uncertainty, and transportation and satellite reliability attributes of the space mission.

Basic input data to a stochastic (Monte Carlo) model consists of both deterministic and probabilistic data. Examples of deterministic data are the number of time periods to be considered, the discount rates, and the number of satellites to be considered and their desired launch schedule. Probabilistic data consist of the "uncertainty profiles" associated with the variables whose values cannot be predicted or known exactly in advance. Typical uncertainty variables include schedule delays, expense items, and capital expenditures, and their specification requires estimates of maximum and minimum values and the identity of the associated uncertainty profile (which then determines the probability distribution of the variable within the range of uncertainty).

The cost and availability analysis is performed by the random sampling of the input data (according to the weighting of the uncertainty profiles), performing computations contained within the simulation model, saving the results, then repeating the process. This process is repeated many times until a reasonable set of histograms can be developed from the saved output. These histograms are worked into the desired form to indicate the variability of performance measures, such as annual cost, net present value of life cycle cost, and sensor availability. Cost measures are summarized in terms of expected values and standard deviations. Also established are the probability density functions of launch and satellite purchase events and sensor availability. It is convenient to display the developed performance measures in the form of "risk profiles," which indicate the chance of a performance measure exceeding specific levels (i.e., the complement of the cumulative probability distribution function).

The life cycle cost model consists of a number of computational procedures or sections (not to be confused with subroutines). These include sections concerned with:

1) Simulating the launch sequence (i.e., operational analysis) whenever a launch is required as determined by desired launch dates, launch and satellite failures, and specified placement and maintenance scenarios

2) The determination of sensor support subsystem failures and the time of failure

3) The determination of sensor failures and the time of failure

4) The estimation of satellite replacement time as determined by considerations of specified launch criteria and a determination of satellite failure time (this feeds results back to the launch sequence simulation)

5) Establishing the number of operational sensors of each type as a function of time

6) Cost analysis that takes into account learning effects and cost spreading

7) Developing total annual and life cycle costs taking into account other costs such as satellite (S/C) control operations and satellite repair. A num-

ber of performance measures are developed with both expected values and standard deviations determined so that risk profiles may be developed

8) Report generation that includes the annual and life cycle cost projections and statistics on launch attempts, satellite purchases, and sensor availability

A simplified computational flow is illustrated in Fig. 6. Within the computational flow there are many levels of disaggregation and hence many computational loops. For the sake of simplification and to illustrate the level of detail often required in space system life cycle cost analysis, dormant spares logic is shown separately in Fig. 7. The dormant spares computation is as follows. When dormant spares are to be considered, it is

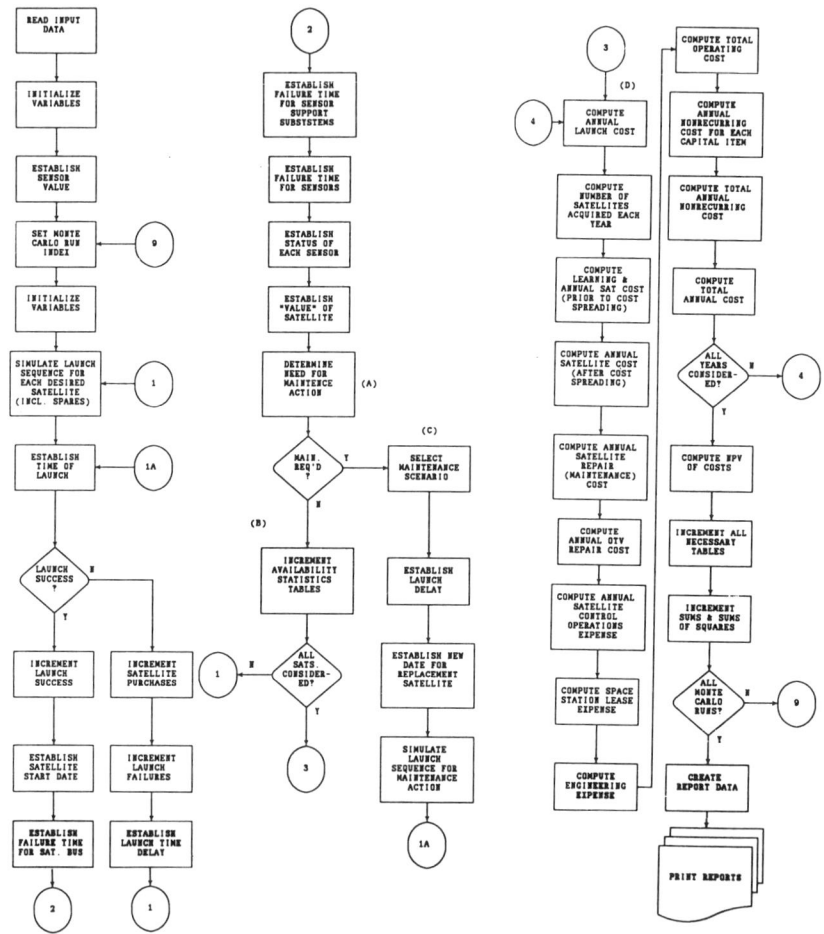

Fig. 6 SATCAV model: computational flow.

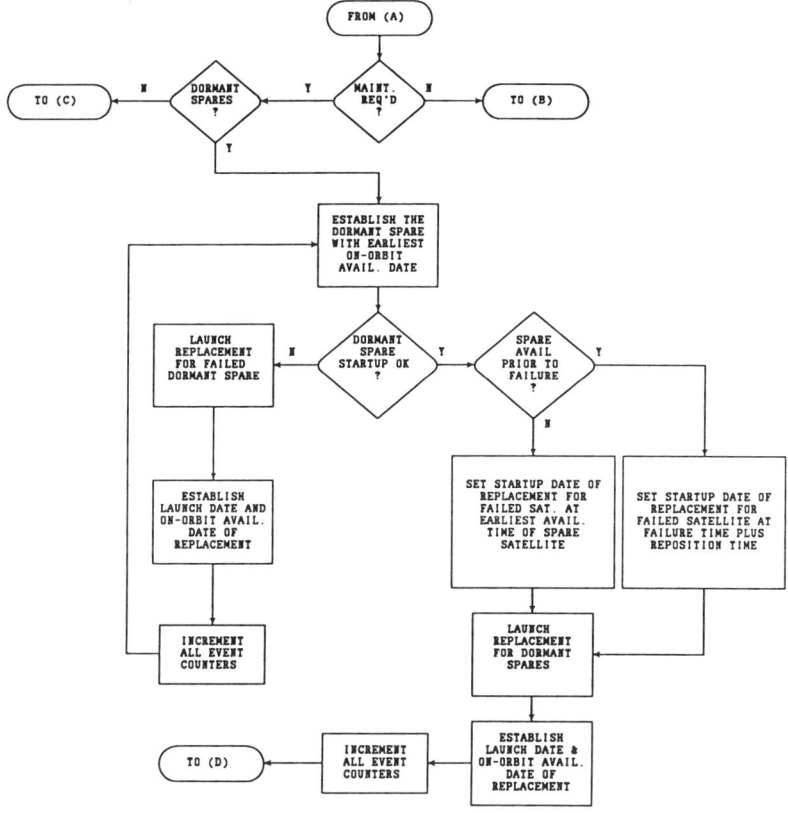

Fig. 7 Consideration of dormant spares. (From Ref. 8.)

necessary to verify the availability of the dormant spares for use when needed. Then:

1) The identity of the dormant spare with the earliest on-orbit availability is established. This dormant spare will be the first utilized (if it operates when turned on) to replace a failed operational satellite.

2) A determination is made as to whether the desired dormant spare operates successfully (the probability of success is specified via input data) when it is turned on (prior to its repositioning).

3) If the dormant spare fails, then the dormant spare is replaced using a placement launch scenario.

4) When a replacement for a failed dormant spare is successfully placed into orbit, its launch date and on-orbit availability date (taking into account all associated launch and on-orbit testing delays) are established.

5) All satellite and launch associated counters are incremented as appropriate and the results later included in the cost computations.

If the dormant spare does not fail, then:

1) When the selected dormant spare starts up, its availability time is compared with the time of failure of the operational satellite it is to replace.

2) If a spare dormant satellite is not available prior to the operational satellite time of failure, then start date is set equal to the earliest availability time of a spare satellite.

3) If a spare dormant satellite is available prior to the operational satellite time of failure, then start date is set equal to failure date plus repositioning time (it is assumed that the dormant spare moves to the position of the failed satellite).

4) A replacement is launched for the dormant spare. The appropriate launch scenario is utilized and all appropriate launch delays and events are taken into account.

5) The launch date and the on-orbit availability date of the new (i.e., replacement) dormant spare are established.

6) All satellite and launch associated counters are incremented as appropriate and the results later included in the cost computations.

To illustrate a user-friendly method for providing data for and the specifics of the data required by a life cycle cost analysis, a set of input data screens (containing illustrative data) are presented in Figs. 8–13. Figure 8 contains the global or overall system data and includes:

1) Number of years to be considered in the analysis

2) Sparing strategy (1 refers to launch on failure, 2 refers to dormant spares, and 3 refers to launch in anticipation of wearout failures)

3) Number of operational satellites and their desired launch dates (actual launch dates depend upon failures and associated time delays)

4) Number of on-orbit (active or dormant) spares and their desired launch dates

5) Number of ground spares and their completion dates

```
MODIFY        PRINT        NEXT         RETURN-MAINMENU
Modify data in this screen

1    [1]  GLOBAL DATA (SYSTEM)          8. MAX. # GROUND SPARES      1
2                                       9. COMPLETION DATE (YR)
3    1. NO. YRS. ANALYZED       10         SATELLITE NO. 1          5.0
4    2. MAX # OF LNCH SCEN'S     9         SATELLITE NO. 2
5    3. SPARING STRATEGY         1         SATELLITE NO. 3
6    4. MAX. # OPER. SATS        3         SATELLITE NO. 4
7    5. DESIRED LNCH DATE (YR)             SATELLITE NO. 5
8         SATELLITE NO. 1       5.0    10. NO. SIMUL. RUNS          100
9         SATELLITE NO. 2       5.0    11. P/L LEARN. RATE (%)      90.0
10        SATELLITE NO. 3       7.0    12. AVAIL. START (YR)        5.0
11        SATELLITE NO. 4              13. ANTICIPATORY TIME
12        SATELLITE NO. 5              TRANSPORTATION SCENARIOS:
13   6. MAX. # ON-ORB SPARES     0     14. INITIAL PLACEMENT         1
14   7. DESIRED LAUNCH DATE (YR)       15. MAINT. & REPAIR           1
15        SATELLITE NO. 1              16. SPACE STATION
16        SATELLITE NO. 2                   STORAGE COST (M$/YR)
17        SATELLITE NO. 3              DORMANT SPARES (SPARING STRAT=2)
18        SATELLITE NO. 4              17. PROB. TURN ON
19        SATELLITE NO. 5              18. DELAY
20   19.DISCOUNT RATE (%)       0.0    5.0   10.0   15.0   20.0
```

Fig. 8   Global data (system).

LIFE CYCLE COST AND AVAILABILITY ANALYSIS

6) Number of Monte Carlo runs to be performed
7) Payload cost learning rate
8) Availability start time (i.e., the point in time at which sensor availability computations start)
9) Probability that a dormant spare will turn on successfully
10) Time delay (fraction of a year) to move the dormant spare into position so as to replace a failed satellite
11) Identification of the transportation scenario to be used for initial placement and maintenance/repair flights
12) Space Station storage cost for those transportation scenarios that utilize the Space Station.

Two transportation scenarios must be identified: one for initial P/L placement operations and the other for P/L maintenance/repair operations. The same scenario may be identified for both initial placement and maintenance/repair operations. It is assumed that a transportation system consists of a generic launch vehicle (LV) that contains two stages (LVS1 and LVS2) and a generic orbital transfer vehicle (OTV). Each of these stages may actually contain other stages, but can be considered from a reliability, cost, and recovery point of view as being lumped into a single stage. The following is a brief description of each of nine transportation scenarios that have been developed and included in the life cycle cost and availability model:

Scenario 1—Consists of a reusable LV (LVS1 may be reusable or expendable) or LV and OTV or expendable LV and OTV for initial P/L placement and/or replacing failed P/Ls. Scenario 1 may be used in conjunction with scenario 4 for the placement portion of a replace/return/repair mission. If reusable launch vehicles are considered, then OTV and P/L checkout failures in LEO may be corrected when and if returned to Earth.

Scenario 2—Consists of a reusable LV (LVS1 may be reusable or expendable) and OTV for placing a P/L into orbit and returning a failed P/L in the same OTV flight. If reusable launch vehicles are considered, then OTV and P/L checkout failures in LEO may be corrected when and if returned to Earth. The returned P/L is repaired and placed into inventory for future use.

Scenario 3—Consists of a reusable LV (LVS1 may be reusable or expendable) or LV and OTV, or expendable LV and OTV for docking a P/L repair kit with a failed P/L and performing on-orbit repair at the P/L location. It is assumed that the OTV is capable of docking with specifically configured P/Ls. If repair cannot be accomplished, then a replacement is performed using the specified initial placement scenario. If reusable launch vehicles are considered, then OTV checkout failures may be corrected when and if returned to Earth.

Scenario 4—Consists of a reusable LV (LVS1 may be reusable or expendable) and OTV for acquiring and returning a failed P/L. It is assumed that the OTV is capable of docking with specifically configured P/Ls. Replacement is performed (prior to returning failed P/L) using an appropriate specified scenario. If reusable launch vehicles are considered, then OTV checkout failures may be corrected when and if returned to Earth.

Scenario 5—Consists of a reusable or expendable LV for providing P/Ls, as required, to a space-based facility such as the Space Station (SS) where on OTV (based at the SS) places the P/L into final P/L orbit. The OTV may be reusable or expendable. It is assumed that P/L and OTV checkout failures can be corrected at the SS.

Scenario 6—Consists of a reusable LV (LVS1 may be reusable or expendable) and OTV for placing a P/L into orbit and returning a failed P/L in the same OTV flight. The reusable LV provides the new P/L to the SS and returns the failed P/L from the SS to Earth. It is assumed that P/L and OTV checkout failures can be corrected at the SS.

Scenario 7—Consists of a reusable LV (LVS1 may be reusable or expendable) or LV and OTV, or expendable LV and OTV for performing on-orbit P/L repair. OTVs are located at the SS and, it is assumed, are capable of docking with specifically configured P/Ls. If repair cannot be accomplished, then replacement will be accomplished either via initial placement scenarios 1 or 5 (as specified). It is assumed that OTV checkout failures can be corrected at the SS. A repair kit is delivered from the Earth to the SS and, upon mission completion, returned to Earth.

Scenario 8—Consists of a reusable OTV for acquiring and returning the failed P/L to the SS. The P/L is then returned to Earth for repair using a reusable LV. It is assumed that the OTV is capable of docking with specifically configured P/Ls. Replacement is performed (prior to returning the failed P/L) using the specified initial placement scenario. It is assumed that OTV checkout failures can be corrected at the SS.

Scenario 9—Consists of a reusable LV and OTV transportation system for placing a P/L into orbit and returning a failed P/L in the same OTV flight. A P/L is stored on the SS and repair is performed on the SS. An initial flight is required to place a P/L into inventory on the SS.

Figure 9 illustrates the transportation system data requirements. These data must be provided for both the placement and maintenance/repair scenarios (if different). The data format is the same for each of the nine scenarios, with the input system designed so that only the required data need be provided. The transportation scenario data are comprised of reliability, delay, and cost data. In order to accomodate improvements in transportation system reliability over time, two sets of reliability data may be provided together with the time of their applicability. The reliability data are as follows and provide the input for the simulation of launch and on-orbit operations:

1) Probability of stage 1 (LVS1) success
2) Probability of stage 2 (LVS2) recovery given a stage 1 failure
3) Probability of stage 2 success given stage 1 success
4) Probability of stage 2 recovery given a stage 2 abort
5) Probability of transfer stage (referred to generically as the OTV) checkout success in LEO
6) Probability of stage 2 recovery given an otherwise successful mission
7) Probability of OTV transfer success
8) Probability that the payload (P/L) checks out successfully in LEO
9) Probability that the payload functions successfully in its final orbit (referred to as GEO)

LIFE CYCLE COST AND AVAILABILITY ANALYSIS

```
MODIFY         PRINT          NEXT          RETURN-MAINMENU
Modify data in this screen

 1    [2]  TRANSPORTATION SCENARIO 1:  DIRECT PLACEMENT (GROUND BASED ASSETS)
 2         ***** Use arrow keys to scroll up and down.*****
 3
 4    RELIABILITY DATA
 5
 6                              EARLY  LATER                        EARLY  LATER
 7    PROBABILITY OF             YEARS  YEARS                        YEARS  YEARS
 8    1. LVS1 SUCCESS            0.980  0.980    7. OTV TRANSFER SUCC 0.970 0.970
 9    2. LVS2 RECV:LVS1 FAIL     0.000  0.000    8. P/L CHECKOUT IN LEO 1.000 1.000
10    3. LVS2 SUCC:LVS1 SUCC     0.980  0.980    9. P/L OK IN GEO      0.960 0.960
11    4. LVS2 RECV:LVS2 ABOR     0.000  0.000   10. OTV REC:OTV SUCC   0.000 0.000
12    5. OTV CHECKOUT SUCC       1.000  1.000   11. OTV RENDEZVOUS     0.000 0.000
13    6. LVS2 RECV:LVS2 SUCC     0.000  0.000   12. YR OF CHANGE         5
14
15    DELAY DATA                                 DELAY TYPE
16                                LVS1   LVS2   OTVC   P/LC   OTV    P/L
17    13. MAXIMUM DELAY (YRS)     0.75   0.50   0.00   0.00   0.00   0.00
18    14. MINIMUM DELAY (YRS)     0.25   0.25   0.00   0.00   0.00   0.00
19    15. DELAY UNCERT PROFILE     15     15     1      1      1      1
20    16. P/L START DELAY (YRS)   0.25
21
22    COST DATA                                        MAX    MIN   PROFILE
23                                                    (M$)   (M$)   ID#
24    17. LAUNCH COST FROM EARTH TO LEO               50.0   50.0    1
25    18. RETURN COST FROM LEO TO EARTH                0.0    0.0    1
26    19. OTV COST FROM LEO TO GEO                    10.0   10.0    1
27    20. OTV COST FROM GEO TO LEO                     0.0    0.0    1
28                                                    MAX    MIN   PROFILE
29                                                    (%)    (%)    ID#
30    21. P/L REPAIR COST FOR CHKOUT FAILURE          10.0   10.0    1
31    22. P/L REPAIR COST FOR P/L FAILURE              0.0    0.0    1
32    23. OTV REPAIR COST FOR CHKOUT FAILURE          10.0   10.0    1
```

Fig. 9 Transportation scenario data.

10) Probability of recovering the OTV given an otherwise successful OTV flight

11) Probability that the OTV rendezvous successfully with a payload (when using a service/repair maintenance scenario)

The above set of probabilities can be used for either expendable or reusable vehicles; an expendable vehicle has recovery probabilities set equal to zero.

The delay data associated with failures are considered as uncertainty variables requiring minimum and maximum values (i.e., the range of uncertainty) and the form of the uncertainty (the identity of one of a large number of probability density functions that may be specified via input data). These delays are a function of the type of failure and include the loss of stages 1 and 2, failure of the OTV to checkout successfully in LEO, failure of the OTV, and failure of the payload to function properly when placed into its final orbit. Finally, a payload start delay may be specified that includes on-orbit checkout and acceptance testing.

All cost data are considered as uncertainty variables. Costs are disaggregated according to flight regime so that the more complex scenarios may be appropriately modeled. Payload repair costs for checkout failures and for satellite failures (when the payload can be returned or is available for repair) are specified as percentages of initial payload cost. OTV repair cost for checkout failures is specified as a percentage of initial OTV cost.

```
MODIFY        PRINT         NEXT         RETURN-MAINMENU
Modify data in this screen

1    [13] SENSOR & SENSOR SUPPORT SUBSYSTEM RELIABILITY DATA
2
3
4                                              SENSOR
5                                     1     2     3     4     5
6       1. MEAN-TIME FAIL (YRS)       25   999   999   999   999
7       2. EXP. WEAROUT (YRS.)        3.0  99.0  99.0  99.0  99.0
8       3. STD.DEV. WEAROUT (YRS.)   0.500 0.100 0.100 0.100 0.100
9                                          SENSOR SUPPORT #1
10                                    1     2     3     4     5
11      1. MEAN-TIME FAIL (YRS)       40   999   999   999   999
12      2. EXP. WEAROUT (YRS.)        5.0  99.0  99.0  99.0  99.0
13      3. STD.DEV. WEAROUT (YRS.)   0.500 0.100 0.100 0.100 0.100
14                                         SENSOR SUPPORT #2
15                                    1     2     3     4     5
16      1. MEAN-TIME FAIL (YRS)      999   999   999   999   999
17      2. EXP. WEAROUT (YRS.)       99.0  99.0  99.0  99.0  99.0
18      3. STD.DEV. WEAROUT (YRS.)   0.100 0.100 0.100 0.100 0.100
```

Fig. 10 Sensor & sensor support subsystem reliability data.

The payload reliability data are provided as illustrated in Figs. 10 and 11. The payload is assumed to consist of multiple sensors, each with multiple support subsystems (the failure of a sensor and/or one or more of its support subsystems causes a sensor failure). These sensor groups are contained on a bus that may consist of multiple support subsystems; the failure of any one causes all of the sensors to fail. Both random and wearout failures are considered for each subsystem with the random failures being characterized by the mean-time-to-failure (i.e., an exponential distribution) and the wearout failures being characterized by an expected wearout time and the standard deviation of wearout time (i.e., a normal distribution).

In addition to the reliability data, information is provided (input form not shown) that indicates the critical sensors, both individual and in combinations, failure of which will trigger a maintenance or repair response.

Figure 12 illustrates the capital expenditure data. A number of specific expenditures may be identified along with associated cost uncertainty data, the year of completion, and the percentage of the expenditure made in

```
MODIFY        PRINT         NEXT         RETURN-MAINMENU
Modify data in this screen

1    [14] BUS SUBSYSTEM RELIABILITY DATA
2
3
4
5                                         BUS SUBSYSTEM
6                                         1     2     3
7       1. MEAN-TIME FAIL (YRS)          20    40   999
8       2. EXP. WEAROUT (YRS.)           5.0   5.0  99.0
9       3. STD.DEV. WEAROUT (YRS.)      0.500 0.500 0.100
```

Fig. 11 Bus subsystem reliability data.

LIFE CYCLE COST AND AVAILABILITY ANALYSIS 139

```
MODIFY         PRINT         NEXT        RETURN-MAINMENU
Modify data in this screen

1    [17] CAPITAL EXPENDITURE DATA
2
3    NON-RECURRING COST                  6. SPREADING
4    ---------------------------------------------------------------
5    1. NAME      2.MAX  3.MIN  4.ID#  5.YR  YEAR1 YEAR2 YEAR3 YEAR4 YEAR5 |TOTAL
6                 (M$)   (M$)                                             |
7    A.XXXX       10.0   10.0     1     5     10    40    50              | 100
8    B.YYYY       20.0   20.0     1     6     10    30    60              | 100
9    C.ZZZZ       30.0   30.0     1     5     20    20    20    20    20  | 100
10   D.                                                                   |   0
11   E.                                                                   |   0
12   F.                                                                   |   0
13   G.                                                                   |   0
14   H.                                                                   |   0
15   I.                                                                   |   0
16   J.                                                                   |   0
```

Fig. 12   Capital expenditure data.

years prior to completion. Payload and repair kit unit recurring costs are also input as uncertainty variables. Figure 13 illustrates the cost-spreading data for launch, payload, and repair kit recurring costs. As with the non-recurring costs, costs are spread backward in time, from when the cost-incurring event was determined to have occurred.

Typical reports resulting from the SATCAV Monte Carlo analysis are illustrated in Figs. 14–17. Figure 14 indicates detailed costs for each year of the time horizon with all results being expected values except those indicated as *ST. DEV.* which represent standard deviations. Also indicated are quantities of satellites and repair kits produced which include all spares and operational (including replacements) equipment. Also indicated is the present value of life cycle cost at the previously specified discount rates.

Figure 15 indicates typical developed event statistics (available for each year of the analysis) in the form of the probability of occurrence of the indicated number of events. Also indicated are expected values and associated standard deviations. Statistics are maintained separately for initial placement flights and for maintenance/repair flights. Because the resulting event data are not normally distributed (an approximation that is reasonable for cost results), the actual event probability distributions require development as is indicated.

```
MODIFY         PRINT         NEXT        RETURN-MAINMENU
Modify data in this screen

1    [18] COST SPREADING DATA FOR P/L AND LAUNCH COST
2
3
4                                     YEAR
5                              1    2    3    4    5   |TOTAL
6                                                      |
7    1. LAUNCH COST........   40   40   20             |  100
8    2. P/L COST..........   30   50   20             |  100
9    3. P/L REPAIR KIT....  100                        |  100
```

Fig. 13   Cost-spreading data for P/L and launch cost.

ANNUAL COST PROJECTION (M$)

YEAR	1	2	3	4	5	6
LAUNCH COST	0.0	0.0	24.8	55.6	68.7	32.8
EARTH TO LEO (SAT)	0.0	0.0	24.8	55.6	68.7	32.8
LEO TO EARTH (SERV)	0.0	0.0	0.0	0.0	0.0	0.0
LEO TO SAT ORBIT	0.0	0.0	0.0	0.0	0.0	0.0
SAT ORB. TO LEO (REP)	0.0	0.0	0.0	0.0	0.0	0.0
LEO TO EARTH (REP)	0.0	0.0	0.0	0.0	0.0	0.0
EARTH TO LEO (REPKIT)	0.0	0.0	0.0	0.0	0.0	0.0
SATELLITE COST	0.0	0.0	80.6	212.4	160.7	60.0
SAT CONTROL OPS	0.0	0.0	0.0	0.0	2.0	4.0
SAT MAINTENANCE	0.0	0.0	0.0	0.0	0.0	0.0
SAT SERV	0.0	0.0	0.0	0.0	0.0	0.0
SAT REPAIR	0.0	0.0	0.0	0.0	0.0	0.0
REPAIR KIT	0.0	0.0	0.0	0.0	0.0	0.0
OTV MAINTENANCE	0.0	0.0	0.0	0.0	0.0	0.0
SPACE STATION LEASE	0.0	0.0	0.0	0.0	0.0	0.0
ENGINEERING EXPENSES	0.0	0.0	0.0	0.0	4.5	5.0
TOTAL OPER. COST	0.0	0.0	105.4	268.0	235.9	101.8
ST. DEV.	0.0	0.0	23.9	58.8	64.9	64.7
EXPENDITURE						
SAT NON-RECURRING	30.0	120.0	120.0	30.0	0.0	0.0
	0.0	0.0	0.0	0.0	0.0	0.0
	0.0	0.0	0.0	0.0	0.0	0.0
	0.0	0.0	0.0	0.0	0.0	0.0
	0.0	0.0	0.0	0.0	0.0	0.0
	0.0	0.0	0.0	0.0	0.0	0.0
	0.0	0.0	0.0	0.0	0.0	0.0
	0.0	0.0	0.0	0.0	0.0	0.0
	0.0	0.0	0.0	0.0	0.0	0.0
	0.0	0.0	0.0	0.0	0.0	0.0
TOTAL NONRECURRING COST	30.0	120.0	120.0	30.0	0.0	0.0
ST. DEV.	0.0	0.0	0.0	0.0	0.0	0.0
TOTAL ANNUAL COST	30.0	120.0	225.4	298.0	235.9	101.8
ST. DEV.	0.0	0.0	23.9	58.8	64.9	64.7
QUANTITY OF SAT. PROD.	0.0	0.0	0.0	0.0	4.5	0.6
ST. DEV.	0.0	0.0	0.0	0.0	0.9	0.8
QUANTITY OF REPAIR KITS	0.0	0.0	0.0	0.0	0.0	0.0
ST. DEV.	0.0	0.0	0.0	0.0	0.0	0.0

	1	2	3	4	5
DISCOUNT RATE (%)	0.0	5.0	7.0	10.0	15.0
PRESENT VALUE	1985.0	1458.0	1305.0	1119.0	889.0
ST. DEV.	262.0	181.0	159.0	133.0	103.0

Fig. 14 Annual cost projection report.

Figure 16 illustrates (for a satellite with a single sensor) developed availability statistics over the planning horizon. Similar reports provide annual availability results. The results indicate the percentage of the time that the indicated number of sensors are available. The availability is measured across all satellites and takes into account all launch and on-orbit delays associated with satellite placement and maintenance/repair actions.

Figure 17 indicates the probability that specific subsystems were the cause of sensor failures. This information is useful in pinpointing specific improvements that may materially influence availability and cost. Also indicated are the expected satellite life and the associated standard deviation. The satellite life statistics are based upon the satellite useful life (the time at which a maintenance action is required) and take into account all random

LIFE CYCLE COST AND AVAILABILITY ANALYSIS

PROBABILITY OF INDICATED EVENTS   YEAR 5

	PLACEMENT FLIGHTS				REPAIR FLIGHTS			
# OF EVENTS	LNCHS	P/LS SERV.	P/LS REPAIR.	OTVS	LNCHS	P/LS SERV.	P/LS REPAIR.	OTVS
10	0	0	0	0	0	0	0	0
9	0	0	0	0	0	0	0	0
8	0	0	0	0	0	0	0	0
7	0	0	0	0	0	0	0	0
6	1	0	0	1	0	0	0	0
5	0	0	0	0	0	0	0	0
4	9	0	0	9	0	0	0	0
3	11	0	0	11	1	0	0	1
2	79	0	0	79	0	0	0	0
1	0	0	0	0	12	0	0	12
0	0	100	100	0	87	100	100	87
AVG. VALUE	2.3	0	0	2.3	0.1	0	0	0.1
ST. DEV.	0.7	0	0	0.7	0.4	0	0	0.4

Fig. 15 Probability of indicated events.

and wearout failures as well as the identified criticality of individual and combinations of sensors.

Typical Application

The results of a life cycle cost and availability analysis, made possible by simulation techniques and models such as the SATCAV Model, include both cost and cost risk information associated with achieving different levels of availability. This information can be used to evaluate spacecraft design alternatives, transportation alternatives, maintenance/repair strategy alternatives, and sparing alternatives and makes possible the explicit and quantitative consideration of risk in their evaluation and comparison. Fig-

PERCENTAGE OF TIME THAT INDICATED NUMBER OF
SENSORS ARE AVAILABLE OVER PLANNING HORIZON

# OF SENSORS	SENSORS				
	1	2	3	4	5
5 OR MORE	0.00				
4	0.00				
3	0.00				
2	69.46				
1	27.80				
0	2.74				

Fig. 16 Percentage of the time that indicated number of sensors are available over planning horizon.

```
PROBABILITY THAT INDICATED SUBSYSTEM WILL CAUSE SENSOR FAILURE
                                    SENSOR
           SUBSYSTEM              1      2      3      4      5
           ---------            -----  -----  -----  -----  -----
        1 SENSOR                0.443
        2 SENSOR SUP. SUBSYS #1 0.240
        3 SENSOR SUP. SUBSYS #2 0.178
        4 BUS SUBSYS #1         0.058
        5 BUS SUBSYS #2         0.041
        6 BUS SUBSYS #3         0.040
                                -----  -----  -----  -----  -----
                                1.000

           SATELLITE LIFE

              EXPECTED..............  1.88
              *STD. DEV.* ..........  1.25
```

Fig. 17 Probability that indicated subsystem will cause sensor failure.

ure 18 illustrates the use of life cycle cost and availability information (as provided by the SATCAV Model) in analyzing alternative sparing strategies.

Figure 18 illustrates the life cycle cost and availability tradeoffs in terms of sparing strategy for a typical mission comprising two operational satellites plus spares with the expected wearout life of each satellite being 7 years. Availability is measured as the chance that two or more sensors will be operational at any point in time. Dormant spares are assumed not to fail while they are in the dormant state, but the probability of startup of a dormant spare may be less than 1.0 (as indicated by the dormant spares

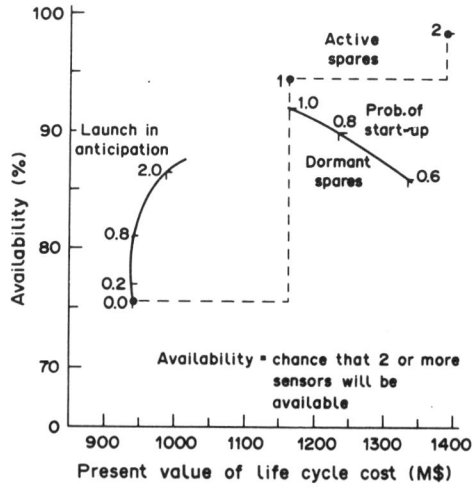

Fig. 18 Life cycle cost & availability tradeoff in terms of sparing strategy: two operational satellites plus spares (7-year expected wearout life). (From Ref. 8.)

curve). Three sparing alternatives are indicated: active spares (with zero, one or two spares), dormant spares (with different probability of turn on), and launch in anticipation (from 0 to 2 years) of expected wearout failure but launch on random failure. It can be seen that the minimum cost approach is a function of the required level of availability. For example, at low level of required availability (i.e., 75–90%) the launch in anticipation strategy is best offering lower expected values of life cycle cost than other alternatives to achieve the same level of availability. At higher required levels of availability the active spares option is best.

Figure 19 illustrates the life cycle cost and availability tradeoffs in terms of sparing strategy for a mission comprising five operational satellites plus spares. Availability is measured as the chance that four or more sensors will be operational at any point in time. As in Fig. 18, three sparing alternatives are indicated. Launch in anticipation of wearout failures is the best strategy except if a very high availability is required. For a very limited set of conditions, dormant spares (two) may be preferred, but at high levels of availability, active sparing is the only viable alternative.

The above is presented for illustrative purposes only and should not be taken as a general rule. The minimum life cycle cost approach within an availability constraint is a function of anticipated stand-down time (given a failure), launch vehicle reliability, and other factors. The above approach demonstrates an important role of life cycle cost analysis relating to the evaluation and comparison of alternatives.

Another application of life cycle cost and availability analysis concerns the setting of satellite reliability requirements in terms of transportation system reliability and satellite sparing and maintenance strategies. As already demonstrated, availability and life cycle cost are a function of the

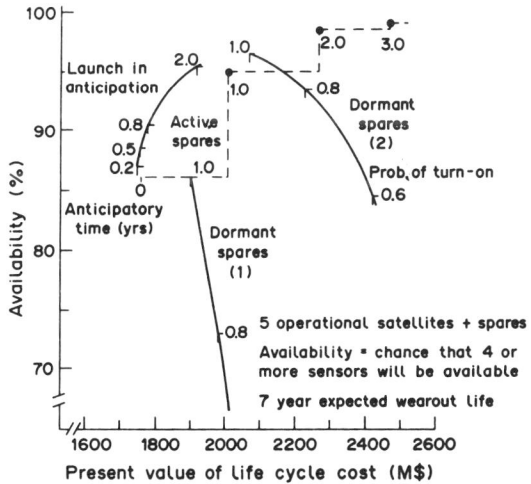

Fig. 19 Life cycle cost & availability tradeoff in terms of sparing strategy: five operational satellites plus spares (7-year expected wearout life). (From Ref. 8.)

combined choice of satellite configuration, launch vehicle and transportation scenarios, and maintenance/sparing strategy. The characteristics of each may, to a certain degree, be traded off with the choice and characteristics of the others. To illustrate these tradeoffs, results of sensitivity analyses[8] for a mission comprising two operational satellites (as indicated in Fig. 18) are presented in Figs. 20–22. These analyses illustrate the sensitivity of availability and life cycle cost of the three considered maintenance/sparing scenarios (active spares, dormant spares, and launch in anticipation of wearout failures) to launch vehicle reliability and satellite expected wearout life. The intent is to illustrate that it is possible to trade off launch vehicle reliability with satellite design life to achieve the desired level of availability but possibly at differing expected life cycle cost.

Figure 20 illustrates the effect of launch vehicle reliability on availability and expected present value of life cycle cost. A 10% (approximately) reduction in launch vehicle reliability results in a 4–6% increase in expected present value of life cycle cost (40–70 M$) with a slight accompanying reduction in availability.

Figures 21 and 22 illustrate the sensitivity of availability and life cycle cost to satellite expected wearout life for launch vehicle reliability of 0.85, when considering launch in anticipation of wearout failures and utilizing active on-orbit spares. It should be noted that much of the increase in life cycle cost resulting from the use of a low reliability launch vehicle may be offset by designing and utilizing a satellite with increased expected wearout life (for the case illustrated, 8 years). (The effects of increased satellite life on satellite cost and mass have not been taken into account. An increase

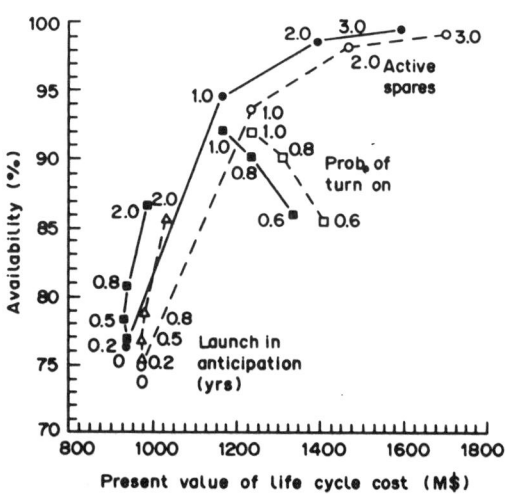

Fig. 20 Life cycle cost and availability tradeoff in terms of sparing strategy: two operational satellites plus spares (7-year expected wearout life). ●, active spares, LV reliability = 0.95; ○, active spares, LV reliability = 0.85; ■, dormant space, LV reliability = 0.95; □, dormant spares, LV reliability = 0.85; △, launch in anticipation (in years). (From Ref 8.)

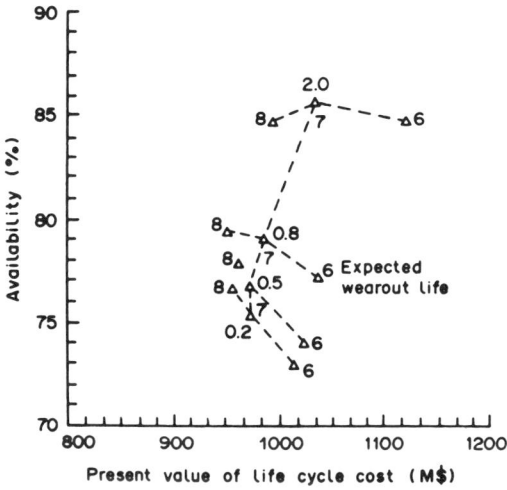

Fig. 21 Life cycle cost and availability tradeoff for zero, one, two, and three active spares sparing strategy: effects of expected wearout life (6, 7, and 8 years; LV reliability = 0.85). △, launch in anticipation (in years). (From Ref. 8.)

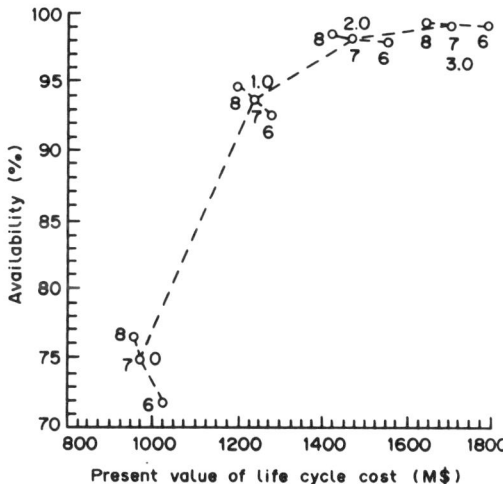

Fig. 22 Life cycle cost and availability tradeoff for launch in anticipation sparing strategy: effects of expected wearout life (6, 7, and 8 years; LV reliability = 0.85). ○, active spares. (From Ref. 8.)

in satellite mass would also likely increase transportation cost. The intent of the above is to demonstrate methodology and not to come to specific numeric values.) Similarly utilizing a higher reliability launch vehicle together with a shorter expected wearout life satellite may result in about the same overall availability and life cycle cost performance as utilizing a lower reliability launch vehicle with a satellite designed for a long wearout life.

Summary

In the space business, annual costs, and therefore the present value of life cycle costs, must be considered as probability distributions. The reasons for this are many but primarily due to uncertainties associated with both nonrecurring and unit recurring cost (in no small part resulting from the use of new technology); less than perfect reliability of launch vehicles, failures of which can significantly affect costs and schedules; schedule uncertainty resulting from both failures and necessary rescheduling to recover from failures and to respond to uncontrollable delays (i.e., weather and stand-down time required to correct previous failures); random and wearout failures of operational satellites which require replacement.

The analysis of life cycle costs is further complicated by a myriad of operational alternatives involving sparing and maintenance/repair strategies. In addition to satellite design considerations relating to sparing and redundancy, there are operational considerations such as the use of on-orbit active or dormant spare satellites, ground spares, and the consideration of on-orbit servicing or repair at a transportation node. Since sensor/transponder availability is a function of overall satellite configuration (including redundancy), sparing and maintenance strategies, and transportation system reliability, it is necessary to consider the many complex interrelationships that exist when configuring a satellite and estimating mission life cycle cost.

The life cycle cost analysis, in addition to the explicit and quantitative consideration of cost and delay uncertainties, requires the consideration of cost learning effects and cost spreading. Since launch and space operations play a large part in both life cycle cost and availability, the specific launch scenarios must be defined for both initial placement and maintenance flights, including reliability or probability of success per major operation, cost (including uncertainties), and delays that may result from different types of failures.

The net result of these considerations is that life cycle cost must be established as a probability distribution, and decisions must be made taking into account quantitative risk measures: the chance that annual costs will exceed annual budgets and the chance that present value of life cycle costs will exceed specified levels. The net result of these considerations is also the determination of quantitative measures of availability. Availability is a function of launch reliability and delays caused by failures, satellite subsystem reliability characteristics, and sparing/maintenance strategies. Different system parameters (ranging from launch system reliability to sparing strategy) will result in different sets of life cycle cost (expected value and standard deviation) data as well as availability data. Thus, it is necessary to compare life cycle costs of alternatives at the same level of availability. In other words, the costs associated with equal capabilities must be compared.

References

[1]Greenberg, J. S., "Reliability, Uncertainty and Risk Assessment of Space Systems. A Methodology for Decision Making," AMS Rept. 1085, Princeton Univ., Dec. 1972.

[2]Greenberg, J. S., and Hazelrigg, G. A., Jr., "Methodology for Reliability-Cost-Risk Analysis of Satellite Networks," *Journal of Spacecraft*, Vol. 11, No. 9, 1974, pp. 650–657.

[3]Greenberg, J. S., "A Simulation Analysis of Space Operations," IAA-87-621, 38th Congress of the International Astronautical Federation, Oct. 1987.

[4]Dole, S. H., et al. "Establishment of a Long-Range Planning Capability," RM-1651-NASA, Rand Corp., Sept. 1969.

[5]Quade, E. S., and Boucher, W. I., eds., *System Analysis and Policy Planning*, Elsevier, New York, 1968.

[6]Greenberg, J. S., *Investment Decisions: The Influence of Risk and Other Factors*, American Management Associations, New York, 1982.

[7]Greenberg, J. S., "Satellite Configuration, Transportation, Maintenance and Sparing Decisions: The SATCAV Model," Third Space Logistics Symposium Colorado Springs, CO, May 1990.

[8]Greenberg, J. S., "Satellite Reliability Requirements: Effects of Transportation System, Sparing, and Maintenance Strategies," *Acta Astronautica*, Vol. 21, No. 6/7, 1990.

[9]Greenberg, J. S., "SATCAV: A Space System Life cycle Cost and Availability Model," 40th Congress of the International Astronautical Federation, IAA-89-694, Oct. 1989.

Chapter 3. Benefit/Cost and Cost Effectiveness Models

Measuring Returns to Space Research and Development

Henry R. Hertzfeld*
HRH Associates, Bethesda, Maryland 20816

Introduction

FOR well over 30 years the United States government has conducted scientific and engineering research and development (R&D) for civilian activities in the exploration and use of outer space. The total expenditure over these years has been over $350 billion (1992 dollars). If space expenditures for national defense and other purposes are added to the National Aeronautics and Space Administration (NASA) R&D, the total expenditures have been over $550 billion (1992 dollars).

The space missions have been spectacular successes. U.S space exploration has generated national and international technological prestige and has captured the imagination of all mankind. Although the annual expenditures on civilian space activities have been large, they still account for well under 1% of the annual U.S. government budget.

Economic benefits to the nation were considered when Congress first created NASA in 1958. So were scientific benefits, engineering, and technological leadership, and the stimulation of the educational system.[1] Nonetheless, economic expectations were secondary.

Technologically. the space "race" is still active. However, it has changed in recent years from a politically motivated race between the United States and the former Soviet Union to an economic and commercial race. This raises a variety of economic issues for the United States and for justifying its investment in space activities. As other nations gain commercial advantages in space technologies, and as the U.S. budget deficit worsens, pressure mounts to orient space activities toward commercial ends and away from scientific and exploratory ends. However, the U.S. space pro-

---

Copyright © 1992 by the American Institute of Aeronautics and Astronautics, Inc. All rights reserved.
*President, HRH Associates, 5208 Baltimore Ave., Bethesda, MD 20816.

gram, more than that of most other nations, remains organized and motivated toward space transportation and infrastructure and less toward the profitable market niches.

Space investments now are unlike those of the 1960s. The Apollo program to put men on the moon was a presidential initiative with virtually as much funding as was needed made available to accomplish a national mission. Today, funds for space vie against many other priorities. With more limited resources, the need to evaluate, justify, and measure economic returns for the investment has taken on higher priority than in the past. This article critically reviews the major studies and methodologies that have been developed to measure the returns to space investments.

Types of Studies Reviewed

Economic returns are best measured for civilian space activities. Because of the nature of defense activities and the difficulty in tracing benefits from defense spending to commercial products, the studies reviewed here focus on civilian space activities. And, because NASA accounts for the predominant civilian space expenditures, most analyses have been limited to evaluating the impact of NASA expenditures.

The easiest to understand measure of benefits is usually expressed as a ratio of returns to expenditures, such as 8:1, 14:1, and so forth. Even though these measures are excellent fodder for speeches, public relations, and congressional testimony, they are inherently inaccurate, misleading, and confusing. And, because such numbers are thrown around somewhat haphazardly, often the results of studies that set out to measure very different economic phenomena are mistakenly compared with one another.

This article focuses on studies that have analyzed the benefits of space R&D to the overall economy. Specific studies that have focused on the impact of space on a particular sector, region, or employment category are noted and in some cases summarized; however, many industry-oriented studies (e.g., for communications, materials, etc.) are not space-specific and incorporate methodologies that cannot be extended toward measuring aggregate benefits.

Also, this review focuses on analyses that attempt to measure the long-term technology-enhancing and economy-wide productivity gains rather than on short-term industry, regional, or employment multiplier effects of government spending. (Employment and income multiplier analyses of government expenditures measure immediate impacts of gains or losses in spending. Spending on the space program will not have a significantly different national multiplier effect from other types of government spending (housing, welfare, environment, etc.) or even from a tax cut. However, the types of jobs and the specific parts of the country affected by space spending will be different.)

It is the lasting and significant economic impacts that arise from the new technologies that are stimulated by the space program that contribute to long-term economic growth and increases in the standard of living that are of most interest to the health of the economy and to furthering U.S. economic competitiveness. Once new knowledge is acquired and used, it cannot be taken away, whereas short-term impacts last only as long as the money is being spent.

Approaches to Measuring Economic Returns to R&D

Three distinct approaches have been used to quantify the economic impacts of space R&D:

1) The adaptation of a macroeconomic production function model to estimate impacts of technological change attributed to R&D spending on the gross national product (GNP) and derivative measures such as employment and earnings. The results of using this type of model can be expressed both as a rate of return to a given investment and as aggregate totals.

2) Microeconomic models that evaluate the returns to specific technologies to the economy through the use of demand and supply curve analyses of consumer and producer surplus. The measures that result from these studies are usually expressed as benefit/cost ratios. Benefits derived from these studies are rarely additive to aggregate benefits across different technologies because of technical incompatibilities in data collection and economic assumptions underlying the models.

3) The examination of actual reported as well as survey data that provide evidence of the direct transfer of technology from federal space R&D programs to the private sector. These measures include: patent and licensing data, contractor reports of new inventions discovered during the performance of the contract, active technology utilization and transfer programs of the government agency, and special surveys of industry. The results of these analyses tend to be reported in actual numbers measured (number of patents or inventions, value of royalties, value of sales, etc.). They are rarely compared to government expenditures because of the difficulty of attributing specific funding to specific products or patents.

Macroeconomic Analysis

The strengths and weaknesses of the macroeconomic models are well documented in the economic literature. Some authors, such as Griliches,[2] feel that the macroeconomic approach has validity and its imperfections can be cured. Others argue that the assumptions behind the model preclude any accurate results (see Denison[3] and Mansfield[4]). Although it is possible to calculate aggregate returns from space R&D through this model, and the results reported below have shown very robust returns, the estimates have not stood up to statistical tests for accuracy or reliability. The value of the aggregate production function approach is more in structuring thought and providing a framework for analysis than in quantifying precise benefits.

The aggregate production function model assumes that a formal relationship exists between R&D expenditures and productivity. Given the many case studies of specific new technological developments and their impacts on the economy, there is no doubt that such a relationship exists.

However, the aggregate production function model skips a number of steps in proving that relationship between R&D and productivity. Expenditures for some R&D (particularly in measurable terms, the research side) may be a stage on the road to productivity. But that stage must be followed by the creation of knowledge, then translated into inventions and, eventually, innovations. Finally, with a lot of additional investment in advertising, marketing, distribution, and quality control, the innovations may become successful commercial products. In other words, the link between formal R&D programs and changes in productivity and GNP is indirect. Further complicating the relationship is that some productivity changes do not come from R&D investments. And some may come from R&D investments that occurred many years ago and are newly applied to unique situations.

Therefore, assuming that there is a direct and measurable link between formal R&D and productivity, as is done with the production function model, is a giant leap of faith. The production function model does not prove the link. However, from documentation arising from numerous case studies, it is irrefutable that government R&D (including NASA R&D) has had a positive impact on long-term productivity. These facts lead to the dilemma economists face in measuring the impact of R&D: we know there is a link, but we can do no better than show a probable statistical relationship.

Another problem with the macroeconomic model as applied to an R&D context is that the sum of all R&D funding is a very small expenditure in the economy. Since 1975 it has varied from a high of nearly 3% of GNP to a low of 2.2%. And, NASA space R&D, while reaching nearly $10 billion per year, in terms of GNP is less than 0.25%. Attempting to empirically estimate the impact of such a relatively small expenditure of funds is very difficult, given that there is a margin for statistical errors in the GNP that is much greater than the NASA component.

Adding to the problem of estimating the impacts of government expenditures on the GNP are the accounting conventions that the U.S. government uses in calculating GNP. Government expenditures, as a class of expenditures, are considered to be spent in the year the outlays are made. There is no capital account, nor is there an investment account for government expenditures. Therefore, there is no rate of return that should be expected or implied from these expenditures. R&D, including government R&D, is really an investment of current dollars with no expected future stream of returns. On the other hand, accounting for private R&D expenditures in the GNP does treat R&D as an investment. Because government R&D is not treated similarly, there is a logical fallacy that develops when macroeconomic models attempt to calculate a future stream of returns to current government R&D outlays. And, an even greater problem occurs when aggregate returns to government R&D are then compared to private R&D returns and benefits. Is it any surprise that the returns from government programs are smaller?

Yet another factor in the unreliability of macroeconomic estimates is the long time frame needed to develop, market, and disseminate new products successfully. Although new technologies may be reported in 2 years or less, it easily can take 20 years for a major product to mature and make an impact on the economy. When most of the macroeconomic analyses reported below were conducted (in the mid-1970s), the available time-series data for space R&D were barely 15 years. This alone should make one question the robust results because the period of space expenditures measured coincided with a generally robust and growing U.S. economy. We now have a longer time series to analyze, but we still lack the sophistication of an economic model sensitive enough to pick up the R&D-to-GNP relationships accurately.

Finally, there are a host of technical assumptions and statistical problems with the macroeconomic approach to estimating current and future impacts. The problems are greatest when dealing with long-term projections of GNP (more than 1 year ahead). These limitations of the models will not be discussed in this review.

Macroeconomic Studies Using Secondary Data Sources

Midwest Research Institute Studies

In 1971 (and repeated again in 1988), the Midwest Research Institute (MRI) conducted a study of the economic impact of NASA R&D.[5,6] Part of these two studies included a subjective review of cases of successful technology transfer, but a major additional task of both studies was an attempt to measure the aggregate benefits of NASA R&D using a macroeconomic production function approach.

These studies (both were done using approximately the same economic techniques) took a national income accounting approach, following the methodology of Solow's pioneering econometric work in analyzing the returns to R&D.[7] The MRI investigators first looked at total national productivity changes in the economy, then subtracted those changes that were attributable directly to capital and labor. The remaining unexplained changes (the "residual") were studied further to find their important components. After accounting for changes in demography, education, health, work-week length, and economies of scale, the investigators attributed the rest of the residual to advances in knowledge, or the R&D component. A least-squares regression then was used to estimate the linear relationship between gains to R&D and a weighted sum of past R&D expenditures.

The results of the 1971 study showed that $207 billion would be returned to the economy through 1987 from the $29 billion NASA outlay between 1959 and 1969. This translated to a 7:1 overall return to NASA, or an estimated 33% discounted rate of return.

The 1988 study found the return to NASA R&D was approximately 9:1. This study used a multifactor productivity analysis and took a more sophisticated approach to measuring the sensitivity of the returns than the 1971 study. They concluded, for instance, that even if the productivity returns to R&D were overstated in their model by 30%, the calculated benefits ratio over the lifetime of the investment would still be 6.5:1. They also tested for varying the lag in years between funding of R&D and measurable results. Again, the return was still robust if the lag was increased from 3 to 5 or even 8 years.

The 1971 MRI study was the first such attempt to measure NASA's returns in the aggregate. Two major assumptions made in that study should be questioned. The first was that NASA R&D was not separated from any other R&D in the economy. In fact, what the MRI did was to calculate the returns for total R&D (federal and private) and assume that space R&D was similar to all other R&D. The 1988 study used the same technique, although MRI in 1988 did try to test the "uniqueness" of NASA R&D and did additional industry analysis on the industries most affected by NASA contracting. However, even though they concluded that NASA R&D was no less productive than other R&D, they did not lay to rest the obvious potential fallacies of this assumption.

The second assumption was that R&D has an 18-year lifetime from outlay to "death" in the economy. After 18 years had elapsed, no returns were measured. Further, each R&D expenditure was considered to be independent of all other R&D expenditures. In terms of the economic lifetime of an innovation, this span (adjusted for lags in implementation) is probably not unrealistic. And, MRI's 1988 analysis tested for changes in the lags and in the economic lifetime of an innovation and found the effects of changes small on the aggregate numbers.

However, knowledge is never lost. It becomes the building block for further technological (and productivity) advances. The MRI studies did not acknowledge the important, but subtle, distinction between a product's economic life cycle and the impact of society's accumulation of knowledge.

Chase Econometric Associates Study

In 1975, Chase Econometrics[8] did a far more sophisticated econometric analysis of NASA spending on space R&D and its impacts on GNP. Chase analysts estimated a production function for the United States based on a potential GNP time series (i.e., GNP adjusted for full employment). As there are several different sources of such estimates, the choice of a particular set of data will influence the results of the analysis significantly. This was one problem with the empirical results of the Chase study, but, as described below, it was not the major problem.

Chase then calculated the residual, the variations in the potential GNP that could not be explained by the capital and labor inputs. This residual was broken down statistically into its components using a regression technique. The independent variables used in this regression to explain changes in the residual included: NASA R&D, other R&D, industry mix, capacity utilization ratios, and demographic factors. These variables were suggested by the earlier work of Denison[3] and others who have done extensive analysis of the determinants of productivity. Chase also built into the model various lag functions to account for the delay between R&D outlays and the introduction of new technology into the economy.

Chase experimented with about 60 variations of the equation and chose the one that appeared "best" to them. One problem is that with only 15 years of time series data on NASA expenditures, the use of many independent variables in the estimating equation restricted the ability of the model to give accurate statistical results. Further, it is not clear that spurious and coincidental business cycle relationships with NASA R&D were not picked up by the short time series. The selection criteria of the "best" equation were not well documented, as an examination of the t ratios for reliability of the regression coefficients even in that best equation revealed that some were only marginally significant. Finally, any equation of this sort is driven not by the R&D variable, but by the capacity utilization variable.

The calculated cumulative "productivity" return to NASA R&D was 14:1. This translated into an annual discounted rate of return of 43% to NASA outlays. From these data on productivity, Chase then ran their macroeconomic simulation model and calculated that a $1 billion increment to NASA expenditure for each of 10 years from 1975 through 1984 would augment GNP by a cumulative $83 billion. In other words, $10 billion of extra investment by NASA in space and related R&D would yield 8.3 times that amount in GNP benefits.

The U.S. General Accounting Office (GAO)[9] performed an extensive review of the Chase study. Not surprisingly, the GAO found a great deal of instability in the equations used by Chase.

In 1980, at NASA's request, Chase replicated the earlier study using a 20-year time series instead of a 15-year time series.[10] The original data and choice of measurement techniques behind the parameters were subjected to new and more careful scrutiny. The results of this update showed that some of the original economic measures were not well constructed and did not measure exactly what they were supposed to measure. Minor modifications were made for this update.

The results of the new Chase analysis showed NASA R&D spending effects to not be significantly different (in formal statistical tests) from zero. This should not be interpreted to mean that there has been no economic return on NASA R&D, but only that this macroeconomic approach to measuring those returns contains many data and theoretical problems that have yet to be solved.

U.S. Department of Labor Study

Over the years, interindustry input–output models have been used to measure the multiplier-type returns to various defense and other major national programs. With this in mind, NASA commissioned the U.S. Bureau of Labor Statistics (BLS)[11] to measure the productivity changes from NASA expenditures at an industry level, using the production function approach and an equation for every industry that is tied closely to aerospace production.

The results were inconclusive, mainly due to the high level of "noise" (i.e., the measurement errors in the data) in the industry-level data. However, there were several interesting general results. First, overall U.S. technology change was found to be labor-saving. Second, the R&D component of the technology change has saved capital. Third, the returns measured to private R&D investments were between 15 and 30%, while returns to government R&D were between 0 and 5%. The equations that isolated NASA expenditures did not show statistically consistent results, probably due to the relative small percent of R&D that NASA accounts for in the economy coupled with the sampling errors contained in the BLS data base. In addition, like the Chase study, they found that there was too short a time period studied to allow for business cycle changes (the beginning of the study was at a peak in the cycle and the end of the study was at a trough), and that the NASA R&D variable may be more representative of the government demand for output than of actual advances in knowledge. Finally, as with the Chase study, the capacity utilization and output variables drove the equation.

Microeconomic Studies: Benefit/Cost Analyses

Microeconomic analysis focuses on the economic theory that explains the behavior of the firm and the supply and demand relationship for individual products. Benefit/cost analysis to evaluate the success or failure of public sector projects is based on this theory. More often than not, individual projects are analyzed to determine their consumer surplus. This "surplus" is the measured value representing the additional "income" to the consumer when the price of a commodity that the consumer pays is less than the consumer is willing to pay. An analysis of the demand curve for a particular commodity provides the input data needed to measure the consumer surplus.

Because demand and supply curves are measured at a given point in time, the savings can be translated into streams of income over time by projecting demand and supply curves into the future. Benefit and cost (from the supply curve) estimates are then discounted back to the present by assigning a time value to money (discount rate). The ratio of benefits to costs provides an estimate of the comparative value of an investment.

Using this methodology to evaluate the impact of R&D and new technologies poses some unique problems. Benefit/cost analysis is predicated on accurate measurement of the demand and supply curves. Often, new technological innovations result in products that are so different from anything on the market that no well-defined demand curves can be constructed. Nor is there a stable supply function (which measures costs of production), as costs and production processes change rapidly as the product evolves.[12] Therefore, using this methodology for evaluating some R&D projects involves little more than advance guesswork.

Measuring appropriate costs can be another serious problem when applying benefit/cost analysis to specific innovations arising from large programs such as those of NASA. Should the denominator reflect only the cost of transferring the technology, the cost of the research project, the cost of the group of related projects, or the entire expenditures of the agency? With some projects that result in unintended (or spin-off) technologies, a good case could be made for any of these choices.

In a large R&D agency as NASA, many new technologies are discovered and developed over time. Due to limited resources, information, and time, only a few technologies can be subjected to benefit/cost analysis. Usually, only the "winners" are selected, which leads to a very unbalanced view of the portfolio of technologies supported by the agency's funding.

Another problem, more general to the methodology than to R&D situations, is the choice of the rate to be applied to the time value of money. For government projects, the most frequently used rate is the one dictated by the Office of Management and Budget (OMB). It will not always reflect the real cost of money in the economy, nor will it frequently be appropriate for the cost of R&D projects .

In spite of all the problems with benefit/cost analysis (the above described problems are not exhaustive of those encountered with this methodology), economists are relatively comfortable with estimates derived from well-executed benefit/cost studies because they are based on traditional economic principles.

Over the years NASA has sponsored numerous benefit/cost and related studies. Many of them have taken a cost-effectiveness approach[14] to the problems, since their major purpose was to choose between alternative methods of achieving a given result. [Cost effectiveness is a modified form of a benefit/cost model. Rather than focusing on the ratio of benefits to costs, this model starts with an objective (e.g., an operational space transportation system, given certain performance requirements) and performs analyses of different production techniques to achieve that objective.] The planning study done in 1970[13] to determine some of the economically desirable characteristics of the (then) proposed Space Shuttle system was an example of the use by NASA of a cost-effectiveness model.

This article is focusing on more aggregate, economy-wide benefits from space investments; therefore two benefit/cost studies are described below. They are representative of the types of analyses conducted using this methodology on NASA-sponsored technologies.

Mathematica Study

In 1975, Mathematica, Inc.,[14] studied four successful technologies that NASA had a role in developing: cryogenic insulation, integrated circuits, gas turbine engines, and NASTRAN (a computer program for analyzing structural properties of large vehicles). Each analysis required a different type of data gathering and statistical modeling. But, the studies used similar methodologies so that the results could be added to get a sum for the benefits from all four innovations.

The benefits measured were not the total impacts on the economy from these innovations. Instead, the study estimated the speed in the introduction and commercial use of these innovations that could be traced to the special program requirements of NASA's space R&D procurement specifications. Mathematica focused on two measurable elements: the introduction of a new commodity and the decrease in the costs of production for an existing good.

The study was not a true benefit/cost analysis, as the costs were not calculated—a handy way of skirting the difficult choice of determining the proper yardstick for costs. It was impossible to attribute or allocate costs of developing the different technologies, particularly when NASA's involvement was not only as a funding source of R&D, but also as a stimulus to private firms already investing in the technology development. And, it would have been impossible to separate the specific NASA funds that went to the technologies, since many of them crossed over various space programs. Still, the methodology used to measure benefits was consistent with traditional benefit/cost analysis.

The results were impressive. The investigators found that, over a 10-year period from 1975 to 1984, the four technologies could be expected to return a discounted total of $7 billion (in constant 1975 dollars) in benefits that were attributable to NASA's involvement in their development. To put that in perspective, NASA's entire budget outlay for 1975 was $3.2 billion.

1977 Mathtech, Inc. Technology Transfer Study

For many years, NASA has had a formal technology transfer program to disseminate information and develop space technology for nonspace uses. In 1977, Mathtech, Inc.,[15] conducted a benefit/cost analysis to measure the economic impact of nine of the more successful commercial innovations transferred from the government to the private domain through this program. In this study, benefits were compared to costs, but only to the costs of transferring the technology, not to the development phase of the underlying NASA work.

The results from the nine innovations showed a sizable variation. In the biomedical field, the cardiac pacemaker had a benefit/cost ratio of 4:1, while the laser cataract tool had a ratio of 41:1. Other biomedical results were: burn diagnosis, 8:1; meal systems, 6:1; and a human tissue stimulator, 10:1. For engineering innovations, the ratios were: nickel-zinc battery, 68:1; zinc-rich coatings, 340:1; track-train dynamics, 3:1; and a firefighter's breathing system, 4:1. The wide range in ratios measured result from the following: 1) the technology is not yet fully ready for the market, 2) the market is primarily government and therefore smaller than private market demand, and 3) nonquantifiable benefits may be larger than those that are easy to measure (particularly in the biomedical field).

The principal advantage of benefit/cost analysis is that similar methods can be applied to all innovations and results can be compared (even if the comparisons are inexact). The major drawback is that these studies are done at one point in time. (If redone today, the benefit/cost ratios would be very different, reflecting changing actual market conditions, changing production techniques, and a very different set of alternative technologies.) A staged series of benefit/cost studies of the same or similar technologies and products would be far more useful in tracking the success (or failure) of NASA technologies and technology transfer mechanisms through the economy.

Other Microeconomic Studies

NASA has commissioned many economic studies. The two described above were selected because they used a formal consumer surplus model to measure benefits and costs, and because they used similar methodologies for different technologies, which enabled the results to be summed together.

Hundreds of technology assessment and technology forecasting studies have been conducted by NASA for specific aerospace technologies. Many are subjective, qualitative evaluations. Because many space R&D outputs are at the cutting edge of technology, subjective methodologies were the only ones that could be applied at the time. It is unfortunate that the results of those analyses cannot be aggregated to give a broad picture of the impact of space-related technology to the economy.

Notably missing from the above discussion are the studies that document the "big" technologies that are a direct or indirect result of the space R&D program. The most obvious examples are satellite communications, weather and remote sensing satellites, and private space launch vehicles. A detailed discussion of the specific impacts of these technologies (particularly studies of the communications industry) is beyond the scope of this article.

Summarized briefly, the telecommunications industry has been altered radically as a result of satellite communications technology. Long-distance communications have become less expensive and new services are being offered. In the 20 years between 1965 and 1985, the cost of living index trebled while the unit service charge on the Intelsat satellites fell 90%.

Not only has the supply side of communications changed, but improved products and services have increased the demand for communications satellites. Associated with space hardware has been the development of a rapid and competitive market for Earth receiving stations, including antennae, sophisticated electronic equipment, and support services.

Although it is likely that economic studies of the impacts and benefits of satellite communications have recorded some of this growth, it is unlikely that they could have accounted fully for its impact. Not only has it changed the industrial structure of the industry and promoted competition where formerly there was none, but it also has created difficult political and social problems. For instance, direct broadcast television, which is capable of broadcasting over wide reception areas, will create sensitive issues for small countries whose governments jealously control scheduling and content of programs within their borders. The same technology that permits coverage of large areas will enable education and information to be made available to remote populations previously inaccessible to TV broadcasts.

Another example of economic analyses of particular NASA R&D programs are those dealing with remote sensing. (See, for example, Ref. 16. During the 1970s, Econ, under contract to NASA, produced a number of studies of remote sensing and agricultural forecasts.) The studies measured the gains from having better and more generally available knowledge of current wheat production. By studying the historical price fluctuations of the crop and by analyzing the potential dampening of those fluctuations due to less uncertainty in the market, the benefits of decreased speculation in the futures market, and, therefore, more stable prices, could be assessed. The results did show the expected trends, but were controversial because of the study's assumptions concerning the industrial structure of the agricultural sector.

Specific applications of the remote sensing satellites led to better information about the spread of crop blight, snow melt, land use, pollution monitoring, and other data previously unobtainable or very expensive. In economics, calculating the value of information is a difficult and not well-developed part of theory. These attempts were exploratory.

The list of new information and data products from satellite technology is extensive. Beyond meteorological information, there have been successful search and rescue satellites, navigation aids, mapping technologies, and so forth. Both new commercial products and improved government capabilities that affect health, safety, and welfare have become available. Specific analyses have measured the impacts of these products. However, the methodologies have varied so greatly that it is impossible to aggregate the results.

Over the years, NASA has conducted numerous cost-effectiveness and engineering studies oriented to detailed decision making. Although these may add to our information about government programs, they tend not to fit into the productivity/impact analyses that this article is reviewing.

Similarly, the various regional and short-term economic multiplier analyses that have been conducted on the space program add little to our information about the major long-term benefits from space R&D that translate eventually into improved products, improved manufacturing processes, and improved productivity.

In recent years, with the advent of private opportunities in space and space-related businesses, there have been numerous business and financial analyses of space technologies with commercial potential. These range from various types of materials that can be produced only in a near-zero-gravity environment to remote sensing products, to actual private space launch vehicles and ground operations. Many of these studies are proprietary. However, there is a growing library of accumulating financial information about the risks, costs, and benefits (payoffs) of space activity.

From a technical economic view, business and financial analyses of space activities are no better or worse than the previously reviewed macroeconomic and microeconomic analyses. But, the business models have their own limitations, due primarily to the short-term focus of the underlying assumptions. Discounted cash flows are extremely sensitive to heavy upfront capital investment requirements, interest rates, and near-term market sales projections. Investments in space business typically involve large upfront costs, and unknown market sales. Both are factors that make it difficult for business to justify investments in space compared to alternative uses of the funds. And, the period of the 1980s was one with high interest rates. This, too, biased financial models away from justifying investments with a long payback.

The results of these studies of commercial space activity also do not lend themselves to evaluating aggregate benefits of government space R&D investments. But, the recent (past 10 years) growth of these studies, of actual private investment, and of a space "industry," provide ample evidence of future economic opportunities and benefits that will translate to real jobs, economic growth, and improvements in the quality of life.

Hybrid Studies

ESA Space Benefits Study

In 1978 and again in 1988, the European Space Agency (ESA) released studies of the economic benefits of its contracts.[17,18] These studies, prepared by the economics faculty at the Louis Pasteur University in Strasbourg, France, took an entirely different approach to measuring overall economic benefits from R&D than the U.S.-sponsored studies.

The study team conducted extensive interviews with the major industrial contractors to ESA aimed at assessing the value to the contractors of doing advanced technological work for the space agency. Benefits were grouped into four major types: technological advantages, commercial advantages, advantages for organization and methods, and advantages for the work factor (labor productivity and skills). The results were presented in aggregate numbers and also broken down by industrial sector.

It is interesting to compare the different results obtained over a 10-year period by ESA. They found:

1) The ratio of benefits to ESA contract expenditures increased to 3.2:1 in 1988 vs 2.9:1 in 1978

2) Indirect benefits to nonspace sectors decreased from 50% to 21.1%, mainly attributed to the growth of the commercial space sector in Europe (i.e., increased opportunities for use of technology)

3) There was a significant change in breakdown of indirect benefits:

Type of benefit	% 1978	% 1988
Technological	24.8	43.3
Commercial	26.9	8.5
Organization and methods	19.4	7.1
Work factor	28.9	41.1

4) The benefits outside the space sector basically did not change in the two samples (after correcting for year differences, etc.). This was attributed to the fact that companies working in space are organizationally not able to switch to nonspace commercial market (i.e., small scale and prototype to large-scale manufacturing).

5) Polarization within companies has contributed to a decrease in benefits from exports to nonmember countries. ("Commercial" space markets are mostly within Europe in the form of national and international programs.)

The studies found some interesting trends of indirect benefits, including a decline in the growth rate of technological benefits, mainly through sales of ESA products. They attribute this to an S-shaped innovation curve at work, showing recent year leveling of forecasts. In the commercial area, spin-off benefits are estimated to drop after 1987, reflecting decreased space flight opportunities.

The ESA studies also measured some geographic differences that mainly reflect types of work done in the various nations. Great Britain, France, and Italy tend to focus on structures and propulsion systems, which are not generic technologies and generate few indirect benefits. Germany has a range of space technologies and specializes in electronics, power supply, conditioning and storage, design, and engineering. Germany's space program has a broader scope and therefore has more direct nonspace sector applications. Sweden's dramatic increase in benefits shows a capability in technology transfer in telecommunications and transports, attributed to the structure of Swedish companies which have a wide range of activities in many sectors and also do fabrication that ranges from small scale to mass production.

The study analyzed benefits by types of contractors. One major conclusion was that the greatest benefits came from equipment developers, and the smallest from service providers. This was attributed to medium-sized high-technology companies not subject to heavy R&D management structure.

Finally, when benefits by industrial sector were analyzed, they found that space spin-offs occur mainly within sectors, rarely among different space sectors. Eighty percent of spin-offs were found in onboard electronics (instrumentation is a large factor here), ground facilities, design, and engineering.

There are significant differences between the European approach and the United States approach to measuring benefits. In the ESA studies, the term "indirect" means anything that is a spin-off of space (ESA) contracts. Successful performance of the contract specifications was not considered a benefit (and would, by default, be their definition of "direct"). In U.S. studies, direct economic effects are used to signify the shorter run employment and income effects (i.e., jobs created and income generated by the expenditure of government space funds). The indirect effects are the longer term productivity impacts as measured in macroeconomic GNP-type data. These include spin-offs as well as productivity from direct space equipment manufacturing (e.g., major contractors, communications, etc.).

In microeconomic U.S. studies of space benefits, direct economic effects are those on the company (contractor) itself (productivity, jobs created, profits generated, etc.), and indirect benefits are the spin-off technologies and new products introduced into the economy. Thus, the ESA study definitions are different. Productivity-enhancing aspects are considered indirect benefits and are spread among the various categories studied. They are measured at a company level, as are all measures in this study. Therefore, the aggregate measures are lower than those calculated for the U.S. economy because social and aggregate spin-off measures are added up from the companies surveyed and their suppliers. They do not include productivity enhancements of purchasers of new products (except where those products are intermediate goods sold to other space industry firms or transferred vertically within a firm), or any other aggregate social welfare improvements.

However, the numbers generated are "believable." And they are based on hard data, not on proxy measures. The geographic and industry detail reinforces the validity of the conclusions concerning organizational and management techniques that work best, as well as the types of activities that are specialities of the firms and industries.

The most interesting policy conclusion that can be drawn from the data is the relative lack of use of the spin-offs outside of the space and defense industries and the various barriers to the successful adaptation and transfer of technologies to other industries. Also, the relative change over time to an inbred commercial space sector of the economy that lowers the ability (and desire?) of firms to find buyers outside of that sector for their products.

A mature internal market implies increased sales, but also less ability to capitalize on state-of-the-art technology outside of that market or place a special advantage on doing work for ESA.

Finally, the major benefits accrued to industries and nations that have done research in space in areas of generic technologies, particularly in those that have specialized in electronics and instrumentation. The ability to apply the results of that type of research to mass markets is superior to other space specialties. Nations and firms specializing in large space structures, space vehicles, and so forth, do not show the same degree of spinoff because of the one-of-a-kind orientation. Some companies with matrix managements and a product line that includes mass production (e.g., auto companies) do show spin-off transfers of technologies.

Other Statistics Showing Impact of Space R&D

Government agencies, including NASA, keep many records of their activities. Some of them include actual counts of the uses and benefits of various space technologies, whereas some reflect results of surveys of users of NASA technology. There are two main sources of this information: the patent office and the Office of Commercial Programs (OCP). The NASA patent office tracks the domestic and foreign use of NASA-held patents, and is also responsible for issuing licenses to private firms for the use of NASA patents. The OCP keeps data on the Technology Utilization program, and on various other NASA programs created to stimulate the development of space technologies that have potential to find their way into the commercial marketplace. The extent of these data are too broad for a comprehensive review in this paper. (A review of patent statistics can be found in Ref. 19. Also, the NASA OCP publishes numerous reports documenting the use of NASA technology in the commercial sector.)

Lessons Learned About Measuring Economic Impacts of Space R&D

Characteristics of Space Research That Can Be Measured

Every age and time has its imaginative and awesome technological achievements. Landing a man on the moon, the most mind capturing of all of our space accomplishments, ranks easily with the building of the Gothic cathedrals of the Middle Ages, the Great Wall of China, and the transcontinental railroad of the 19th century. In their time, each of the projects was without precedent. The investment capital had to come from a central governmental (or religious) source with a surplus of funds. New technology had to be developed and perfected. And, partly as a result of their investments, the nations that were able to support those activities maintained and increased their political and economic leadership.

All of the great engineering projects had major economic impacts: The money to build the structures was diverted from other uses; the wages paid to those employed by the projects had direct and indirect income and employment effects; and the learning and new technological capabilities enhanced trade and national prestige. But, like the space program, most of those great feats had to amass their support from a coalition of political, social, and economic institutions. Economic impact arguments alone would not have been enough of a stimulus to see them through to completion. Even projects like the railroads had social and defense objectives. Government subsidies built and kept such major projects going. Because of the subsidies and monopoly grants, the operating companies realized profits. Without support from public sources, the time required to recover investments would have been long, and the economic profits, if they materialized, would not have been realized quickly.

Outer space is a national and international resource of great potential value. Rather than concentrate on the economic impacts that are identifiable directly, one can view space as a national laboratory. Rarely does our system question the value of having government research devoted to national needs.

The Space Shuttle as well as future space-based facilities are similar to existing government laboratories. We tend to justify them with projections of economic benefits. But that is not the real reason to build and operate such laboratories. The main reason is political and cultural. We must preserve our technological leadership in order to be among the industrial and military leaders of the world. As a society, we have a cultural need to devote resources to exploring the unknown. The short-term benefits may or may not outweigh the longer term economic returns. We do not and cannot know what those economic returns will be.

The returns may well be in new products and improved processes. Current space projects may create an infrastructure that will lead to other space projects and improvements in our economic system not even thought of today. The railroads and interstate highway system created new delivery methods and new land-use patterns in the United States. While they made the transport of materials and products cheaper and faster, they also changed the shape of our cities, the way we do business, and the way we live. And, these structural changes in our society from those investments are not clearly separated and measured in our standard economic yardsticks of GNP and related statistics. About all that can accurately be said of space R&D investments is that economic impacts will occur, and those impacts in the long run will be sizable.

Standard economic models are devised to quantify what *can* be measured. What *should* be measured is entirely different. We need to develop a measure of the value of man-made national resources. Such a value must integrate easily understood and standard economic measures with political and social values.

Economic Measures of Space R&D to Be Avoided

First, no economic study should attempt to calculate a "bottom line" return on space R&D investments. There is no such number that has any realistic meaning—it lives only in the uncharted world of general economic equilibrium theory. To date, all such numbers that have been published and used have, at the very best, only measured partial returns. And, all such numbers are products of economic models with many limiting assumptions.

Even when these assumptions and qualifications have been laid out carefully, the mere existence of the number is an attractive bait to those politicians who have a pressing need to find a justification for space R&D. Once a bottom line economic ratio of returns to R&D has been calculated, it quickly finds its way into misuse.

Second, another measure to be avoided is one that contrasts returns from government investments with returns from private investments. There is no a priori reason why government R&D must have a measurable GNP or productivity return. For technical measurement reasons—such as the fact that government accounting standards treat all government expenditures, including R&D, as current spending with no imputed investment returns—the measured returns to government R&D should be smaller than those to private R&D. And, for the obvious reason that investment of federal funds for R&D is undertaken because of a national public need, the returns may reflect the completion of a mission and not be translatable into GNP effects. Therefore, it should be no surprise that comparative studies show economic returns to government R&D much lower than returns to private R&D investments.

Third, studies that have poorly defined objectives are particularly susceptible to error when dealing with R&D. Research and development is a very broad term, covering many activities. Often economic models of R&D may not fit the questions really being asked, and are really answering other questions. Research activities generally have knowledge as their output. Development activities have products or processes as their output. The same economic model is inappropriate for analyzing both ends of the R&D spectrum. Government R&D is different from private R&D. Again, using the same model to calculate returns is not good use of economic theory. Yet, time and time again, these mistakes are made.

Fourth, new studies should not repeat the mistakes of earlier analyses. Over the years economists have begun to fine-tune both the models and the data applied to measuring the returns from R&D. Although the profession is still far from attaining even "good" results, the cumulative body of literature has added to our understanding of the R&D process and the economic underpinnings of the system. Continued research and improvements will be forthcoming.

Conclusion

The reasons for performing economic studies of government space R&D investments vary. However, the most common purpose is to justify continuing R&D programs or to gain support for new R&D initiatives. Unfortunately, the currently available economic models imperfectly measure returns from past investments. There is no accurate method available to forecast future returns on incremental space investments.

Therefore, the main role of macroeconomic benefit studies has been for public relations purposes rather than as a real tool for decision-making purposes. In order to improve the economic state of the art and to make the models more relevant for policymakers, a number of actions should be taken.

First, a balance must be reached between economic and political objectives. Models and measures should be constructed to meet this balance, and the nonquantifiable returns should be recognized. Understanding the different economic and political objectives will begin to put space R&D in perspective. Since almost all space R&D is government supported, the traditional economic models built on the operating assumption of a freely competitive market must be modified. In current practice, this is rarely done. And, the right models to answer the specific question being asked must be used.

Business planning for space products and markets has been taking place with the rising interest in various commercial space projects. Traditional discounted cash flow models rarely show space R&D investments as advantageous over alternative uses of the funds because of the high up-front costs of space, the long pay-back time frame, and the comparatively high risk of failure. However, there are business models that refine the discounted cash flow methodology to take into account the value of the technology as it is being developed, and the possible decision points where the technology could have market value. These business models are most often proprietary and do not lend themselves to macroeconomic impact evaluations. However, a developing literature and compendium of space business activity models and data will show true economic (i.e., GNP) effects of private R&D that bootstraps on and adds to government R&D infrastructure investments.

Finally, the most promising type of economic study for measuring returns to space R&D is the documentation of actual cases based on direct survey information. Case studies provide relatively clear examples of the returns, both theoretical and actual, to space R&D investments. A well-structured series of case studies coupled with good theoretical modeling aimed at integrating the data and incorporating those data with more general economic measures of benefits may well be the way to establish reliable and "believable" economic measures of the returns to space R&D.

References

[1]U.S. Senate, Committee on Commerce, Science, and Transportation, "Space Law," Selected Basic Documents, 2nd ed., U.S. Government Printing Office, Washington, DC, Dec. 1978.

[2]Griliches, Z., "Issues in Assessing the Contribution of Research and Development to Productivity Growth," *Bell Journal of Economics*, Vol. 10, No. 1, 1979, pp. 92–116.

[3]Denison, E. F., *Accounting for United States Economic Growth, 1929–1969*, Brookings Institution, Washington, DC, 1974.

[4]Mansfield, E., "Contribution of R&D to Economic Growth in the United States," *Science*, Vol. 175, No. 4021, 1972, p. 477.

[5]Midwest Research Institute, "Economic Impact of Stimulated Technological Activity," NASA NASW-2030, Nov. 1971.

[6]Midwest Research Institute, "Economic Impact and Technological Progress of NASA Research and Development Expenditures," National Academy of Public Administration, Sept. 20, 1988.

[7]Solow, R. M., "Technical Change and the Aggregate Production Function," *Review of Economics and Statistics*, Vol. 38, Aug. 1957, p. 312

[8]Evans, M. K., "The Economic Impact of NASA R&D Spending," NASA NASW-2741, Chase Econometric Associates, Philadelphia, April 1976.

[9]U.S. Congress, General Accounting Office, "NASA Report May Overstate the Benefits of Research and Development Spending," Rept. PAD-78-18, Comptroller General, Oct. 1977.

[10]Cross, D. M., "The Economic Impact of NASA R&D Spending: An Update," NASA NASW-3345, Chase Econometric Associates, Philadelphia, March 1980.

[11]U.S. Department of Labor, Bureau of Labor Statistics, "Impact of Government and Private R&D Spending on Factor Productivity in Space Manufacturing," Washington, DC, July 1980.

[12]Kochanowski, P., and Hertzfeld, H., "Often Overlooked Factors in Measuring the Rate of Return to Government R&D Expenditures," *Policy Analysis*, Vol. 7, No. 2, 1981, pp. 153–167.

[13]Heiss, K., and Morgenstern, O., "Economic Analysis of New Space Transportation Systems," NASA NASW-2081, Mathematica, Princeton, May 1971.

[14]Mathematica, Inc., "Quantifying the Benefits to the National Economy from Secondary Applicatins of NASA Technology," NASA CR-2674, Princeton, March 1976.

[15]Mathtech, Inc., "A Cost-Benefit Analysis of Selected Technology Utilization Office Programs," NASA NASW-2731, Princeton, Nov. 1977.

[16]Bradford, D. V., and Kelejian, H. H., "The Value of Information for Crop Forecasting with Bayesian Speculators: Theory and Empirical Results," NASA NASW-2558, Econ, Inc., Princeton, 1974.

[17]Fitussi, F. P., "Economic Benefits of ESA Contacts," European Space Agency, Paris, July 1978.

[18]BETA, "Study of the Economic Effects of European Space Expenditure," Faculte des Sciences Economiques, Universite Louis Pasteur, Strasbourg, Oct. 1988.

[19]Hertzfeld, H. R., "Measuring the Economic Impact of Federal Research and Development Investment in Civilian Space Activities," Committee on Science, Engineering, and Public Policy, National Academy of Sciences, Washington, DC, Nov. 1985.

Measuring and Managing Spinoffs: The Case of the Spinoffs Generated by ESA programs

L. Bach,* P. Cohendet,* G. Lambert,* and M.J. Ledoux*
Louis Pasteur University of Strasbourg, France

Introduction

SINCE the 1960s, economists have tried to measure the economic impact of space programs with a variety of different tools. The level of expenditure these programs involve is so high that public opinion is increasingly demanding an assessment of the tangible benefits to the economy in return for the considerable sums that have been invested. Thus, as it is largely described by the article of Hertzfeld in the present book ("Measuring Returns to Space R&D), macroeconomic analysis combined with econometric tools have been applied to assess the global impact of space expenditures [macroeconomic modeling, influence of research and development (R&D) expenditures on a macroeconomic production function, etc.]. A different approach was used to evaluate the economic activity and the employment directly induced by space programs in the space industry and its suppliers (input-output analysis, use of economic multiplier). Other studies focused on the impact of the use of meteorological or communication satellites on weather forecast or activities related to telecommunication, as well as on the evaluation of space technology transfer policy (by the analysis of some of the markets created around or "fertilized" by space technologies).

This article is part of the methodological debate on the economic effects of large space programs. It aims to present a methodology designed by the Bureau d'Economie Théorique et Appliquée of the University of Strasbourg (BETA) in order to evaluate what we consider to be the most specific economic effects of those programs: the *spinoff effects*.

Copyright © 1992 by the American Institute of Aeronautics and Astronautics, Inc. All rights reserved.
*Bureau d'Economic Théorique et Appliquée.

As a practical example of the use of this method, the measure of the economic impact of projects implemented by the European Space Agency (ESA) is described. This evaluation integrates two kinds of objectives. The first consists of obtaining discriminant quantitative figures that can be used to test the effectiveness of a program, so as to justify the financial commitments made from the public authorities by giving a minimum approximation of what is called the indirect industrial effects of ESA contracts. The second is of an informative and prescriptive kind which does not question the established status of the program, but attempts to improve its effectiveness by analyzing all its economic, scientific, and organizational impacts on those who contribute to it and on their corporate environments. In other words, it depicts the behavior and requirement of the industrial side concerning the management of the diffusion of technology and know-how.

In the first section we will define more precisely what comes under the all-purpose term of spinoff. The second section is devoted to the problem of measurement, and especially to the studies carried out by the BETA in this field. Finally some of the factors playing a part in the spinoff generation as well as some issues of spinoff policy are reviewed in the last section.

Definition of Spinoff

The term *spinoff* is very often understood as defining the cases in which technologies developed in the framework of space programs are used in nonspace activities. Space technologies are thus transferred and allow firms to make profits by helping them to design and then sell new products or services or to modify their production processes in order to enhance their efficiency. These effects, spreading over all the economy through sales of goods and services, purchase of licenses, imitation, technical or scientific documents, and so forth, constitute the basis of what is commonly called the long-term economic effects of space programs.

However, in a much broader sense, the term spinoff covers all the ways in which what has been learned during one activity of a firm, here the space program, is used by it or by another organization in another context. In that way, spinoff should not be restricted to technology transfers, the introduction of new methods of management, the change of organizational structures, the strengthening of collaboration between firms, the use of having worked for space applications as a marketing reference, the improvement of employees' know-how, and so on, could also be considered as spinoffs.

Thus a clear understanding of what is and what is not a spinoff is needed. For this purpose, we first compare them to the other types of economic impacts of space progams. This leads us to introduce the typology of spinoffs used by the BETA in its studies. Some examples as well as some qualitative dimensions are mentioned in order to emphasize the variety of cases that come under spinoff phenomena.

Spinoffs and Economic Impact of Space Programs

Distinction between short-term and long-term effects, or between macro and micro effects, are well known in the literature. However, it is possible to propose another approach in many ways more adapted to the specific characteristic of space programs.

To make it clear, let us observe that a large-scale technological development program such as a space program, with a significant financial involvement compared to the private R&D expenditure of the sector, generates two kinds of economic effects on the industrial structure: *direct* and *indirect effects*. The former are those which arise out of contracts performed within the set framework of the program (designers, constructors, suppliers of services, and end-operators). The latter are different in that they go beyond the scope of the contract objectives and subsequently spread throughout the economy as a whole. (In this respect, it is important to note that the definition proposed here of what are direct and indirect effects is slightly different from definitions according to which direct effects are those affecting the participants to the space program and indirect effects are those affecting the other organizations.)

However, when attempting to define the nature of the full range of direct and indirect effects, we have to consider a specific characteristic of large-scale technological programs. These usually depend on a twofold contractual relationship: on the one hand, between a government of a state (or of a number of states) and an agency; on the other hand, between that agency and a group of business contractors. Furthermore, each type of contractual relationship has its own sets of direct as well as indirect effects. It is important to distinguish between these two types for evaluating purposes: for any given contractual relationship, the related direct effects can be seen in terms of the specific objectives agreed upon between the parties, whereas the indirect effects correspond to general objectives (e.g., improvement of scientific knowledge, social equity, macroeconomic equilibrium, etc.). In the case of the ESA, the related economic benefits are as follows (see Fig. 1):

1. The contractual relationship between the Member States (European countries participating in ESA) and the Agency (ESA) provides that the latter shall coordinate space activities with a view to establishing the operational facilities (launchers, satellites, and ground control) needed to attain given political, scientific, and economic objectives. In the economic sphere, the Agency is required, for example, to make the meteorological satellites operational, which, by enabling more accurate weather forecasting, will lead to benefits affecting a large number of business sectors, such as agriculture, construction industry, transport, and so on. Other economic objectives are clearly designated in connection with the implementation of telecommunication, remote sensing, and earth sciences satellites. On the basis of economic objectives of this kind, stipulated in the contractual arrangements between the Member States and the Agency, we can identify a first category of direct economic effects corresponding to the benefits obtained by users of the services provided using the space infrastructure:

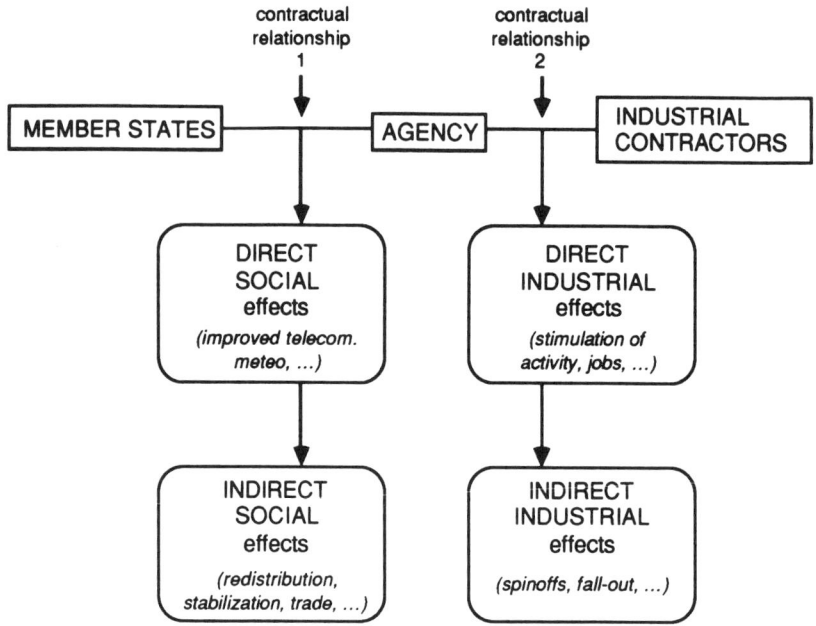

Fig. 1   Economic impact of space programs.

direct effects on the social community, such as benefits derived from more efficient telecommunications systems, more accurate weather information, or extended knowledge of the Earth.

Beyond these direct effects, which are to be evaluated in terms of the specific provisions set out in contractual instruments, we can identify a whole range of indirect effects on the social community (or indirect social effects) that are also generated by the program but which correspond to economic phenomena of a more general kind (cost redistribution effect in the case of structurally influential projects, possible environmental nuisance, income redistribution effects, etc.).

2. The contractual relationship between the Agency and the group of project contractors requires the latter to carry out—generally according to very stringent technical and quality specifications—the industrial projects laid down by the Agency. We can link to that relationship a set of direct industrial effects arising out of the establishing and operating of an industrial infrastructure, mainly on account of stimulation of activity (measured in terms of production level and net job creation) by the orders for construction of launchers, satellites, or ground control centres. (These effects are sometimes called short-term economic effects in U.S. studies.) The measuring of these direct industrial effects is often based on objective factors corresponding to marketable services on fully known markets.

The indirect industrial effects (often collectively described as "spinoffs" or "fall-out" effects) include all the benefits in terms of technology, know-how, corporate image, or business contracts, which Member States companies derive from their participation in ESA programs and are able to deploy elsewhere (this constitutes a "first circle" of effects). The process then expands beyond these firms, spreading first to customers and suppliers of the contracting companies and subsequently throughout the economy by many varied ways.

An Extensive Typology of Spinoff

In economics, spinoffs are traditionally compared with externalities, and more precisely technological externalities. According to Griliches[1,2] there are two kinds of externalities.

In the first case, technologies developed or enhanced in a sector of activity are embodied in marketable products, and the economic advantages related to this kind of externality appear in the sale and purchase of these products on markets. Firms which sell or use these products are thus able to increase their incomes, while consumers benefit from new or better and more efficient products.

The second kind of technological externality corresponds to the spreading of knowledge and its impact on the research endeavors and more generally the activities of other sectors. The knowledge can be transferred without direct links between sectors, and there are many ways by which it is conveyed (movement of personnel, reverse engineering, printed articles, news release, patents, licenses, colloquia, mergers and acquisition of firms, etc.). This was the basis of the argument put forward in their seminal papers by Nelson[2] and Arrow[4] to justify public R&D expenditure: Because of these externalities, and despite the patent system, firms cannot appropriate all the benefits of their in-house research, thus their incentive to innovate is not strong enough; as a consequence, the national R&D effort may be nonoptimal without the support of public funds.

There is no doubt that such technological spinoffs are central in the case of space programs which precisely assume the role of leader in the technological development of an industry and even of a country as a whole. Besides, NASA has been trying to promote and develop these spinoffs for years, through the Technology Utilization programs, and similar initiatives have been taken more recently in Europe, at an international level through ESA pilot projects as well as at national level, for instance, through the creation of the Novespace company in France.

But as underlined above, the spinoff phenomenon can be seen as much broader than the technological transfers. The BETA has proposed a typology that takes into account the different forms that spinoff can take. Before presenting it, two of its characteristics must be pointed out. First, only spinoffs affecting contractors of the space agency in charge of the program(s) studied are taken into account. [As described later in the paper, this typology was designed for the purpose of evaluations based on direct interviews with firms in one way or another statistically representative of the size of the space programs (e.g., the space agency contractors). For this reason, spinoffs affecting the whole economy (spill-over effects) cannot be taken into account since it is by definition not possible to form a statistically significant sample of organizations having benefited from these "second-order" spinoffs (only case studies can be provided about this more global phenomenon, unless econometric tools using statistical data are used).] Second, spinoffs concern nonspace-related activities of these contractors as well as space-related activities, provided that these latter are not carried out for the space agency in question.

The typology is implicitly based on the analytical framework proposed by Schumpeter (linear model of innovation). According to this author, new economic configurations have an impact on products, production and sales techniques, the market, and company organization and methods. Referring to this theoretical background, but needing to preserve an operational character, the BETA classification distinguishes four categories of effects: the technological, commercial, managerial, and work factor effects.

Technological Effects

The fundamental—and even more applied—research work carried out in the framework of the space programs gives rise to technological innovations leading to the emergence of new product generations and subsystems subsequently deployed by other space programs. It also enables a technology developed in the space sector to be applied to other industrial sectors, resulting in the creation of new products—sometimes leading to a diversification of activity—and improved characteristics (quality, performance) of existing products.

These are the classical spinoffs that were referred to above. From Teflon materials or miniaturization of electronic components for Apollo to ceramic materials for the coating of the Space Shuttle, NASTRAN computer software for structural analysis, programmable implantable medication systems, or power controller for energy savings in engines, one can find numerous examples of such spinoffs in U.S. industry (see also the annual Spinoffs reports from NASA or the qualitative part of the Midwest Research Institute (MRI) study[5]). In Europe, there are no systematic surveys of these technology transfers; air-bag security systems for cars derived from gas generators technology, remote-control systems for professional TV cameras, or different specialized electronic devices as hybrid components are cases in point.

Commercial Effects

Commercial effects basically take the form of increased sales of products or services that do not incorporate significant technological innovation. The space agency contractors are able to take advantage of new market areas opened up as a follow-up of those space programs, for instance at a national level (e.g., ground control stations). Many of them have furthermore acquired a quality label associated with space activities, which is likely to give them considerable competitive leverage. On that commercial level, ESA programs—more likely than other space programs—also enable some contracting companies to form closer business ties which are then extended to foster joint activities outside the space agency framework. For instance a company operating in the space market for connector technologies was in a position to join forces with Belgian and Swedish companies to bid for a Eureka contract in order to solve problems of connectors operating in a hostile environment (automated station for North Sea oil rig).

Effects on Organization and Methods

Another important contribution of the space programs is in the managerial and production methods innovations they have inspired, for instance in terms of quality control, production techniques, and project management. These innovations result from the high standards imposed by space performance and reliability specifications (principle of zero-fault in a hostile environment). Laser technology for cutting and welding electronic control units, control of EMI/EMC and ESD problems in electronic components production, or design review methods are example of techniques and methods developed or learned by firms in space programs and then applied elsewhere.

These effects are also the consequence of the particular form of the industrial network set up for space programs, joining together at different levels of responsibility many firms originating from very diverse industrial sectors (the generalization of the PERT method initiated by the U.S. Polaris program is a good example of this effect, although not originating from space application). In the case of the ESA programs, the competence in project management is perhaps even more necessary since the production is less concentrated than in the United States and is shared between firms from different countries.

Work Factor Effects

The economic effects induced by ESA programs are to a large extent connected with the "men." Space departments are often regarded as training schools for personnel as well as for managers. The induced work factor effects are related in particular to the heightened qualifications and skills acquired by the personnel employed in those programs, which enable them to feed expertise into the company departments not directly concerned by space activities. For instance the technical staff responsible for maintaining fluids and mechanical systems, UHF radio links, and so forth, on the Kourou site are trained to fit into a highly disciplined framework working to stringent standards. They are later employed on oils rigs, chemical production plants, or nuclear power stations and prove to be more aware of the importance of quality and control.

In addition to promoting this permanent enhancement of skills, in certain firms space programs support the creation, maintenance, or growth of well-structured teams of specialists, scientists, engineers, and technicians that constitute what can be called the "critical mass" of the firm. The technological potential which this critical mass represents is a decisive qualification for securing contracts relating to the increasingly complex systems in all sectors of industry.

For a major prime contractor in the European space industry, ESA programs were the catalyst that enabled it to bring together within a single team the technical skills that had previously been scattered through the different departments of the company. Another prime contractor freely confessed that one of its main considerations was to reach a critical size, through its contracts with the Agency, so as to be able to compete with American firms. Similarly, space firms in the smaller countries, by working for the Agency, are able to keep certain specialists in the space industry and even in the country itself, and thus form national centers of advanced-technology skills.

Some Dimensions of the Spinoff Phenomenon

If some spinoffs are "spontaneous" (e.g., skills improvement) or quasi-immediate (use of space as a marketing reference), most of them constitute processes that require a deliberate policy of the space firm (set up before its participation in a space project or once the results of this latter are known), time (several years may elapse before tangible results are observed), and carry costs for the adapation of the knowledge to its new environment (typically the case of technology transfers). These costs include, in particular:

1) The cost of acquiring new knowledge about market needs, opportunities, existing and potential competitors

2) The cost of adapting the technology to its new conditions of use, i.e., to the industrial and market requirements of recipient sectors

3) The cost of adapting the firm itself to this new technology or products (for instance education and training of production and marketing personnel)

4) The cost of giving up existing products that are replaced by new ones

5) The cost of giving up or not being able to discover alternative ways of research (opportunity cost)

6) The transaction costs between the space firm and the recipient one(s) in the case of external transfer

As far as technologies are concerned, the diversity that characterizes transfers derives first from the type of technology involved: technologies relating to products, production processes, or methods and procedures. In practice, transfers often concern more than one of these three aspects (for example, certain technologies cannot be used for a product unless a special manufacturing process is also used).

Another element in this diversity is the extent to which the transferred technology is formalized or codified, i.e., the precision with which it has been possible to define its characteristics and its conditions of use. This leads to another definition of "types of technology," that can be classified according to their degrees (or levels) of formalization. To each type of technology associated to a level of characterization correspond sets of possible forms of transfer and of modes of appropriability.

By way of example, the characteristics of a product or process covered by the sale of a license are "frozen" and clearly defined: the license provides a definition of the conditions of use of the product or process, it is thus an explicit knowledge. Its transfer can be regulated by market mechanisms, even if it is not always the case in reality. This also applies to all products (including software) or processes which can be used more or less as they stand by nonspace sectors. In contrast, the know-how possessed by specialists is a relatively indeterminate combination of scientific and technical knowledge, work habits, and experience. By definition, this know-how is in the head and in the hands of specialists, and is very often tacit or uncodified. For instance, in a firm specializing in the design and building of electronic tubes for space applications, the ability to shape the special glass for given applications lies almost exclusively in the skills of a very limited number of specialists. It is then dramatically difficult for these specialists to transfer their knowledge. Sometimes, different pieces of knowledge are in the mind of different specialists, and only the combination of these different skills allows the firm to design or produce products. (This phenomenon forms the basis of the concept of "critical mass effect.") In almost every case, the only way to transfer this type of technology is to transfer the specialists themselves.

Between these two extremes range a large variety of "types" of technologies: precise and specific technical information on a particular aspect of a technology; management procedures and production or quality assurance methods, the main features of which may or may not be easy to identify or which may offer a valid methodology; algorithms or procedures used in computer programs; technologies whose conditions of use are fully understood by the space industry but the limits of whose applicability have not yet been established. It must be underlined that a significant proportion of the technologies generated by the space sector relates to know-how and technical expertise.

Spinoffs may involve different actors interconnected according to different patterns of agreements. These forms of spinoffs are determined by the extent to which the technology is formalized—this conditions the scope for its transfer (for example, know-how realized only by a transfer of staff, obviously more easily done within a company)—and by the firm's technological, productive, and commercial capabilities and strategic choices.

The following forms of spinoffs can be identified:
1) Transfer within firms, between two departments or divisions
2) The creation within a firm of a new department or division
3) The creation of a new firm, for example, a subsidiary
4) Transfer between a space firm and a firm in the recipient sector (In the case of granting of a licence or patent, the market is sometimes divided up geographically or is shared on the basis of industrial sectors and/or of the size of the customer's orders.)
5) The creation of a new firm in conjunction with a firm specializing in the recipient sector (joint-venture)
6) Technical assistance by the space firm in product development by a nonspace firm

In all such cases, a consultancy firm or organization may be called in. Such firms may identify technological or commercial opportunities on behalf of the space firm or the transferee, liaising between them (technology brokers) or taking part in the transfer itself (contract research organizations).

Finally, it must be remembered that the actual transfer from the space to another sector is very seldom a "pure one-way" spinoff, from space to nonspace application; on the contrary, it is most of the time one step in the overall process of technology development. For instance, a technology is first developed in a nonspace sector, then used in the space field where some of its characteristics are modified; then it is transferred back to the nonspace originating sector or another one. There are also cases of synergies in which each sector is fertilized by the other. Combinations of these basic patterns are obviously common in reality.

Measurement of Spinoffs

This section will be mainly devoted to the presentation of the methodology designed by BETA and its applications to the case of the ESA programs. But first we will emphasize some of the main issues related to the evaluation of spinoffs and to the problems encountered in such an exercise.

Issues in Measuring Spinoffs from R&D and Space Programs

As was underlined in the first section of this paper, in economics spinoffs are usually compared with the two kinds of technological externalities defined by Griliches.[1,2] The first occurs when technology is fully embodied in products, and is related to the price mechanisms as they are taken into account by the theory of the firm. It can theoretically be measured in terms of producer and consumer surplus generated by it, and derived from the supply and demand curves' representation. The producer surplus is basically equal to profit, while the consumer surplus is the difference between what the consumer is willing to pay for the product (represented by its demand curve) and the price the purchaser actually has to pay for it. Moreover, if the innovative product is, for instance, a machine used by another firm, it may allow this latter to increase its profits by lowering its production costs, and by the same mechanism it entails increased profits for the downstream firms and finally in the surplus of the final consumer. The difficulty of measuring these effects will then depend on at least three parameters: the complexity of the relations between suppliers at each step of the production process of the innnovative product; the ability of price indexes used by evaluators to reflect the change in the quality of the product; and the competitive structure of the industry determining the distribution between buyers' and suppliers' surplus.

The general diffusion of knowledge from one sector to the others, and the impact on the latter is the second kind of spinoff or technological externality. To evaluate them directly, and apart from the technical problems of measurement, one has first to identify either the firms or the sectors where they are localized and the features of the phenomenon itself (channels, direction, and path).

In practice, it is often very difficult to identify separately the two kinds of externalities. Basically, two types of evaluation are to be found in the literature, using "classic" tools of economists: estimates of private and social return limited to a particular industry or sector, and regression based estimates of the impact of R&D expenditures on the economic activity. But other more qualitative approaches have also been used in this field.

The first classic approach is based on the theory of the firm and the consumer/producer surplus concept mentioned above (cost/benefit approach). Apart from some earlier studies in agriculture, one of the main applications is the work perfomed by Mansfield and his team[6] on 17 cases of innovation. [The private return takes into account the surplus of the producer (income from the innovation less costs of producing and marketing the new products as well as costs of carrying out the innovation) and the profits that would have been made if the innovation had not occurred; the social return takes into account the consumer surplus, the research expenditures of related unsuccessful innovators, and the profits (losses) of imitators (unsuccesful competitors).]

These approaches have been extensively reviewed and criticized. We will not attempt to review them all. Two points must nevertheless be stressed. In the case of new products, the proven and stable enough demand and supply curves required for quantification do not often exist. On the other hand, it only draws the attention to social and private rate of returns for "successful" innovations, and thus may not be "representative."

The second approach is based on the use of the production function, linking output (for instance, Gross National Product in studies at the macroeconomic level) to different production factors, basically capital, labor, and R&D (the contribution of Solow[7] can be considered as the pioneering work in the field; Dension[8] and the works performed by the team of Griliches are among the most representative references). Most studies attempt to estimate the contribution of R&D expenditure to economic growth, by measuring the contribution of the other factors and affecting the remaining influence to a "technical progress" factor, R&D then representing part or all of this factor ("residual factor models"). Some other works directly link the total factor productivity (excluding R&D) to the intensity of R&D investment (typically R&D to sales or added value ratios); the coefficient of regression between the two elements can be explained as the rate of return of R&D expenditures (see, e.g., Ref. 9 for a good survey on this). Regression-based estimates raise a lot of problems: measurement of output, capital and labor factors and especially R&D "capital," short time frame of available data on R&D, "scope" of the effect of R&D actually captured by the measure (for instance, quality changes are difficult to take into account), assumption of separability of the influence of the different inputs, lack of understanding of the innovation and diffusion processes, and so forth. (See Ref. 1 for discussion on some of these points.) In the basic specification of these models, and insofar as spinoffs are to be evaluated this way, one must note that if the study is carried out at a sector level, the interactions between sectors are not taken into account, whereas if the study is carried out at national level, it is not possible to distinguish what is a spinoff of what.

Thus more recently, numerous studies have attempted to specify the transmission channels of externalities between sectors, trying to identify the "suppliers" and the "receivers" as well as the nature of the links between them. In other words, the problem is to evaluate the influence of the R&D of one sector on the activity of another. Different approaches are proposed, some considering the influence on a given sector of the R&D

of all other sectors, some others considering the influence of a weighted amount of R&D of the other sectors. In this second case, the weighting function is based on different assumptions (proportional to the input-output flows of intermediate consumptions between sectors, to the flows of patents, to empirically determined flows of "innovations," to a "technological distance," for instance based in the United States on Standard Industrial Classification (SIC) or National Science Foundation (NSF) classifications, etc.). The results of these analyses are sometimes used in more sophisticated specifications of the production function for the same purposes as mentioned above, and some others stand alone, as patent statistics, bibliometrics, reviews on interindustrial flows of innovations, and so on (see Ref. 10). The latter are by far much closer to the very large family of approaches not strictly based on consumer/producer surplus theory or production function analysis, which very often put more emphasis on qualitative aspects of spinoffs evaluation (including also case studies, financial investment models, studies on skills and competences, etc.).

In this very large family, one growing stream follows the more radical criticism of the basic assumptions underlying classic tools of economists such as the concept of production function on its own, questioning, for instance, the hypothesis of perfect competition, rationality of choices, technology akin to information, nonincreasing returns, and so on. In particular, attention is drawn to the tacit, localized, and path-dependent characters of technologies (with a fundamental role played by learning processes), and on the interdependence between technological development and organizational forms (internal to firms, interactions between administrations and firms, network of firms, etc.) in shaping the evolution of the economy. The dynamic analysis of the processes of wealth creation is thus emphasized instead of the problematics of resource allocation in a fixed context implicit to mainstream economics (for recent synthetic works of authors representing this "nonorthodox" view sometimes known as evolutionary economics, see Dosi et al.[11]). From this standpoint, the scope of spinoffs or indirect effects of R&D expenditures and especially R&D public programs, such as space programs, must be enlarged; as a matter of fact, the BETA typology of indirect effects provided in the first part of this article is in some way close to this kind of "evolutionary" approach. Nevertheless, in this field the diversity of the methods of evaluation put forward has so far prevented the development of a standard methodology that could lead to providing a tool for universal application. (Pieces of the puzzle can be found in Irvine and Martin,[12] Callon et al.,[13] and David et al.[14]).

In the field of space programs, different estimates have been made of the spinoff phenomenon, resorting to the methodologies briefly described above. Most of them were conducted in the United States. The production function approach was applied by MRI[5,15] to NASA programs. (Results obtained with this approach were also used in macroeconomic models in Evans,[16] Cross,[17] and Econ,[18] in order to estimate the impact of space industries on aggregate economic indicators such as prices, employment, or balance trade.) Cost/benefit calculations were completed on some secondary applications of NASA programs[19] and on the NASA Technology Utilization Program (see, e.g., Refs. 20 and 21). A lot of studies actually focused on this NASA TU program, using miscellaneous indicators (cost/benefit-related approaches, sales and cost reductions for users of NASA technologies, statistics on commercialization of patents or licenses, etc.) (see Refs. 21 and 22). Most of them are largely described in the contribution of Hertzfeld ("Measuring Returns to Space R&D") as well as in an earlier paper of this author.[23] We will therefore come to the presentation of the works completed by the BETA in this field.[24-26]

BETA Methodology

General Presentation

The main features of the methodology designed by the BETA can be summarized as follows:

1) The evaluation is limited to the indirect effects/spinoffs affecting the ESA contractors.

2) It is based on first-hand data, since they are obtained by direct interviews with the managers of the ESA contractors, and carried out by BETA members.

3) The inventory of indirect effects is thus of a microeconomic type, but since the sample of irms can be considered as statistically significant, the result may be extrapolated to the whole set of ESA contractors. (Note that the result that may be obtained with this calculation is still different from macroeconomic evaluation.)

4) The objective of the evaluation is twofold: 1) qualitative, since it aims to describe in more details the spinoff phenomenon; 2) quantitative, since it aims to provide a minimal estimation of its importance.

5) The scope of the spinoff phenomenon studied corresponds to the typology proposed in the first section.

The economic indirect effects studied by BETA correspond to the different learning processes tried by firms during their work for ESA, affecting them in many varied ways (widening of scientific, technical, and "organizational" knowledge; innovation in products and procedures; new links with new external organizations), and applied to other activities than ESA contracts (space or nonspace-related activities). In fact, if the economic effect of large R&D programs are likely to spread to the whole of the economy, it seems clear that the phenomenon of "wealth creation" first appears in the organizations contracting with ESA, where they obviously have their origin and first concrete use in economic terms. Such a choice implies that the BETA methodology does not make it possible to know what are the long-term effects of ESA programs on the whole economy.

The procedure followed was to make as exhaustive an inventory as possible of indirect effects resulting from ESA programs among the ESA contractors, and to identify the various forms they may take. For this purpose the typology of indirect effects presented in the first section has been refined and gave birth to the classification presented in Table 1).

Quantification of Effects

The final unit of measurement used to express indirect effects on a firm is the added value (the sum of the firm's wages and profits), together with

Table 1 BETA classification of spinoffs

TECHNOLOGICAL EFFECTS	COMMERCIAL EFFECTS
Derivatives form ESA products New products Diversification Product improvement	International cooperation New sales networks Use of ESA as marketing reference
EFFECTS ON ORGANIZATION AND METHODS	**WORK-FACTOR RELATED EFFECTS**
Quality control Project management Production techniques	Formation of critical mass of specialists Improvement of workforce skills

the estimated value that results from setting up and keeping highly skilled design and production teams (what was defined above as the "critical mass"). The quantification exercise thus consists in determining how the work carried out for ESA programs affects these two parameters; the process is illustrated in Fig. 2. The contracts that firms get from ESA, like all their other activities, affect the four basic factors corresponding to the four types of effects described earlier (technological, commercial, organization and methods, and work factor-related effects). These in turn contribute to increasing the volume of sales and reducing costs and thus, under some circumstances, to increasing the firm's added value. The work factor also specifically affects the critical mass, which is estimated in a broad fashion on the basis of the payroll of the staff concerned.

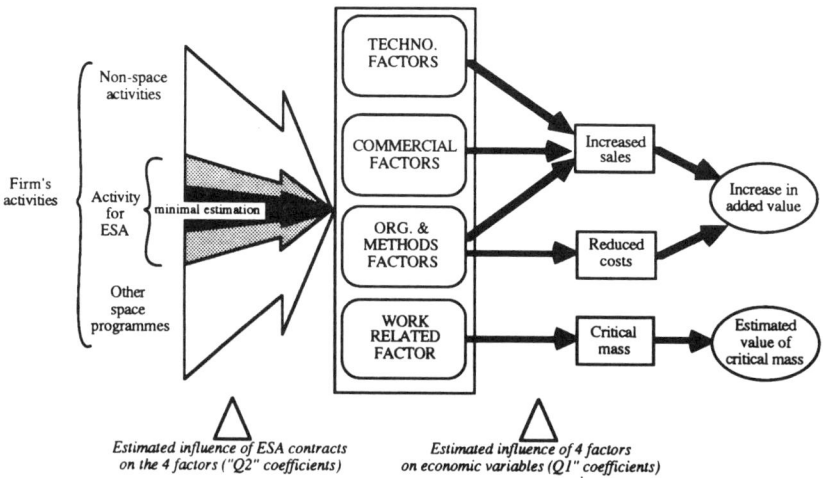

Fig. 2 Principle of quantification of indirect effects.

1) In the case of *quantification by sales*, the managers interviewed are thus asked to estimate, as a percentage, two sets of coefficients:

a) The ones (Q1) accounting for the parts played by the three factors, Technology (Q1T), Commercial (Q1C), and Organization and Methods (Q1OM), in influencing sales; their sum must be equal to 100%. Q1 does not therefore refer exclusively to the firm's ESA activity.

b) The ones (Q2) accounting for the parts played by ESA contract work in each of the three factors above (Q2T, Q2C, and Q2OM); they must be between 0 and 100%. They are very often based on objective data like the share of ESA funding in the development of the product in question. The industry representatives also specify the exact nature of the influence of ESA contracts expressed by Q2 in each of the three categories.

These figures are, of course, relative to the sales of the product which constitutes the indirect effect in question. The final result is obtained by multiplying these two sets of coefficients by the increase in added value caused by the increase in sales:

Technological Effect: Sales × rate of added value × Q1T × Q2T

Commercial Effect: Sales × rate of added value × Q1C × Q2C

Organization and Methods Effect: Sales × rate of added value
× Q1OM × Q2OM

It is also possible to extend the scope of evaluation by including those suppliers of ESA contractors who helped to make products among the items which constitute indirect effects as defined (in this case, the complementary of added value to sales is discounted). But the distinction between technological, commercial, and organization and methods effects is no longer relevant, for the suppliers do not benefit from an ESA experience, but only from sales opportunity. It gives:

Effect for Suppliers: Σ[Sales × Q1 × Q2 × (1 − rate of added value)]

Example

Firm X has developed new composite high-pressure tanks. The sales of this product came up to U.S. $ 10 M (60% of which were export sales), and the purchase of raw materials and components came up to U.S. $ 4 M, of which U.S. $ 2 M went to foreign suppliers. The representatives interviewed estimated that 50% of the sales were due to the technological performance of the product and its degree of innovation, as compared to competing products, while 30% were due to the commercial strategy of the firm and 20% to the quality, efficiency, and reliability of its production process. Seventy percent of the product technology is based on technologies developed under ESA contracts. Moreover, these contracts helped firm X to get to know a European firm and, together with it, to obtain contracts on the European market representing 30% of the worldwide sales of this product; this collaboration was the centrepiece of firm X's commercial strategy in Europe and helped to open up the European markets to it; and it accounted for 20–30% of the success of the firm's worldwide commercial strategy. Fifty percent of the production of the product in question is based on methods and techniques mastered under

ESA contracts, half of which corresponds to quality insurance and the other half to production methods.

Numerical data:
Sales U.S. $ 10 M, of which 6 M in export sales
Value added   60%
European inputs   50%

Coefficients:
Q1T   50%
Q1C   30%
Q1OM   20%
Q2T   70%, new product
Q2C   20%, international collaboration
Q2OM   50%, of which 25% for quality management and 25% for production methods

Quantification:
Technological Effect, new product:
  U.S. $ 10 M × 0.6 × 0.5 × 0.7 = U.S. $ 2.1 M
Commercial Effect, international collaboration:
  U.S. $ 10 M × 0.6 × 0.3 × 0.2 = U.S. $ 0.36 M
Organization/Method Effect:
Quality control:
  U.S. $ 10 M × 0.6 × 0.2 × 0.25 = U.S. $ 0.3 M
Production methods:
  U.S. $ 10 M × 0.6 × 0.2 × 0.25 = U.S. $ 0.3 M
Effect for Supplies:
  U.S. $ 10 M × 0.4 × [(0.5 × 0.7) + (0.3 × 0.2) + (0.2 × 0.5)] × 0.5 = U.S. $ 1.02 M

2) In the case of *quantification by cost reduction*, the data are quantified using savings on inputs, lower reject rates or savings in production time. It is done:

a) Directly, by adding up the savings realized due to methods acquired under ESA contracts

b) Indirectly, by multiplying the following data: amount of savings realized as a result of a particular method and percentage of influence of ESA experience in implementing that method (Q1).

Example
To produce a line of hybrid high-performance components, firm Y has set up a new quality control, which it was therefore able to test before adopting them for the whole of the firm's production. At first, these quality controls entailed a slight increase in production costs, because of the need to adapt the methods to the characteristics of the product and the know-how of the work force. But in the end the firm's production costs decreased, and the net savings came up to U.S. $ 100,000. The firm estimated that 50–60% of the newly introduced quality control methods had been acquired under ESA contracts, notably certain documentation techniques.

Numerical data:
Amount of cost reduction   US $ 100,000
Coefficient Q1   50%

Quantification:
Organization/Method Effect, Quality management:
   U.S. $ 100,000 × 0.5 = U.S. $ 50,000

3) In the case of *quantification of the critical mass*, for reasons of homogeneity, the quantification is made in monetary terms by taking into account the average cost of an engineer working in the space division. The effect thus measures the minimum cost the company would have to bear for qualifying for space contracts if it had not been able to benefit from ESA contracts.

The data are quantified by the representatives interviewed in three stages:

   a) Estimating the firm's critical mass in terms of the number of people in the space sector; managers often provide at the same time a distribution of the number of specialists by field of technology.

   b) Estimating the share of the critical mass which is created or maintained by ESA contracts; industry representatives either give overall estimates in percentage terms or examine the given fields of technology one by one.

   c) Multiplying the number of people making up the critical mass by the average cost of an engineer for the company.

   Example
   Firm Z employs 650 people, 125 of them in the Space Division, which specializes in on-board telecommunication equipments. These 125 people work under ESA contracts as well as for national or bilateral programs and commercial export programs. The representatives interviewed estimated that if their firm was to maintain its technological capacity and hence its ability to obtain contracts at least equivalent in terms of technical specifications, level of responsibility, and monetary value, it must absolutely preserve a critical mass (or "nucleus") of specialists. This critical mass is composed of 70 specialists, as follows:
      28 for signal and data processing
      15 for computing (software and hardware)
      15 for system design
      5 for quality control

   ESA contracts employ an average of 30 people, but only a high-frequency multiplex study program really helps to maintain part of the critical mass, estimated at 11 specialists, distributed as follows:
      5 for signal and data processing
      3 for computing
      2 for system design
      1 for quality control

   The annual average salary cost of an engineer is about U.S. $ 150,000.

   Numerical data:
   Critical mass in Space Division   70
   ESA critical mass   11
   Annual average salary cost   U.S. $ 150,000

   Quantification:
   Work-Related Effects, critical mass:
      11 × U.S. $ 150,000 = U.S. $ 1,650,000

Finally, additional information relevant to each indirect effect identified is gathered (technological areas giving rise to it, application areas in which it occurred, time lag, etc.), so that the results could be analyzed in detail.

This relatively complex procedure is designed to meet two essential requirements. First, it must make it possible to isolate the specific contribution made by ESA contracts as against the firm's other activities, so as to allow for the fact that technologies or production methods often stem from developments made in a number of different programs over a period of time. [Other studies used sales and costs reduction for assessing some indirect effects of NASA programs, but without such a fine search for the "fatherhood" of the effects (see, e.g., Ref. 22).] Second, the purpose of the study is to provide a minimum estimate of the volume of indirect effects rather than to set a precise value on them (given that some items may be overlooked). Consequently, the corporate managers taking part in the survey were asked to assess the influence of ESA contracts in terms of an estimated range, of which only the lower figure was used for the final calculation.

Findings of ESA Program Evaluation

Overall Results

Under this heading, we describe the results of different evaluations performed since the late 1970s by BETA, bearing on the indirect economic effects of the ESA programs. The overall results are formed by adding the effects observed with respect to the contractors and their suppliers (see explanation above): it is the total value of indirect effects identified by BETA. They are shown in Table 2.

This result can be expressed in the form of an overall economic spinoff coefficient, representing the ratio between the total value of the indirect effects generated by the ESA contractors studied and the total payments made by ESA to those contractors during the period covered by the study. It means that on average, for the sample of firms studied, every 100 units paid by ESA to industry results in a minimum indirect economic benefit of around 300 units via the ESA contractors forming the sample.

Moreover, this figure must be seen as a conservative estimate, at least for three reasons. First, as mentioned above, the study takes no account of long-term effects on society as a whole. Second, some effects inevitably escape the interviewers while some others are impossible to quantify (only 60–70% of the identified effects were quantified). Third, the option taken for the quantification exercise was always to retain the lower boundary of the figures provided by the managers interviewed.

This coefficient may nevertheless be a little confusing, because one figure cannot adequately describe the large variety of cases of indirect effects studied, and may also give the false feeling that all the economic effects of space programs are covered. (This variety of cases is also linked to the variety of the nature of the effects; for instance, critical mass effects may be considered as of a different nature than other effects. The reader can easily take into account this remark by computing differently than is done

Table 2 Overall results of BETA studies

	ESA 1980	ESA 1988	Canada 1989
Period covered	64-82	77-91	79-93
Number of firms in the panel	128	67	10
Total indirect effects	7 551 (MAU 86)	12 680 (MAU 86)	256 (MAU 89)
• among ESA contractors	6 023 (MAU 86)	9 214 (MAU 86)	189 (MAU 89)
Ratio effects / contracts	≥ 2.9	≥ 3.2	≥ 3.5
Indirect effects outside space sector	50 %	21.1 %	24.4 %
Indirect effects on exports	28.2 %	12.8 % (out of ESA Member States)	66.4 %
Nature of the effects (% of contractors' effects)			
— Technological	25	32	40
— Commercial	27	8	18
— Org. & Methods	19	6	18
— Work factor	29	54	24

here the figures provided.) Another shortcoming of this kind of presentation is that readers could be tempted to compare it with results from other studies based on completely different approaches, as, for instance, those based on the use of the production function analysis. For these reasons, the more detailed analysis that will be provided below is undoubtedly much more interesting. However, it could be noted that the overall results obtained from the three studies carried out by BETA are quite similar; it suggests that there is a certain homogeneity between the European and the Canadian "performances" as regards indirect effects, despite the different characteristics of the space industries successively studied. Incidentally, it also proves to a certain extent that the method developed by BETA is both repeatable and transferable.

The results set out in the rest of the article are expressed in terms of added value, given that they correspond to the fraction of the effects observed at the level of the contractors. We will also focus on the 1988 ESA study results. (Results from the Canadian study are extensively analyzed in Ref. 26.)

Breakdown by Type of Indirect Effect

The indirect effects for the contractors can be broken down into different categories according to their nature; and inside each broad category, "subcategories" can also be distinguished.

The main part of the indirect economic benefits from the ESA programs relates to product technology, together with the enhanced potential of the design and production departments of the companies involved. By contrast, the commercial and organization and method effects are relatively slight, whereas the four types of effects carried more or less equal weight at the time of the 1980 study. It should be noted that in the case of Canada, the breakdown shows that technological benefits are clearly dominant, the three other effects being close to one another in size (the "Canadian critical mass" being supported by Canadian or United States programs).

The *technological effects* generated by the European space programs were considerably on the increase during the period considered, confirming the trend observed at the time of the previous BETA study. We also found a time lag of about 5 years between the marketing of a product and the ESA program from which its technology was wholly or partially derived ("incubation" phase for know-how applied to new products).

The *commercial* and *managerial effects* increased very slightly during 1977–1991, but were in sharp decline as compared with the period from 1964 to 1982. There are a number of explanations for this twofold trend: fading novelty of the ESA connection; stability of the network of companies working in the space sector up to 1986/1987 (restricting opportunities for fresh contacts); emergence of new programs tending to reinforce cooperation at European level (EEC programs, growing internalization of the aerospace industry); production methods imposed by ESA already common and not being renewed; and so on.

The *work factor effects* increased in step with an evolutionary trend in the space industry. In most of the companies surveyed, we found an ongoing process of structural expansion of the space activity, with the original space project team becoming a "Space Department," then a "Space Division," and in some cases, a self-contained subsidiary. This work factor effect is of course linked to ESA expenditures and it can reasonably be expected to continue growing if the three major programs scheduled by the ESA are completed (Ariane 5, Columbus, and Hermes), since they should enable the European space industry to cross an important technological threshold.

Further Analysis

Thanks to the different qualitative information received from the managers interviewed in connection with each case, it is possible to make a further analysis of the results from different standpoints. One consists in classifying the induced effects according to the space technological specialty from which it is derived, and the space technological specialty (if the effect remains in the space sector) or the industrial sector where it takes shape. In other words, a distinction is made between indirect effects within and outside of the space sector. Another view is brought by a classification of the indirect effect according to the type of firm in which it was generated. We will examine this second type of results in the section "Managing Spinoffs: Key Factors and Policy Issues."

Effects Inside and Outside the Space Sector

The results of this analysis are summarized in Table 3. Remaining constant in value terms, the effects recorded outside the space sector accounted for a smaller proportion of the total indirect effects than was found under the 1980 study (20% vs 50%). This trend seems to correspond to the process of building up a major European space industry during the last 10 to 15 years, which caters mainly for the "commercial" space market—as it is reflected in the increase in the size of the highly skilled work force and in the impact of the technological effect (sales to the private sector of systems developed in the course of ESA programs).

The indirect effects observed inside the space sector reveal that there is very little synergy between the different technology areas. Most of the effects are concentrated in areas where there is a "commercial" space market, satellites, or launchers ordered by Arianespace (telecommunication and propulsion technologies and the like). Outside the space sector, the

Table 3 Indirect effects outside the space sector / 1988 ESA study (21.2% of total indirect effects)

ORIGIN		FALLOUT APPLICATIONS	
Space technology area	% total effect	Industrial activities	% total effect
On-board equipment	31.1	Aeronautics	31.3
Production & testing equipment	19.6	Defense	29.5
Power supply & storage	11.7	Data processing	8.1
Ground equipment	9.6	Electronic equipment	7.8
Design & methods	9.0	Telecommunications	6.5
Telecoms systems	6.5	Medical equipment	5.8
Structures and mechanisms	5.9	Transport	4.5
Propulsion	3.8	Energy	2.8
Thermal control	1.6	Design engineering	1.8
Attitude & orbit control	0.9	Others	1.9
Optics	0.3	------	
	------		100
	100		

main part of the indirect effects are generated in the aircraft and defense construction divisions of ESA contractors. In fact, many of the companies in the space industry, especially the largest ones, are usually active in both sectors. The technology areas most apt to generate indirect effects outside the space sector are those whose applications cover several industrial sectors and can more readily be transferred. By contrast, those whose applications are more specifically within the space sector generate fewer indirect effects. These findings confirm the existence of a synergy effect between sectors which have similarities on both the technological expertise and the organizational levels ("technological clusters" analysis). It also appears that some companies encounter organizational obstacles to converting from an innovation-oriented space activity characterized by increasingly complex systems and very small scale of production, to a commercial nonspace type of production based on standardized products to be produced in large series. These difficulties are mostly related to readapting a corporate culture ruled by the observance of quality standards that are often too strict for direct appplication to a nonspace activity (see "Managing Spinoffs").

Technology Transfers (or "Classic" Spinoffs)

It was possible to extract from the data base on indirect effects those which are more traditionally called spinoff, i.e., the technology transfer. They are clearly set apart from other indirect effects by the fact that they require technological content, a nonspace recipient sector, and a deliberate policy on the part of the firms making the transfers. The analysis, undertaken solely from the point of view of the transferring party (the ESA contractors), leads to the identification of 133 cases of technology transfer based at least in part on knowledge acquired by firms working on ESA programs. The results are shown in Table 4.

The transfers represent 17.2% of the total indirect effects, and the bulk of them involve product-linked technology (61.2%), with or without adaptation of the technology concerned, and project or quality control procedures (20.8%). So space transfers give rise above all to new products, while on the other hand few production processes developed for space purposes are transferred between firms.

The space technology fields that generate most transfers are those relating to onboard equipment (32%) and to production and test equipment (24.5%). The main recipient sectors are again aeronautics and the defense industry.

A very large proportion of space technology transfers are internal (85% of the total), i.e., in the direction of other activities of firms working in the space field. The majority of external transfers (the remaining 15% of the total) are towards sectors further removed from the space industry (transport, electronics), sometimes in the context of international cooperative projects.

Managing Spinoffs: Key Factors and Policy Issues

This section looks closely at the mechanisms of spinoffs and at the factors determining their success or failure. We will mainly focus on the factors

Table 4 Some results on technology transfers (1988 ESA study)

Number of transfers	133
Total value of transfers	2 179 MAU 86
to contractors	1 345 MAU 86
Transfer coefficient (transfers/estimated payments)	0.6
Technology transfers as a % of total of indirect effects	17.2 %
Internal transfers	84.8 %
External transfers	15.2 %
Product technologies	61.2 %
Process technologies	10.3 %
Procedures	20.8 %
Others	7.8 %

affecting "classic" spinoffs (transfers of technology), since the other effects are not specific to the spinoff phenomenon. Effects such as cooperation strengthening, opening up of new markets, or formation of critical mass can be analyzed on the basis of the theory of the strategic management of firms, theory of organizations, or theory of R&D management which go beyond the frame of this article.

If we leave aside important but rather obvious factors such as the need for the "existence of a market" or an "efficient management of technology transfer projects," it appears that the elements presented here play an important part in the existence and the success of spinoffs. Corroborating evidence of the significance of these factors was obtained by correlating empirical studies, especially BETA studies carried out for ESA, and the replies to a questionnaire sent to European space firms (this analysis is detailed in Ref. 27). But we will not classify these factors by order of importance, according to recent developments in the Contingency Theory of Organizational Innovation showing that the performance of an organization in terms of innovation, covering differents aspects such as administrative and technical issues,[28] product and process,[29] or radical vs incremental innovations,[30] depends more on the conjunction of several factors (relation of "congruence") than on each type separately (see, for example, Ref. 31 for the analysis of innovation as a multidimensional phenomenon).

Some factors have a general influence on the transfer of space technologies to other sectors, and mainly concern *structural* elements of the industrial network potentially involved in such a phenomenon. Some others specifically affect internal spinoff (within one single firm or company) or external spinoff (between firms), and are more linked with the *behavioral* dimension of the transfer inside and between the industrial structures at stake, emphasizing the role of the individual capability and the communication between them. Their importance varies according to the different types of spinoffs as they are defined under "Definition of Spinoff." In relation to these key factors, we present some policy actions taken at the micro or macro level in order to provide the conditions for successful spinoffs.

Structural Factors

Technological Complexity of Space Activities

The argument here is based on a "congruent" property according to which the higher the level of complexity of the technology, the more important the potentiality of transfers. A complex technology and all the R&D efforts associated with it should generate more technical ideas, and consequently more potential sources of innovation. And from this point of view, technologically and organizationally, space industry ranks among the most complex areas of activity, whatever the indicator used to express this complexity (number of elements and linking as in Ref. 32, R&D intensity representing the effort required to reduce the uncertainty in technological development as in Ref. 33 or 34). Thus, space technologies theoretically have the greatest potential for solving the less complex organizational and technical problems encountered in other sectors. Nevertheless, if this assertion may sometimes be accepted when thinking of technology, it cannot as easily be so if we think in terms of products and processes. Space technologies and products are very often seen as too sophisticated and overqualified for commercial applications, and even space staff suffer from the same criticism, being considered as unable to design commercial products using current technologies.

Technological Proximity of Receiving Sector

The adaptation work, and associated costs, required by transfers from the space industry will be less (and the volume of transfers that much greater) where transfer is towards sectors with technologies having features in common with space technologies. Two aspects should be distinguished: the generic (common to several industrial activities) or specific nature of the space technologies concerned and the technological proximity of the space and recipient sectors.

The two arguments, and mainly the second, lead to the application of technological bunching strategy (also called "technological cluster") generally defined as a systematic search for combinations between different sectors of technological activity. At least two phases have to be distinguished for the control of the combined feature of the different technologies.[35] One is the management of a minimal scope of know-how during the constitution of the space technology; the other corresponds to the exploitation of the technological similarities between the space sector and the recipient needs. Several studies show that the larger the span of technological specialties controlled by the firm, the higher the control level of a transfer involving different sources of technology. Furthermore, all along the technological transfer process, Teece[36] shows that, beyond the complementary feature of the different shapes of know-how in the firm, some more logistic and downstream skills like marketing and distribution problems are important. The situation of the space activity is unusual and covers the main fields of scientific knowledge on which industrial activities are founded. From this point of view, space industry is privileged in terms of strategic situation for transfers of technology.

For instance, it is interesting to note that space activities have grown more frequently among big companies first involved in aeronautics and defense systems. A lot of similarities exist between these activities and the space department. Manifold technical solutions included in the first generation of launchers in Europe came from the aeronautic and defense knowledge. The quantitative results of the BETA studies on performance of the European space industry gives a significant illustration of this phenomenon as shown under "Measurement of Spinoffs."

Nature of the Firm

Two interconnected aspects are mainly covered here, the size and the position of the firm inside the network of the R&D programs from which the technology is coming. We emphasize that these two structural dimensions influence the industrial learning process of the firm, and hence the type and amount of spinoffs that it is able to generate.

Former outlines based on the combined feature of the technologies leads to believe that big companies are in a better position than small ones because of their larger scope of know-how. However, the highest capacity to generate technological synergies is certainly reinforced by the financial capabilities for new ventures. Financial aspects could be determining for a transfer strategy when the latter implies high costs of technological adaptation for the user. For Porter,[37] for example, the size of the firm is important in the achievement of an R&D program where large scales of production are at stake.

The results of the BETA studies provide partial and somewhat contradictory elements on this point. The BETA team studied how much the size of firms affected their likelihood of generating indirect effects, and took particular notice of the results of small and medium-sized firms, divided into two types: "general" firms, with a staff of under 1000 employees on all types of work (25 firms in our sample), and "space" firms with a staff of under 100 engaged in space work (30 firms in our sample).

It is clear from the study that these two classes of firms generate proportionately more indirect effects than the overall sample of firms (coefficient of 3.5 and 4.1, respectively, and that these effects are generated largely outside the space sector (34 and 61.2%, respectively). The firms also produce more commercial effects, but tend not to form a critical mass of employees. It should be noted that the "space" firms include a number of large firms engaged in space work on a relatively modest scale. These appear, however, to generate the most indirect effects, particularly outside the space sector, no doubt because of the interaction of technological and organizational factors and because they have the money to finance space technology transfers.

Then it seems that to find a significant relation between size and technological efficiency, a more "contingency" vision is required pooling different influencing factors, such as the size of the company, the existence of a scope of activities, and the firm's internal communication system. Anyway, interpretations can be drawn in this direction from our empirical results. An interesting complementary point of view on this question is given by the second feature determining the nature of the firm, that is its position in the industrial network setup for space programs (here ESA programs).

For this purpose, the correlation between the level of responsibility of ESA contractors and the indirect effects they generated was studied, to see whether there was a link between the various functions perfomed by firms and the indirect effects on them (see Table 5). ESA contractors are divided into four categories: prime contractors, system developers, equipment developers, and service providers.

The prime contractors, and to a lesser extent the subsystem developers, tend to concentrate their efforts on the space market and have to maintain a highly skilled workforce. They also gained experience in managing complex international projects that can subsequently be put to good use in other programs. Prime contractors tend to diversify more (creation of new activities or new division), no doubt because their size and financial position allow them to do so. The firms generating the greatest indirects effects, especially outside the space sector, are the equipment developers. They are generally innovative, medium-sized firms or large firms with a small space department, using generic technologies to manufacture components and they are quite capable of moving on to mass production. They are in "direct contact" with the technologies, and most of the indirect effects they generate therefore have to do with technology or production processes. If we observe in more detail the nature of those latter firms, it is interesting to note that, again, most of the time they are large companies with a "small" space activity or small firms integrated in a large network, thus corroborating the importance of factors of size and variety of know-how for a strategy of transfer. Finally, few indirect effects are observed among service providers, because they are usually making use, in the context of ESA programs, of skills they have already developed elsewhere. The type of transfers realized by the service companies confirms the importance of the position in the network, whereas approximatively 80% of their transfers are linked to administrative innovations as methods and quality control procedures in relation with their ESA participation work (studies, consultancy, assistance, and maintenance).

Table 5 Analysis by contractors' level of responsibility
(1988 ESA study)

	INDIRECT EFFECTS (% of total)	RATIO indirect effects / contracts
PRIME CONTRACTORS	36.6	2.0
SYSTEM DEVELOPERS	36.1	2.3
EQUIPMENT DEVELOPERS	22.5	3.9
SERVICE PROVIDERS	4.8	1.8

Decision-Making Procedure and Financial Criteria

The last feature of the structural factors having some impact on the transfer of technological knowledge concerns the usual framework structuring the decision in the company and the place that a transfer can take in it. The traditional framework for decision making is based on the financial analysis, i.e., the comparison of flows of returns and flows of expenses through such criteria as return on investment or net present value. While the larger part of projects of investment is analyzed inside the company on this basis, transfer of technologies is often perceived as an astonishing project. Its rentability being not immediate and seldom easy to express in terms of financial profit, such a "nonorthodox" project can be viewed as not adapted to the usual framework of decision of the firm. Thus, a strategy of transfer could be a source of problems vis-à-vis the financial authority in the firm; this handicaps and finally restrains the development of such a project.

According to the managers interviewed, the more the technology is formalized, or specified, the easier it will be to determine its impact on the technological choices of the receiver, and then to give figures for potential markets. In the case of purchase of patent on a product, evaluation appears easier in the sense that classic parameters for criteria can be deduced (production cost, market size, expected profit, etc.). But if the technology is less formalized, containing some form of tacit knowledge, the real impact of its transfer on the reconception of the receiver's products, on the resources required for the production, and on markets it allows to reach, will be difficult to determine. More generally, it seems that for the largest proportion of cases of transfer a strong uncertainty bears on the financial gains. Several market studies are sometimes required to demonstrate the commercial interest of a transfer. Such steps need approval at the highest level of the hierarchy, and very often the R&D department or engineers-scientists initiating the project are not able to provide evidence for a nonroutinized activity.

The rationality of the manager justifying the relevance of these criteria corresponds to a substantial rationality, i.e., emphasizing the financial results of a choice. In fact, another representation of the rationality could provide a more appropriate framework for taking into account the positive aspects of a transfer within the decision-making procedure. This is what Simon calls procedural rationality in which elements of appreciation of a project go beyond strict financial dimensions to incorporate some qualitative features of the investment (such as higher flexiblity, organizational learning, new technological opportunities, etc.). We can observe that most of the time, the existence of a policy of technology transfers in the company depends on the consideration of qualitative aspects in the decision making, i.e., the introduction of procedural elements in the traditional financial framework. Briefly summarized, spinoffs will then depend on the firm's ability to reconcile, in determining its transfer investment policy, conventional cost-benefit criteria (expected profits, time taken to achieve a return on investment, etc.) and more qualitative criteria (new technological openings, acquisition of expertise, company image, etc.).

This last category of factors, linked with the usual framework of decision making in the firm, denotes how individual and behavioral dimensions are necessary to go beyond what is commonly allowed by the structure. Two different aspects in the problem of transfer exhibit strong interactions with human factors. One concerns the notion of technology itself and emphasizes its tacit dimensions having some impact on a process of transfer of knowledge. The other is connected with the role of communication, formal and informal, in the development of new ideas within the organizational framework.

Individual and Behavioral Factors

Problem of Transaction Costs

According to the theory of transaction costs developed by Coase[38] and Williamson, collaboration between different organizations induces some costs generally related to the meetings required for the negotiation and the fulfilment of contractual forms and leads to lengthening the delay of reaction required for elaborating a decision.[39]

One direct and important application of this development to the problem of technological transfer is due to the fact that the transmission of the information related to the technology and the body of knowledge "all around it" is often the more costly operation. As mentioned in a new development of this theory by Teece,[40,41] transfer of technology is not only an exchange of "commodity" or of a codified information, but includes also a large proportion of nonexplicit know-how and knowledge.

In this respect, a serious barrier seems to be the problem of the translation of the technological know-how into a language understandable by the technology user. Therefore considering the tacit or nonspecified part of the technology, a strategy of transfer between two organizations will require either a similar learning process for the user to build up the same information, or important efforts for the supplier to specify formally the expertise embodied in the technology, and to make it explicit and comprehensible for external organizations.

This process of making the technology explicit is certainly an important source of transaction costs and can be avoided in big companies by organizing internal transfers just by moving people or, as we have observed in the European industry, by diversification corresponding to the creation of a new department with employees of the organization, indeed by the creation of a new company by engineers. A "mixing" of these three different solutions is often observed when specific task-force teams are created involving both people from the supplying and receiving companies. The most advanced form of such a collaboration is probably a joint-venture strategy.

Organizational Dimension

Some of the arguments described above point at several strong difficulties encountered when implementing a strategy of technological transfer from one organization to another. It comes out that a company will sometimes first attempt to realize an internal transfer, in order to avoid some costs like transaction costs. This choice is corroborated by our evaluation in the space industry showing that 85% of transfers remain internal to the companies. However, if different barriers to transfer can be overcome this way, some barriers persist due to the organizational dimension. The causality link between technology development, to which spinoffs contribute, and organizational structure has been the subject of a huge literature. An evolutionary perspective on spinoffs and more generally on the diffusion of the technologies exhibits two kinds of barrier for transfers.

The first concerns the degree of decentralization of an organization required in order to provide new issues in the utilization of the technology. This argument emphasizes a relation where the organization has an impact on technological performances. Thus, in the phase corresponding to the *development of the technology* in the industrial organization, it appears that some properties in the organizational structure (existence of vertical links, degree of decentralization for decision making, etc.) condition not only the stimulation for new ideas by cross-fertilization between the different fields of activity of the firm, but also have an impact on the transaction costs mentioned above.

In particular, *mutual adjustments*, meaning informal communications between people in the organization, seem to play an important role in the dynamics of new technological developments.[42] On the other hand, a multiproduct company will generate *economies of scope*, which are savings due to "shareable inputs." In particular, intangible inputs, such as knowledge, are common to several activities and can be employed in different projects.[40,41,43] Organizations such as matrix structures, often seen as the attribute of innovative organizations, combine informal relations and economies of scope. Studying European firms in the space sector seems to confirm that such a phenomenon appears when these firms follow a matrix organization as opposed to independent departments. The MBB-

ERNO company considers that thus the total critical mass was reduced by one third. But it is interesting to note that according to the majority of the managers interviewed, the matrix shape is not necessary and other types of structure have the properties described above.

A second type of barrier exhibits an inversed relation between technology development and organizational structure: new technical features need organizational modifications where tighter couplings are required (dynamics of standardization). Thus, in the phase of the *application growth* of the technology, what is central is the ability of the space firm that created the technology to adapt to the industrial environment in the recipient sector (e.g., mass production, quality requirements, marketing strategies). In other words, for a successful transfer, the organization must be adapted to more commercial features in terms of quantity, price, and timing, and be able this way to move in the expertise from complex products to production programs. This often results in a shift from the aim of maximizing the technical performance characteristics of a product to that of holding costs down.

As a consequence, a firm based on a small series of complex products (e.g., space and avionics) is constrained to introduce standardization rules to diffuse technical progress in a large production series in order to reduce production cost. However, some phenomena of irreversibility prevent such a diffusion from happening without major changes in the firm. The justification of the irreversibility comes both from technical and commercial aspects. On the one hand, the knowledge necessary to design industrial prototypes does not correspond to a continuous search for optimization in order to reduce time and consumption of input in the process of production. On the other hand, the high cost of qualified employees, as well as the existing commercialization structures, is inconsistent with the manufacturing rule of a large series for which a commercial valorization of innovation is at stake instead of permanent innovation per se. For according to Mintzberg,[41] an organization based on complex mechanisms between people, using mutual adjustment more than standardization rules (autocracy form), will become a structure based on standardization of the process of production (divizionalized form) in order to commercially exploit a technological success. But the dynamics between different types of structure conditioning technological transfer encounters several types of inertia.

Management Policies for Spinoffs

Obviously, there is no "recipe" to make spinoffs, and the following comments are more conceived to provide a guideline enriched by some quantitative results coming from an empirical test on the European space industry. On this basis, several management decisions can be taken at the microlevel in the company in order to stimulate the spinoff phenomenon. From each of the key factors described previously, some actions can be derived, although in a set of possibilities more or less bounded by structural elements. It is, for instance, difficult to change rapidly the nature of the firm, its size, or its scope of activities. But it seems, according to experiences of transfer in the realm of industry, that the first step is settled by the willingness to improve linkages in order to initiate such a phenomenon.

Different ways are available for that without disrupting the structural features on which the firm is based.

Some Examples of Firms' Policies

Firms' policies can take the shape of an individual role (opinion leader, product champion, or gatekeeper), or at a higher level, a task force, a project team, or a matrix structure. Generally, the objective is to place the company in an environment more open to new opportunities and stimulate new uses for the technology created. Some examples of microeconomic decisions to begin an active policy on transfers have been observed in the European space industry. Several interesting examples of such a strategy are provided by the German industry. In order to stimulate the use of the technology, companies like MBB-ERNO or Dornier have created a "transfer unit," or simply a special team within the staff responsible for systematically identifying the potential applications of space technologies. MBB-ERNO started a technology application division in 1989, where approximatively 50% of the development projects are coming from space. Dornier, in order to promote cross-fertilization between different technologies developed inside the company, gathers every month members of the staff in a "synergy board meeting" to identify internal technological opportunities and to see what is needed on the demand side of the technology in the different departments of the company. (Dornier and MBB-ERNO are now part of the Daimler-Benz group.)

With a higher level of implication of the organization, the French company Aerospatiale has settled a systematic "swarming" strategy in creating a "New Products" department with the objective of maximizing the valorization of the technologies partly born in the space sector. The specificity of the microstrategy regarding German companies is the opening of the organization as a whole on the environment, considering also, and mainly, external transfers. After an identification of technological opportunities, an assessment of the market for potential applications is needed.

Some cases of industrial organization entirely devoted to the realization of transfers can be observed. ELAB in Norway provides a good example of such a company. Belonging to the SINTEF Foundation (Engineering Research Foundation), ELAB is a laboratory doing research in electronics and computer science for the rest of the industry. The different bodies of knowledge cover realms of acoustics, telecommunications, telematics, and physical electronics, all organized in a matrix structure crossing these home based scientific fields with several research projects like satellite, communication, and environmental protection systems. In accordance with the matrix organization,[44,45] the concept of matrix swing characterizing a moving role of authority is illustrated in this company. Indeed, at the beginning and at the end of a project, the authority is mainly carried out by the general manager due to the functional priorities, and it moves to the project manager during the realization phase.

Finally, an interesting illustration of how internal transfer can be improved by organizational structure mutation is provided by the Swedish Ericsson company. In a first phase, a matrix organization was adopted inside the Ericsson Radar Electronics department in order to start the development of two new activities (antennas and hybrid electronics), particularly due to the growing space activity in the company. But, the separate evolution of these two home-based technologies has led to an organizational mutation. Indeed, while for the antenna activity the technological challenge has remained unchanged over time (prototype or small series), the hybrid function has been more and more standardized in all of the different project applications in the company. In order to achieve the process of standardization coming from an internal transfer of the hybrid technology, but also because of an intensive strategy of diffusion outside the company, the hybrid activity became a new department of the company with its own organization and hierarchical structure. This autonomization process by leaving the matrix structure was mainly guided by scale effect in order to provide a successful transfer of the technology.

Role of Public Spinoff Policies

A lot of interesting microexperiences could be noted in the European space industry, but they are most of the time isolated types of action, not sufficient to provide an optimal rate of diffusion for the space technology. Help is often needed on both the supply and the demand sides. On the supply side, in addition to the creation of a special unit, a study by external experts of the potential applications of the firm's technologies could have a substantial impact. The same is true for the demand side: persons familiar with space technologies could examine with potential users whether their technical problems could be solved by technologies from the space sector. One justification for the intervention of a neutral and external entity comes from the "asymmetric" nature of the technical information to be transferred. According to the paradox of information pointed out by Arrow,[4] a situation of transfer can be characterized as follows: In some cases the recipient has to purchase information for which the real will not become apparent until acquired, whereas the proposer must possess clear evidence of ownership of the technology (patents, licences, commercial agreements) to agree to divulge it.

A solution to this paradox, coming from the diverging interests of, on the one hand, the social optimum of a systematic diffusion, and, on the other hand, the private optimum of the firm, has to be found by adding a third organization making the balance. A policy of technological transfers will have the tricky task to make these two objectives converge, turning the private technology into a quasi-public good, without colliding with the private objective of the providing company. Beyond the problem of protection for the supplier, the role of this third organization is also to translate the objective characteristics of the technology in terms of the perception of the receiver having sometimes a totally different technical environment.

Following this argument, it seems clear that there is space for an involvement of the public sector in order to improve linkages between potential providers and receivers, to relieve costs of transaction by financial support, and even to give some guarantees during the realization for the protection of the technological advance (property rights, policy of patent).

But, as firms wish above all to keep total control over their "home" technologies, the span of action for a public policy of transfers can meet strict limits due to some resistances from private companies in sharing responsibility for the technology.

The NASA TU program is a well-known example of such a policy, although it partly aims at supporting spinoffs from NASA technologies. In Europe, different initiatives have been taken. Apart from the current ESA pilot project which should help the Agency to move from a rather "passive" attitude vis-à-vis spinoffs to the design and adoption of a clearly established strategy, two actions are worth mentioning here because they represent two different approaches. An example of "classic" approach (using technology brokers) is provided by Novespace, a subsidiary of the French Centre National d'Etudes Spatiales (CNES). This organization releases a newspaper reviewing space technologies available at CNES or in French space firms; they are thus acting only as an intermediate organization putting in contact supply and demand sides, without being really involved in the spinoff project by sharing risk or acting as technical supporting organization.

A typical example of sharing responsibilities is provided by the Swedish Board for Space Activity. This organization, created in 1986, was based on the idea of setting up a fund to stimulate new technological developments or commercial applications. This organization was partly funded by the government (for 40%) and partly by the three biggest companies in space (Saab–Ericsson–Volvo) with an equal participation (20% each). In the early 1990s the budget amounted to some 100 million Skr (Swedish Kroner) and was used for providing subsidies to each of the three companies in order to a promote their positioning on future markets. Each proposal application from the companies is submitted to an executive committee composed both of government and firm members. Different technological projects like spinrock for Saab, microwave equipment communication for Ericsson, and propulsion system for Volvo have been developed, funded by this Swedish Board for Space Activity.

Conclusion

Using different methodologies, BETA studies reviewed here and other estimates performed in the United States exhibit optimistic conclusions regarding the existence and the importance of spinoff from space activities. Some authors comparing the benefits from pure R&D activity with transfer projects even emphasize the interest for the latter strategy. But we have to be careful about studies leading to the conclusion that a program of transfer is more profitable than current R&D activities: one is the consequence of the other and, indeed, before exploring some directions for transfer, the original program of R&D has to be performed. More than an alternative program, the transfer strategy leads to increasing the valorization of the knowledge accumulated by the firms in the R&D departments. From this standpoint, the spinoff phenomenon is a very interesting field of observation and research for economics and management specialists, because public and private interests are mixed, and sometimes are conflicting ("socialization" of the technologies for the whole economy vs protection of information on technology to ensure leading corporate position). One essential challenge of the private as well as public management of spinoffs is to make these interests compatible.

Many research projects must still be done in order to develop an analytical framework able to take into account the variety of spinoff phenomena and the complexity of the channels by which they have an impact on the economic activity, and on this basis to design methodologies that can provide accurate measurements. In this respect, recent advances in evolutionary economics could help to shed a new light on spinoffs, by considering them as part of the factors shaping the technico-economic development instead of treating them as isolated phenomena.

Such a change in the research perspective could contribute to overcoming two types of criticism to which studies on spinoffs are very often exposed. The first is the tendency to justify space activities by its spinoffs, whereas the growing importance of space activities as an autonomous economic sector leads more and more to find such a justification in what we defined as its direct effects. On the other hand, studies on spinoffs often give the impression that the space sector is seen as the only innovator of new technologies which are later used in the rest of the economy. Experience shows in fact that the space industry is rather the place in which technologies developed elsewhere are assembled and improved. This is the reason why we may perhaps also consider the spinoff of the space sector in terms of complementarity and interactivity with other sectors rather than in terms of impact that would justify space programs.

Acknowledgment

The authors gratefully acknowledge the help of Monique Flasaquier during the translation of this text in English.

References

[1]Griliches, Z., "Issues in Assessing the Contribution of R&D to Productivity Growth," *Bell Journal of Economics*, Vol. 10, No. 1, 1979, pp. 92–116.

[2]Griliches, Z., "The Search for R&D Spillovers," Working Paper, Harvard University, Cambridge, MA, 1990.

[3]Nelson, R., "The Simple Economics of Basic Scientific Research," *Journal of Political Economy*, Vol. 67, 1959, pp. 297–306.

[4]Arrow, K., "Economic Welfare and the Allocation of Resources for Invention," *The Rate and Direction of Inventive Activity: Economic and Social Factors*, edited by R. Nelson, 1962. Reprinted in Arrow, K., (1985), *Collected Essays*, Vol. 5, Production and Capital, Blackwell, Oxford, 1985.

[5]Midwest Research Institute (MRI), "Economic Impact of Stimulated Technology Activity," Rept. for NASA, 1971.

[6]Mansfield, E., Rapoport, J., Romeo, A., Wagner, S., and Beardsley, G., "Social and Private Return from Industrial Innovations," *Quarterly Journal of Economics*, Vol. 77, 1977, pp. 221–240.

[7]Solow, R., "Technical Change and the Aggregate Production Function," *Review of Economics and Statistics*, Vol. 57, 1957, pp. 312–320.

[8]Denison, E., *Trends in American Economic Growth, 1929-1982*, Brookings Institution, Washington, DC, 1985.

[9]Mairesse, J., and Sassenou, M., "R&D Productivity: A Survey of Econometric Studies at the Firm Level," *STI Review*, No. 8, 1991, pp. 9-43, OECD, Paris.

[10]Mohnen, P., "New Technologies and Interindustrial Spillovers," Presented at the OECD International Seminar on Science, Technology and Economic Growth, Paris, France, June 5-8, 1989.

[11]Dosi, G., Freeman, C., Nelson, R., Silverberg, G., and Soete, L., eds., *Technical Change and Economic Theory*, Pinter Publishers, New York, 1988.

[12]Irvine, J., and Martin, B.R., "The Economic Effects of 'Big Science': the Case of Radio-Astronomy," *Economic Effects of Space & Other Advanced Technologies*," edited by T.D. Guyenne and G. Lévy, European Space Agency, 1980.

[13]Callon, M., Laredo, P., and Rabeharisoa, V., "Des Instruments pour la Gestion et l'Évaluation des Programmes Technologiques: le Cas de l'AFME," *L'Évaluation Économique de la Recherche et du Changement Technique*, edited by J. De Bandt and D. Foray, Presses du CNRS, Paris, 1991.

[14]David, P., Mowery, D.C., and Steinmueller, W.E., "The Economic Analysis of Payoff from Basic Research: an Examination of the Case of Particle Physics Research," CEPR Publ. 122, Stanford University, Stanford, CA, 1988.

[15]Midwest Research Institute (MRI), "Economic Impact and Technological Progress of NASA Research and Development Expenditures," Rept. for the National Academy of Public Administration, Washington, DC, 1988.

[16]Evans, M.K., "Economic Impact of NASA R&D Spending," Chase Econometric Associates, Inc., Bala Cynwyd, PA, 1976.

[17]Cross, D.M., "The Economic Impact of NASA R&D Spending, An Update," Rept. for NASA-NASW-3345, Chase Econometrics Assn., Philadelphia, 1980.

[18]ECON, Inc., "Assessment of the Economic Impacts of the Space Station Program," NASA Space Station Task Force, 1983.

[19]Mathematica, Inc., "Quantifying the Benefits to the National Economy from Secondary Application of NASA Technology," NASA, Washington, DC, 1975.

[20]Mathtech, Inc., "A Cost Benefit Analysis of Selected Technology Utilization Office Programs," NASA, Washington, DC, 1977.

[21]Johnson, F.D., and Kokus, M., "NASA TU Program—A Summary of Cost-Benefit Studies," Denver Research Institute, Denver, CO, 1977.

[22]Chapman, R.L., et al., "An Exploration of Benefits from NASA 'Spin-off'," Chapman Research Group Inc., Littleton, CO, 1989.

[23]Hertzfeld, H.R., "Measuring the Economic Impact of Federal R&D Investment in Civilian Space Activities," Workshop on The Federal Role in R&D, National Academy Press, Washington, DC, 1985.

[24]BETA, "Economic Benefits of ESA Contracts," Final Rept. for ESA, June 1980.

[25]BETA, "Study of the Economic Effects of European Space Expenditure," *Results* (Vol. 1) and *Report on Investigation Theory and Methodology* (Vol. 2), Repts. for the European Space Agency, ESA contract 7062/87/F/RD/(SC), 1988.

[26]BETA/HEC Montréal, "The Indirect Economic Effects of ESA Contracts on the Canadian Economy," Final rept. for the Canadian Space Agency, Contract 67SPS-9-0001/01-SS, 1989.

[27]BETA, "Analyse des Mécanismes de Transfert de Technologies Spatiales: le Rôle de l'Agence Spatiale Européenne," Final rept. for l'Agence Spatiale Européenne, Paris, 1989.

[28]Mintzberg, H., *The Structure of Organizations*, Prentice-Hall, Englewood Cliffs, NJ, 1979.

[29]Utterback, J.M., and Abernathy, W., "A Dynamic Model of Product and Process Innovation," Omega, Vol. 3, No. 3, 1975.

[30]Schumpeter, J.A., "The Theory of Economic Development," *Trans Redvers Opie*, Harvard University Press, Cambridge, MA, 1934.

[31]Damanpour, F., and Evan, W.M., "Organizational Innovation and Performance: The Problem of Organizational Lag," *Administrative Science Quarterly*, Vol. 29, 1984, pp. 392–409.

[32]Ayres, R.U., "Complexity, Reliability and Design: Manufacturing Implication," Working Paper, IIASA, Laxenburg, Austria, 1987.

[33]Hughes, K., "The Interpretation and Measurement of R&D Intensity—a Note," Research Policy, No. 17, 1988.

[34]Lambert, G., "Complexité Microéconomique et Diffusion Technologique à Partir d'un Grand Programme de R&D," *Evaluation de la Recherche et du Changement Technique*, edited by J. De Bandt and D. Foray, Presses du CNRS, Paris, 1991.

[35]Zimmermann, J.B., "Les Stratégies d'Accords Inter-Industriels," *Les Stratégies d'Accord des Groupes de la CEE, Integration ou Éclatement de l'Espace Industriel Européen*, LAREA, "Europe industrielle et technologie" program of the Commissariat Général au Plan, Paris, pp. 4–35, 1986.

[36]Teece, D.J., "Capturing Value from Technological Innovation: Integration, Strategic Partnering, and Licensing Decision," Conference on Innovation Diffusion, Venice, March 1986.

[37]Porter, M.E., *Competitive Strategy*, Free Press, New York, 1980.

[38]Coase, R., "The Problem of Social Cost," *Journal of Law & Economics*, Oct. 1960.

[39]De Jong, H.W., *The Structure of European Industry*, 2nd ed., Kluwer Academic Publishers, 1988.

[40]Teece, D.J., "Economies of Scope and the Scope of the Enterprise," *Journal of Economic Behavior and Organization*, Vol. 1, 1980, pp. 223–247.

[41]Teece, D.J., "Towards an Economic Theory of the Multiproduct Firm," *Journal of Economic Behavior and Organization*, Vol. 3, 1982, pp. 39–63.

[42]Mintzberg, H., *Structure et Dynamique des Organisations*, Les Editions d'Organisation, Paris, 1982.

[43]Levy, D.T., and Haber, L.J., "An Advantage of Multiproduct for the Transferability of Firm-Specific Capital," *Journal of Economic Behavior and Organization*, Vol. 7, 1986, pp. 291–302.

[44]Galbraith, J., *Organization Design*, Addison-Wesley, Reading, MA, 1977.

[45]Davies, M., and Lawrence, P.R., "Problems of Matrix Organizations," *Harvard Business Review*, May–June, 1978, pp. 131–142.

Economics and Regulation of Space Activities

Carissa Bryce Christensen*
Princeton Synergetics, Inc., Princeton, New Jersey 08540

Overview

FEDERAL agencies are required to assess the economic impacts of regulations they impose. Such an assessment is designed to serve as an aid to decision making—the agency's responsibility is to achieve its objectives in the most cost-effective manner possible and to ensure that its regulations are in the public interest. When economic benefits exceed economic costs, the public interest is served.

This chapter addresses the practicalities of performing economic impact assessments of regulations covering the new field of commercial launch operations. Many of the issues raised are likely to pertain to other new commercial space activities that may emerge, such as materials processing and Space Station-related commercial operations. Regulation of existing space commerce—specifically communications and (to a lesser degree) Earth sensing satellites—is not covered, because these industries are well established and do not present the same analytic difficulties as new space industries.

This chapter is intended more or less as a brief "how-to" guide. It identifies the requirements for economic impact analysis of federal regulations, describes some problems that have emerged in performing such analyses for several regulations, and presents the approaches that were used to solve these problems. The examples that follow are based on the activities of the U.S. Department of Transportation (DOT), which regulates commercial space transportation activities through the Office of Commercial Space Transportation (OCST). Among its regulatory activities over the past several years, OCST has investigated two regulations, one dealing with commercial launch insurance requirements and one levying

---

Copyright © 1992 by C. B. Christensen. Published by the American Institute of Aeronautics and Astronautics, Inc. with permission.
*Vice President.

user fees. (It should be noted that the interpretation, analysis, recommendations, and conclusions presented in this chapter are the responsibility of the author only, and that none of the contents of this chapter should be regarded as expressing the official views or plans of the U.S. Department of Transportation.)

This chapter first explains the requirements for economic analysis of regulations promulgated by federal agencies, and then describes the two rules used as examples. The specific analytic requirements and in particular the difficulties associated with the required economic analysis are discussed, and the resolution of these difficulties is explained.

Requirements for Economic Analysis of Federal Regulations

A federal agency issuing a regulation must comply with several legislative and regulatory directives requiring analysis of the economic effects of the regulation.

Executive Order No. 12291

Executive Order No. 12291 prohibits agencies from issuing regulations unless the potential benefits outweigh the potential costs to society. The Order requires that a Regulatory Impact Analysis (RIA) be prepared for every major rule that is to be enacted, defining a major rule as a regulation likely to result in an annual effect on the economy of $100 million or more, or expected to have significant effects on employment, inflation, or industry viability. [The Regulatory Impact Analysis is prescribed by the Office of Management and Budget (OMB) in the "Regulatory Program of the United States Government."]

The Executive Order does not require a full Regulatory Impact Analysis for a non-major rule, but does require that agencies ensure that the regulation is consistent with the principles of the Executive Order. Included among these principles are that potential benefits must outweigh potential costs to society, net benefits to society must be maximized, and the action must represent the least cost alternative. Typically the analysis for a major rule provides much more detailed analysis of the quantitative value of its benefits than that for a nonmajor rule, where difficult-to-quantify benefits may be simply described in qualitative terms.

Department of Transportation Order "Policies and Procedures for Simplification, Analysis, and Review of Regulations", DOT 2100.5, May 22, 1980

DOT 2100.5 requires that a Regulatory Analysis be prepared for any proposed regulation that will result in an annual effect on the economy of $100 million or more; that will result in a major effect on the general economy in terms of costs, consumer prices, or production; that will result in a major increase in costs or prices for individual industries, levels of government, or geographic regions; that will have a substantial impact on the United States's balance of trade; or that the Secretary determines deserves such analysis.

The DOT directive also requires that a determination be made as to whether a rule is "significant." This determination is based on factors such as whether the regulation is of substantial public interest or controversy or will involve other government entities (such as state governments).

Most other regulatory agencies have similar implementing guidelines for regulatory analyses. These may simply ratify the requirements of the Executive Order, may levy new requirements (such as the need to determine whether a rule is significant, as above), or may provide information on the appropriate interpretation of certain requirements in the Executive Order.

Regulatory Flexibility Analysis

The Regulatory Flexibility Act requires federal agencies to determine whether their actions will have a significant impact on a substantial number of small entities. These terms are not defined specifically in the Act.

Background

There is currently only one agency with specific authority to regulate commercial space transportation activities—the U.S. Department of Transportation. Other agencies, such as NASA and the U.S. Air Force, may impose certain requirements on commercial space firms, but these requirements occur in conjunction with the use of the resources of those agencies (such as launch ranges, the Shuttle, and so on) rather than as the result of a regulatory mandate. In addition, regulatory agencies with general authority over industry (such as the Occupational Safety and Health Administration) of course encompass commercial space firms, but these agencies are not specifically aimed at space. DOT's legislative mandate for regulating commercial space transportation is the Commercial Space Launch Act.

Commercial Space Launch Act

In October of 1984, Congress passed the Commercial Space Launch Act, with three purposes:

1) To promote economic growth and entrepreneurial activity through utilization of the space environment for peaceful purposes

2) To encourage the United States private sector to provide launch vehicles and associated launch services by simplifying and expediting the issuance and transfer of commercial launch licenses and by facilitating and encouraging the utilization of government-developed space technology

3) To designate an executive department to oversee and coordinate the conduct of commercial launch operation, to issue and transfer commercial launch licenses authorizing such activities, and to protect the public health and safety, safety of property, and national security interests and foreign policy interests of the United States.

The Act built on earlier federal actions and increasing Presidential support of a commercial launch industry, which resulted in the 1984 Executive Order (12465) designating DOT as the federal government's lead agency for commercial launch activity. Congress intended the 1984 Act to consolidate and build on earlier actions, providing a stable regulatory framework not subject to de-emphasis or modifications in policy as administrations changed. The Act also enabled Congress to provide direction and specify its own policy preferences.

In 1986, the Space Shuttle Challenger accident resulted in a shift in federal policy from sole reliance on the Space Transportation System for government launches to a policy of relying on a mixed domestic fleet. Further, the Shuttle (with certain limited exceptions) was prohibited from launching routine commercial and foreign satellites.

This shift in policy increased the importance to the nation of a commercial launch industry, and resulted in Administration and Congressional actions to foster the emergence of a strong domestic commercial launch industry.

The Commercial Space Launch Act was amended in 1988. The amendments both clarified certain provisions of the 1984 Act, and responded to dramatic changes in federal policy toward commercial space.

Description of the Office of Commercial Space Transportation

The examples of economic impact assessments used in this chapter are drawn from past and current analyses conducted for the Department of Transportation's OCST. OCST has three divisions: Industry Policy and Planning, Program Affairs, and Licensing Programs. The regulations promulgated by or under consideration by OCST, for which these analyses were performed, were the primary responsibility of the Licensing Programs Division of OCST. The responsibilities of this Office are briefly described as follows. (It should be noted that in each of these cases there was also significant input and guidance provided by the Industry Policy and Planning Division.)

The Licensing Programs Division has seven primary responsibilities:

1) *Regulation*—OCST develops standards and procedures related to safety.

2) *Licensing*—OCST performs mission review analysis (which covers aspects of proposed activities, such as the launch range to be used, the purpose of the mission, the nature of the launch or other activity, the flight plan, and the type of vehicle that may affect national interests and treaty obligations) and safety review analysis (which covers safety aspects of proposed activities, such as site evaluation, safety processes and procedures, range safety expertise, tracking and instrumentation, the flight termination system, and the launch vehicle) in association with licensing activities. OCST is also developing procedures for licensing commercial launch sites. For nongovernment payloads and those not licensed by the Federal Communications Commission (FCC) or National Oceanic and Atmospheric Administration (NOAA), OCST determines whether the launch of such payloads would jeopardize public health and safety, safety of property, or U.S. national security. OCST performs maximum probable loss determinations for all licensed vehicles, to aid in determining financial responsibility (insurance) requirements.

3) *Safety Enforcement*—OCST ensures that license conditions and financial responsibility requirements have been met, and evaluates safety

waivers and inspections of safety processes throughout the term of a license, investigates commercial space transportation-related accidents, and enforces its requirements.

4) *Safety Research*—OCST conducts substantial research to aid in determining standards, setting insurance requirements, developing and implementing licensing procedures, and providing technical assistance.

5) *Environmental Studies*—Environmental Impact Studies are performed for launch sites and vehicles.

6) *Intergovernmental Coordination*—OCST coordinates various activities with the USAF Space Command and with federal launch range organizations, and also acts to obtain necessary access to government facilities for the commercial space transportation industry.

7) *Technical Assistance*—OCST provides preapplication consulting to potential licensees, to aid in their planning and preparation. Preapplication consulting may apply to launch licenses, to licenses for nonfederal space launch facilities, and to safety aspects of certain other types of facilities.

Examples Used

Financial Responsibility Requirements

OCST is investigating a rulemaking that would specify the financial responsibility requirements imposed on firms offering commercial space transportation services seeking licenses from DOT. The purpose of this regulation would be to codify and standardize the financial responsibility requirements that such companies must meet as a condition of their licenses. These financial responsibility requirements are determined based on a number of factors, including the maximum probable loss (MPL) associated with a license activity. MPL encompasses damage to third parties (third-party liability, TPL) and damage to government property. A draft report on the economic impacts of such a rule was prepared; its structure and contents are described in Fig. 1.

User Fees

OCST investigated levying user fees in association with its activities relating to commercial space transportation. Anticipated revenue from user fees was incorporated by Congress into the FY91 budget, in an amount up to $300,000. An economic impact assessment was conducted examining alternative methods for assessing user fees. The purpose of the study was to provide DOT with information to aid in its decision making about the type of user charge regime it might implement.

The economic impact assessment examined current DOT programs pertaining to commercial space transportation, and evaluated user fees in this context. User fee policy and practice in the federal government was investigated, and the potential impacts of user fees on the commercial space transportation industry were assessed. (This study, in a slightly modified form, served as the regulatory evaluation required in association with the DOT Notice of Proposed Rulemaking titled "Commercial Space Trans-

---

CONTENTS OF REPORT—FINANCIAL RESPONSIBILITY REQUIREMENTS

Executive Summary (Section 1): Provides a brief overview of the legislative history and statutory authority for the regulation, states the method of analysis, and summarizes the economic impacts identified.

Introduction (Section 2): Provides background information, including an extensive discussion of the provisions of the 1984 Commercial Space Launch Act, and the manner in which it was implemented by DOT, and the provisions of the 1988 Amendments to the Act. Discusses the need for the proposed regulation, and lists the requirements imposed by law and policy on economic analysis associated with a rulemaking action. Also details the methodology and assumptions of the economic analysis contained in this report.

Economic Analysis (Section 3): Identifies parties affected, types of effects, and provides qualitative and, where possible, quantitative descriptions of the economic impacts of the financial responsibility provisions of the 1988 Amendments to the Commercial Space Launch Act of 1984, using the Act and DOT implementation of the Act as the baseline case.

Economic Analysis (Section 4): Describes the differences between the proposed rulemaking and no rulemaking by DOT in the area of financial responsibility requirements for licensees.

Regulatory Flexibility Analysis (Section 5): Assesses the impacts of the 1988 Amendments to the Commercial Space Launch Act and the regulation proposed by DOT small entities.

Conclusions and Recommendations (Section 6): Provides conclusions and recommendations based on the economic analysis performed.

Fig. 1 Financial responsibility requirements report.

portation; User Fees," published in the February 28, 1991 *Federal Register*.) The structure and contents of the report on the economic impacts of this rule are shown in Fig. 2.

Main Steps in Performing an Economic Impact Assessment

The main steps in performing an economic impact assessment are: specifying the baseline case, obtaining data, defining alternatives, structuring analysis, and quantifying effects. Each of these steps, with examples, is discussed below.

Specifying the Baseline Case

Any analysis of the economic impacts of an action requires that the situation that would exist in the absence of that action be defined. This is generally referred to as the baseline case. Defining the baseline may simply require describing the status quo, but is not always so simple or straight-

---

CONTENTS OF REPORT—USER FEES

Introduction.

DOT Programs: Describes DOT's role in regulating the commercial space transportation industry and the programs through which it fulfills this function.

User Fees: Defines user fees and describes existing federal policy and practice with regard to user fees.

Applying User Fees to OCST Activities: Defines and analyzes the user fee alternatives under consideration by DOT.

Appendix A: Background: Provides additional information on user fees in theory and practice.

Appendix B: Detailed Economic Analysis: Provides a full description of the data, method of analysis, and results of the economic impact analysis described and summarized in the body of the report.

Appendix C: Federal User Fee Programs: Discusses user fee programs currently implemented some by federal agencies.

Fig. 2 User fees report.

forward. In one of the examples used here, user fees, the baseline case is simply the status quo—licensing launches free of charge. In the other example defining the baseline was more difficult, and two baselines were ultimately used.

The financial responsibility regulation discussed here is almost of necessity primarily procedural. The financial responsibility requirements the OCST levies on licensees are specified in the 1988 Amendments to the Commercial Space Launch Act of 1984. The contents of a codifying regulation are to a large degree dictated by the contents of the legislation.

Consequently, an economic analysis of the proposed regulatory action which measured the impacts only of a codifying regulation, using the existing legislative requirements as a baseline, would provide little insight into the true economic effects of the financial responsibility requirements levied on commercial firms. The analysis, therefore, considered two types of economic impact—essentially using two different baselines. First, the analysis identified and, to the extent possible, quantified the economic costs and benefits of the 1988 Amendments to the Commercial Space Launch Act. The Amendments impose specific financial responsibility requirements, including important provisions for indemnification of licensees above a certain threshold, the specification of the maximum probable loss associated with a licensed activity as the level of required financial responsibility or insurance, the requirement that DOT not set financial responsibility requirements at a level greater than what can be procured on the world market at a reasonable cost, and requirements for interparty waivers of claims.

Second, the analysis considered the economic impacts of codifying financial responsibility requirements through a rulemaking, relative to a baseline of no regulation. The costs and benefits of such a rulemaking were quantified to the extent possible.

This dual approach—an analysis of the effects of the imposition of the statutory requirements and an analysis of the impacts of a regulatory action building on those statutory requirements—was also followed in the regulatory flexibility analysis.

Obtaining Data

Obtaining data is often a challenge in performing an assessment of economic impacts, for any field. It can be particularly difficult in assessments of impacts on the commercial space transportation industry for several reasons:

1) There is a lack of historical data on which to base projections.

2) There have only been about 1000 U.S. orbital launches in the history of the space program, and because of factors such as vehicle modifications and improvements and differences in military and civil launches, relatively few launches of any particular vehicle have occurred.

3) There are a small number of firms in the industry.

4) It is difficult to determine economic or financial impacts for the industry in such a way that preserves proprietary information of a particular firm. While it is appropriate for a government study to report that a number of firms in an industry are in financial straits, it is not appropriate, generally, to make such an assertion about a single firm.

5) Government plays multiple roles in commercial space transportation. Federal policy has increasingly emphasized commercial development of space, and the role of government has expanded to include acting as a venture capitalist, providing physical resources (such as launch infrastructure), serving as a customer for space transportation services, and acting as a regulator. As a result, it is sometimes difficult to determine clearly the aggregate effects of a proposed regulation on the government, because many different facets of government are involved or affected.

There is also the commonality of industry practice and government requirements. The years of government operation of commercial space activities through contractors and the continuing reliance of commercial operators on government-run infrastructure make it difficult to separate industry practice from government requirements imposed in the past—there is a limited historical information base and a relatively small number of firms. This leads to a lack of statistically meaningful data from which to extrapolate and to occasional problems with the use of company proprietary data (such data normally would be aggregated by industry segment, but with a small number of firms such aggregation may not adequately protect proprietary data).

In general, flawed data are best dealt with by providing sufficient information about the effects of the flaws and by enabling a reader to replicate the analysis with better data if they are available. Specifically, this implies the need to explain in full both the limitations of the data and every step of the analysis. It is particularly important to specify any methodological assumptions that are made, and it may even be advisable to conduct sensitivity analyses to show that the impact of data limitations on the results of the analysis is not extensive.

For example, the data available on the multiple attributes of launch companies were in some cases poor. The economic impact analysis showed all of the data used and explained how the calculations were performed. For the purposes of the user fee analysis, the commercial space transportation industry was categorized on the basis of launch vehicle capability. Specifically, vehicles with a delivery capability to low earth orbit (LEO) of more than 6500 lb were categorized as *large orbital vehicles*, vehicles with a LEO delivery capability of less than 6500 lb were categorized as *small orbital vehicles*, and vehicles with a suborbital delivery capability were categorized as *suborbital vehicles*. (A nominal LEO of 150 n.mi., at an orbital inclination of 28.5 deg, was used. In some cases, fully consistent data were not available, and substitute data were used.) The vehicles in each category are shown in Fig. 3.

For purposes of the analysis, launch rates and OCST licensure rates were assumed to occur at a specified rate. The estimated launch rates were included to clarify the impacts of the proposed user fee programs, but were described (specifically due to the difficulty of accurately projecting commercial space launches) as mainly an illustrative case rather than a firm forecast of expected levels of activity.

OCST analyzed four alternative user fee programs, each representing a two-tier user fee. The first tier consisted of a flat fee levied on all license applications. These flat fees were the same in each alternative: $2500 for either the issuance of a license or license renewal. The second tier of the user fee program was a variable per-launch fee. This fee was calculated

---

Large Orbitals

Martin Marietta Commercial Titan, EPAC S-4, S-3, S-2, General Dynamics Atlas I, Atlas II, Atlas IIA, Atlas IIAS, McDonnell Douglas Delta II (6925), Delta II (7925)

Small Orbitals

SSI Conestoga 2, 4, W, AMROC ILV-1, EPAC S-1, AMROC ILV-S, ILV-1, SMLV, OSC Taurus, Pegasus, LTV Scout (WSMC)

Suborbitals

OSC Prospector, OSC Aries, SSI Starfire, AMROC SMLV, Conatec C-317, C-567, C-568, C-569

Fig. 3 Launch vehicles by category.

on a different basis under each option. The bases for calculating the per-launch fee under each option were referred to in this analysis as the *proxy attributes*. The values associated with these proxy attributes were the main data used in the user fee analysis, and were provided in full, with detailed source information. The information used is shown in Fig. 4.

The calculation of user fee revenues under each option was described precisely, as follows:

> Fee levels were calculated so that, based on an adjusted FY91 Commercial Manifest, FY91 revenues would be constant across Options. [The adjusted manifest included seven manifested large orbital launches, eight mani-

DATA ON PROXY ATTRIBUTES BY VEHICLE

VEHICLE	ATTRIBUTE			
	LEO Delivery Capability (lbs.)	Max LEO Launch Mass (lbs.)	Third Party (Casualty) MPL (millions)	Estimated Launch Price (millions)
Large-Orbital				
Commercial Titan	32000	1531376	$135	$120
EPAC S-4	20300			$49
Atlas I	13000	378000	$84	$50
Atlas II	14950	427950	$84	$65
Atlas IIA	15700	428700	$84	$65
Atlas IIAS	19000	543500	$84	$80
EPAC S-3	15800			$42
Delta II-6925,1-stg	8763	492463	$84	$45
Delta II-6925,2-stg	15224	510224	$84	$45
EPAC S-2	6500			$28
MAX	32000	1531376	$135	$120
MIN	6500	378000	$84	$28
Small Orbital				
Taurus	3400	183400		$15
SSI Conestoga 4	3500			$20
AMROC ILV-1	5390			$12
EPAC S-1	2500			$18
AMROC ILV-S	1210			$8
Pegasus	950	41450	$12	$6
LTV Scout (WSMC)	400	47600	$18	$12
NASA COMET	1800			
MAX	5390	183400	$18	$20
MIN	400	41450	$12	$6
Suborbital				
Prospector		25525	$5	$1
SSI Starfire		5000	$1	$15
DARPA/SDC		30000	$5	$1
AMROC SMLV			$12	$10
Conatec 3-317		5000	$10	
OSC LPX		30539	$5	$1
MAX		30539	$12	$15
MIN		5000	$1	$1

Fig. 4 Launch vehicle information used in calculating user fee impacts.

fested suborbital launches, and two added small orbital launches (not manifested). Two small orbital launches were included in the fee calculation to better reflect the expected vehicle mix in the next few years. These two small orbital launches were assigned attribute values equal to the average attribute values of all vehicles in the small orbital class. These additional launches were *not* used in calculating FY91 user fee revenues.] To determine the fee levels for each Option, the values associated with each proxy attribute were normalized by dividing each attribute value by the maximum value for that attribute. The normalized attribute values for each vehicle on the adjusted manifest were then multiplied by the number of manifested launches. This calculation provided the total normalized units associated with the adjusted FY91 manifest, i.e. the fee basis for FY91....On the basis of a target revenue level of $300,000, the required normalized unit cost to achieve that target was calculated....(It should be noted that the target revenue level was adjusted for the Preferred Option, to reflect revenues from suborbital launches, which will be assessed a flat fee. The target was reduced from $300,000 to $292,000 to reflect suborbital fees which would not be calculated based on delivery capability to LEO, but would be a flat fee.) Finally, this normalized unit cost was converted into a fee per attribute unit by dividing the normalized unit cost by the normalization factor (i.e. the maximum attribute value).[1]

The data on launch vehicle attributes used for this analysis are shown in Fig. 4. The calculations and resulting fee levels are shown in Fig. 5.

Defining Alternatives

The definition of alternative approaches may be relatively informal, in cases where the regulation in question is largely procedural or specified extensively by legislation. In other cases, where the agency promulgating the regulation has a wide range of options, the definition and analysis of alternative approaches must be structured and complete. The examples here varied significantly. The brief statement that follows constituted the complete consideration given to alternative approaches in the financial responsibility analysis. On the other hand, the user fee analysis detailed four alternative approaches—using different bases for determining fee values—to assessing user fees and provided a full analysis of the economic impacts of each approach. The advantages and disadvantages of each approach were discussed. This information is summarized below.

The alternatives available to DOT with regard to the proposed rulemaking were to either continue existing practice, applied on a case-by-case basis, or to undertake a rulemaking that codifies existing practice. The discussion of these alternatives in the draft analysis was:

Codifying Existing Practice

DOT believes that a rulemaking codifying its practices has value to the commercial launch industry. A rulemaking clarifying DOT practice and committing to these practices would provide important information in advance to licensees, and would to a degree reduce the business risk that they face. Increased certainty helps firms to plan better, and increase their competitiveness.

FY 91 MANIFEST: NORMALIZED ATTRIBUTE LEVELS

		Normalized Attribute Values X # Launches (FY91)			
		Preferred Option	Option 1	Option 2	Option 3
Vehicle	# FY91 Launches (Adjusted)	LEO Delivery Capability	Launch Mass	TPL · MPL	Launch Price
Large Orbital					
Atlas	4	1.96	1.16	2.49	2.17
Commercial Titan	0	0.00	0.00	0.00	0.00
Delta II	5	1.87	1.64	3.11	1.88
Small Orbital*					
(Avg. Used, 2/yr)	2	0.15	0.12	0.27	0.22
Suborbital					
SSI Starfire?	1		0.00	0.01	0.13
OSC Suborbitals	6		0.12	0.44	0.10
Prospector	1		0.02	0.04	0.01
TOTAL	19	3.98	3.06	6.36	4.49

SETTING FEE LEVELS USING NORMALIZED VALUES: NORMALIZED UNIT COST

	Total Normalized Units During FY1991			
	Preferred Option	Option 1	Option 2	Option 3
Vehicle Class	LEO Delivery Capability	Launch Mass	TPL MPL	Launch Price
Large Orbital	3.832	2.798	5.600	4.042
Small Orbital*	0.150	0.119	0.267	0.216
Suborbital	0.000	0.140	0.489	0.233
Total	3.981	3.056	6.356	4.491
Annual Variable Revenue Target:	$300,000			
Normalized Unit Cost to Achieve Annual Target Level	$75,351	$98,160	$47,203	$66,797

SETTING FEE LEVELS USING NORMALIZED VALUES: FEES PER ATTRIBUTE UNIT

	Average Normalized Units Per Year			
	Pref.Opt.*	Option 1	Option 2	Option 3
Normalized Unit Cost to Achieve Annual Target Level	$67,351	$98,160	$47,203	$66,797
Fee Calculation	LEO Delivery Capability (lbs)	Launch Mass (1000 lb)	TPL MPL ($ mil)	Launch Price ($ mil)
Normalization Factor	32000	1531	135	120
Unit Fee to Achieve Target	$2.10	$64.10	$349.65	$556.65

Fig. 5 Calculations of user fee levels.

In considering such a rulemaking, DOT has examined its economic impacts. The effects on DOT (as the implementing agency) and on licensees are described in each case. Effects on other parties (specifically other private party participants to licensed activities than the licensee, the U.S. Treasury, the government as owner of launch facilities, and third parties) are noted when they are present.

No Regulation

The option of no regulation would entail the same provisions as described under *codifying existing practice*, applied on a case by case basis. DOT has not been informed of any problems created for industry by the application of these provisions and procedures on a case by case basis, and there is no specific statutory requirement for a regulation. DOT believes that a regulation would be of benefit to industry, by providing clear information in advance about the financial responsibility requirements of the launch licensing process, and by providing an assurance that DOT is committed to a certain set of rules, not to be changed unexpectedly.[2]

The discussion of alternatives in the user fee analysis was significantly more detailed. The economic analysis included discussion of two types of alternative approaches. The first was the use of a completely different approach to assessing user fees than the approach recommended by DOT — a cost-accounting basis vs the preferred approach of using a proxy value to set fees. The second type of alternative was a group of options that used different bases (different proxy values) for assessing a user fee, but the same general structure. The reasons for rejecting the cost-accounting approach were discussed qualitatively. The impacts of the use of different proxies were analyzed quantitatively as well as addressed qualitatively. This former analysis (of a cost-accounting-based user fee program and its relationship to the recommended program) read:

A Cost Accounting Based User Fee Program

One way of implementing a user fee system to recover costs in the manner described above would be to track the costs incurred by the Office in processing, issuing, and maintaining in force each license, and to charge each licensee in such a way as to recover costs incurred by OCST to administer and issue such a license. However, such a tracking system would have a number of disadvantages. First it would require a complex cost accounting system, where OCST staff would be required to log-in and log-out to register the time required to perform the numerous review and processing steps associated with each license application. The additional costs of developing and implementing such a system would have to be spread across the nine or ten firms likely to seek launch licenses during the next few years, and would, accordingly, result in a higher fee in the final analysis than an alternative system that does not rely on a precise cost-accounting system.

Another disadvantage of the cost-accounting approach to user fees is that it would discourage development and use of innovative technologies. One of OCST's missions is to encourage the growth of the commercial space transportation industry and the use of innovative technologies. License applications that incorporate the use of innovative technologies require extensive OCST staff time to review and process because the risks of the new technology must be fully understood and evaluated. Because high fees would discourage innovation, OCST has decided to exclude costs from the user fee system that are directly applicable to the additional effort associated with reviewing innovative technologies.

A third disadvantage of the cost-accounting approach to user fees is due to the fact that certain types of licenses—namely those that involve new elements affecting public safety or U.S. interests, and those that incorporate a new OCST launch license format—require considerable one-time administrative effort for processing an application. Once the initial work is completed, subsequent applications with similar elements or which support previously-issued launch license formats require considerably less effort. An example is the first use of a launch site from which many future licensed launches will occur. The effect of the cost-accounting approach would be to pass these one-time costs to the first of many potential licensees proposing activities reflecting similar elements, or acquiring similar types of launch licenses.

Recommended OCST User Fee Program

OCST has developed a user fee system that uses a proxy to calculate and recover the majority of appropriately recoverable costs incurred by the Office in producing benefits to licensees through the processing of applications and issuance and administration of licenses. The Office's recommended user fee program also assesses each firm a lesser, fixed fee for the issuance of a license and for annual license renewal, which captures the remaining recoverable costs. The total amount of costs directly related to licensing is estimated under this program using the budget line item "Licensing Operations" (which represents contract funds), plus personnel costs based on an estimate of 3 to 3.5 labor years per year. The proposed user fee program does not address OCST services still under study, such as the development and application of hardware standards to the licensing process and licenses for operators of commercial or other non-federal launch sites, where it would be premature to assess user fees.

The proposed user fee program has several advantages:

- Fixed license issuance and renewal fees will generate a predictable revenue stream to cover routine administrative activities associated with licensing. A fixed fee is appropriate in recovering these costs because the level of routine administrative activity associated with issuing and keeping in force different licenses tends to be relatively stable. (Note, this assertion applies to routine administrative activities, as distinct from special activities required to process a first time or new technology license.)
- The proxy used to calculate the per launch fee (payload delivery capability to low earth orbit (LEO)) is easily calculated, and tends to correlate with the cost to issue and keep in force a license. The number of launches also tends to correlate with the on-going costs incurred by OCST in administering a license. A per launch fee better accommodates the uncertainty in assessing the full cost of administering a license that results from uncertainty about the total number of launches that will be conducted under the license.
- The majority of user fees under this program are assessed in association with launches by licensees, which are in turn associated with income to licensees. A per launch fee is less onerous for fee payers

than a lump sum fee which attempts to recover total costs at the time a license is issued.
- This method of calculating fees excludes special costs associated with first time and innovative technology licenses.[1]

In addition, the user fee economic analysis specified in detail the levels of user fee for different classes of vehicles that the four discrete options would generate. Again, these options differed in the proxy attribute used to assess fees for a particular vehicle. For each option, the positive and negative aspects of the approach were discussed (including a specific analysis of the administrative simplicity of the option), the impacts of each option relative to the others were noted. Examples of this analysis are given below. In addition, for each option, the resulting fee levels were specified in terms of the average fee level for launches of suborbital, small, and large vehicles, and the revenues generated annually by each of these types of launches.

Preferred Option—LEO Delivery Capability

Justification: As noted above, delivery capability in general correlates the cost of licensing with the fee assessed for launch activities. Under this Option, suborbital vehicle launches, the benefits of which are relatively modest are assessed at a flat fee of $1,000 to reflect the fact that applying a delivery capability fee scale to suborbital vehicles requires addressing multiple variables—payload weight, altitude, and duration.

Administrative Simplicity: LEO delivery capability is an easily quantifiable attribute, and is largely invariant from launch to launch. It is administratively a convenient user fee basis for both licensees and OCST.

Relative Impacts: Delivery capability is generally a good indicator of the relative license-related costs of different types of vehicles.

Option 1—Launch Mass

Justification: Mass is often used as a proxy for costs incurred, upon which fee schedules are based.

Administrative Simplicity: Launch mass is an easily quantifiable attribute, and is largely invariant from launch to launch. It is administratively a convenient user fee basis for both licensees and OCST.

Although there is some relationship between launch mass and costs incurred, there are anomalies in attempting to correlate weight, productivity, and benefit conferred on a licensee for launching a particular vehicle in comparison with other vehicles. Vehicle technology may be such that vehicles with similar capabilities pay different per launch fees.

Relative Impacts: Launch mass is generally a good indicator of the relative license-related costs of different types of vehicles. In some cases, however, vehicle technology may be such that vehicles with similar capabilities pay different per launch fees. This problem especially applies to large orbital vehicles.[1]

The key factor in defining alternatives for an economic impact analysis is to provide sufficient information to allow a reader to assess the recommended program in context. The discussion of alternatives serves as a forum to more clearly identify the objectives the regulation is to meet and the obstacles that must be overcome to do so.

Structuring Analysis

The basic questions addressed by an economic impact analysis are:
1) Who is affected by the proposed regulation?
2) How are they affected?
3) What are the magnitudes of the effects?

Implied in this approach is, of course, the need for information about the status of affected parties in the absence of the regulation—the description of the baseline.

An accurate characterization of affected parties of course must be based on the specifics of the rule under consideration. However, it can be said that in general affected parties are likely to be firms in the industry at which a regulation is aimed, the suppliers and contractors to those firms, the customers of those firms, the implementing agency (which will incur implementation and enforcement costs), and the general public (as taxpayers or consumers). A complicating factor in defining parties affected by commercial space regulations is the government's many layers of interaction with commercial space firms. For space transportation providers, the U.S. government is an important customer, a supplier (of launch infrastructure), and a regulator. It is important to show these different types of effects on the government.

Both of the examples used here stated explicitly the parties affected, the types of effects, and, to the extent possible, the magnitude of the effects. Quantifying effects are discussed further in the next section. The parties and types of effects identified for the two examples are shown below.

The analysis of financial responsibility requirements examined the economic impacts of the 1988 Amendments to the Commercial Space Launch Act of 1984 on the government and private sector, compared to the impacts of the Commercial Space Act of 1984 (prior to the Amendments), DOT practice prior to the Amendments, and the provisions of the National Space Policy of 1988. There are three major areas that change under the Amendments from the baseline case: provisions to protect participants in licensed activities against claims by third parties, provisions requiring participants in licensed activities to waive claims against one another, and provisions to protect the government against loss or damage to property. The first of these changes is discussed in detail below.

The parties affected by this provision were identified as:
- Government as implementer of regulations
- Licensee (generally launching company)
- Customer(s)
- Insurance company
- Other users of launch range—commercial
- Other users of launch range—government
- Treasury
- Government employees and contractors involved in licensed activities

- Government as owner of property not covered by property insurance
- On-range personnel not involved in licensed activities—government and government contractors
- On-range personnel not involved in licensed activities—e.g. other launch and payload companies
- Owners of on-range property (parties not involved in licensed activities—government and government contractors—and other launch and payload companies)
- Owners of off-range property identifiably at risk (pre-orbit)
- Owners of off-range property not identifiably at risk (pre-orbit)
- Off-range populations—identifiably at risk (pre-orbit)
- Off-range populations—not identifiably at risk (pre-orbit)
- Owners of on-orbit property
- Owners of property—post-orbit re-entry
- Populations—post-orbit re-entry

The types of potential effects of this provision were identified as:

- Bearing Costs of Injury or Loss of Life
- Bearing Costs of Damage to Property
- Paperwork Costs
- Change in Business Risk/Change in Opportunity Costs

Specifically, proposed provisions to protect participants in licensed activities against claims by third parties were found to have the following impacts:

- eliminating liability of licensees and other private party participants for third party claims in the range from the insurance requirement to $1.5 billion beyond that amount (this affects the allocation of expected cost of risk among the affected parties),
- increasing liability of U.S. Treasury for third party claims in the range from the insurance requirement to $1.5 billion beyond that amount,
- decreasing risk to third parties of bearing damages because of provision for government payment of $1.5 billion above the required level of insurance,
- decreasing business risk because of statutorily defined insurance requirements.[2]

Detailed matrices of the parties affected and the types of effects were prepared for each of the major provisions of the proposed regulation. The matrix for this particular provision is shown in Fig. 6.

The analysis of the effects of user fees also specified affected parties and types of effects in detail. Five groups were identified that might be directly affected by user fees: the providers of commercial space transportation services, payload owners, users of payload services, U.S. Treasury, and the general public. Impacts from a general user fee policy on each of these groups were summarized for nine different response scenarios that described different financial decisions on the parts of those charged user fees. These scenarios are listed in Fig. 7. Parties that might be secondarily affected were also identified: the insurance industry, the financial community, general suppliers, and launch ranges. A general description of each of the potentially affected groups was provided.

	TYPES OF EFFECTS			
PARTIES AFFECTED	Bearing costs of injury or loss of life	Bearing costs of damage to property	Paperwork costs	Change in business risk/Change in opportunity costs
Government as implementer of regulations	NA	NA	None.	NA
Licensee	Decreased likelihood of bearing costs because of government payment of claims and cap on insurance requirement. (With small offset due to additional individuals in category of "third party"—government employees.)	Decreased likelihood of bearing costs because of government payment of claims and cap on insurance requirement.	None.	Decreased because firms know method used to determine insurance requirement. Also decreased because of government commitment to industry implied by government payment of claims.
Customer(s)	Same as above, depending on extent risks/costs are passed through by licensee company.	Same as above, depending on extent risks/costs are passed through by licensee company.	None.	Same as above, depending on extent risks/costs are passed through by licensee company.
Insurance company	NA	NA	NA	None.
Treasury	Increased because of payment of claims up to $1.5 billion.	Increased because of payment of claims up to $1.5 billion.	None.	NA
Third parties	Decreased, because of government payment of claims up to $1.5 billion above requirement (higher likelihood of recovering damages from government without excessive litigation.)	Decreased, because of government payment of claims up to $1.5 billion above requirement (higher likelihood of recovering damages from government without excessive litigation.)	NA	NA

SETTING OF FINANCIAL RESPONSIBILITY REQUIREMENTS FOR DAMAGE OR INJURY TO THIRD PARTIES AT THE LESSER OF MAXIMUM PROBABLE LOSS, $500 MILLION, AND INSURANCE AVAILABLE ON WORLD MARKETS AT A REASONABLE COST, WITH GOVERNMENT PAYMENT OF CLAIMS FOR DAMAGES OR INJURY TO THIRD PARTIES ABOVE REQUIRED LEVEL OF FINANCIAL RESPONSIBILITY, UP TO $1.5 BILLION

Fig. 6 Qualitative description of economic impacts of 1988 Amendments to the Commercial Space Launch Act—provisions to protect participants in licensed activities against claims by third parties.

Scenario 1:	User Fee Absorbed By Transportation Provider
Scenario 2:	User Fee Passed on to Payload Owner U.S. Government Agency Payload U.S. Government Agency Budget Held Constant
Scenario 3:	User Fee Passed on to Payload Owner U.S. Government Agency Payload U.S. Government Agency Budget Increased to Cover User Fees
Scenario 4:	User Fee Passed on to and Absorbed by Payload Owner Commercial U.S. Payload
Scenario 5:	User Fee Passed on to and Absorbed by Payload Owner Foreign Payload
Scenario 6:	User Fee Passed on to Users of Payload Services Commercial U.S. Payload
Scenario 7:	User Fee Passed on to Users of Payload Services Foreign Payload
Scenario 8:	User Fee Passed on to Payload Owner Commercial Payload Decision Not to Utilize U.S. Launch Vehicle
Scenario 9:	User Fee Passed on to Payload Owner Commercial Payload Decision to Terminate Payload Program

Fig. 7 **Identification of considered scenarios.**

Quantifying Effects

The full range of methodological challenges associated with cost–benefit analysis is clearly beyond the scope of this article. The techniques used in an economic impact analysis can range from mainly data gathering, to financial analysis of affected firms or industries, and even to macroeconomic modeling (for regulations with major impacts). There will be, for any economic impact analysis, the need to identify and apply appropriate analytic techniques and the ability to apply them appropriately.

An important part of this process is the recognition that certain effects are simply too costly to quantify. Benefits are generally more difficult to quantify than costs. Costs are usually incurred in dollars, while benefits may be in terms of saved lives, improved environment, better health, and so on. (This does not always hold true; costs can be imposed through negative externalities, which may be extremely difficult to quantify.) Quality of life improvements (as opposed to productivity improvements) are notoriously difficult to quantify. Particularly in cases where the costs of a regulation are limited, the quantification of benefits may be unnecessary; a qualitative statement may demostrate sufficiently that benefits are likely to exceed costs.

In many cases, it will be appropriate to quantify only the primary impacts of a rule, and to partially quantify or even simply describe the lesser impacts. Such a description should clearly identify the parties affected and the types of effects, and include a statement explaining why quantification is too costly to undertake or otherwise unnecessary.

Perhaps the most important technique for quantifying effects given limited data is placing bounds on potential effects. It may not be possible to state what the magnitude of an effect will be, but it is often possible to determine that the magnitude will not exceed (or will not be less than) a certain level. If the upper or lower limit of an effect can be shown to be relatively insignificant, then it may not be necessary to attempt to quantify the effect more precisely.

A good example of the bounding process is one from the financial responsibility requirement analysis. In that case, it was necessary to show the impact of a commitment by government to pay successful claims against launching companies by injured third parties, when those claims exceeded the level of required insurance. (This commitment covers claims in an amount from the level of required insurance to $1.5 billion above that level.)

The analysis identified the maximum value of this commitment by using the probability of an accident that would result in damages in the amount of the required level of financial responsibility. This probability (referred to as the threshold probability) was by definition higher than the probability of an accident resulting in damages above that amount, and the use of this probability in calculating expected value resulted in a maximum bound on the expected value of the government commitment. The analysis is explained below.

> For each launch undertaken, the exposure of the licensee is the financial responsibility requirement, plus amounts above $1.5 billion plus the requirement. Prior to the provision for government payment of claims in the Amendments, this exposure would have been from the first dollar, comprising the financial responsibility requirement, and liability for all damages to third parties above that requirement.
>
> The savings to licensees can be calculated by determining the expected value of the change in exposure. Similarly, the cost to the Treasury can be calculated. (This cost is equal to the savings of the licensee.)
>
> To calculate the expected value of exposure, the threshold probability used in determining maximum probable loss (10^{-7}) has been used. The threshold probability is the probability that losses in excess of the maximum probable loss (MPL) will occur.
>
> The reduction in exposure of the licensees for third party damage occurs in the damage range from required level of insurance (assumed to be equal to MPL) to $1.5 billion plus the required level of insurance. It is this range of damage for which the government now (via the Amendments) assumes responsibility. It is assumed that the probability of being in this range of damage is the threshold probability (10^{-7}). This is an approximation which assumes that there is little or no chance that damage levels will exceed the value of $1.5 billion plus the level of required insurance.

Table 1 Change in allocation of expected costs of claims by third parties (1990–1994)

Bearer of cost	Change in exposure[a]		Total difference in		Change in expected cost per launch[b]	Projected number of launches	Total expected cost
	Amendments	Baseline	Exposure	Threshold probability			
Licensee	>(MPL+$1.5B)	>MPL	−$1.5B	10^{-7}	−$150	320	−$48000
Insurance Co.	MPL	MPL	$0	NA	$0	320	$0
Treasury	$1.5B	$0	+$1.5B	10^{-7}	+$150	320	+$48000

[a]Licensees assumed to purchase insurance to fulfill financial responsibility requirements.
[b]Exposure multiplied by threshold probability.

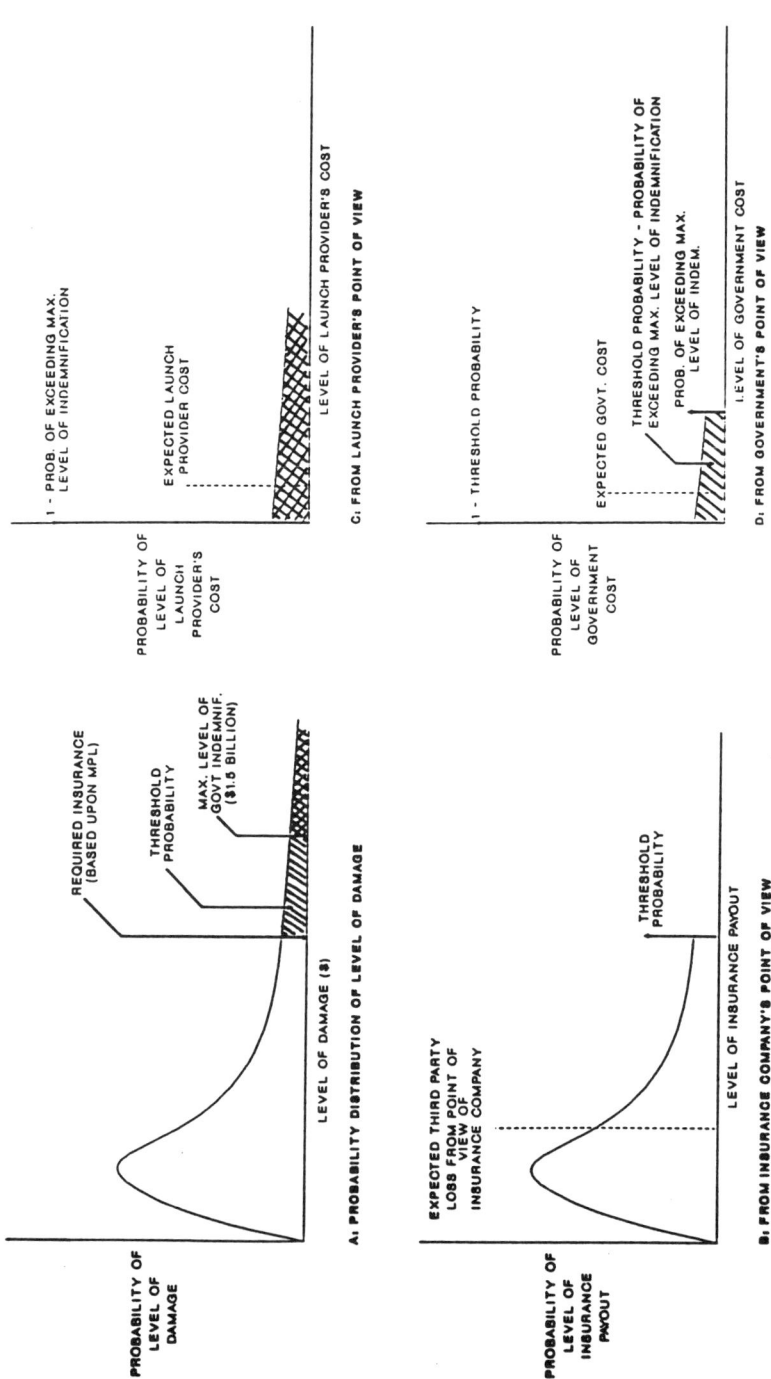

Fig. 8 Partial quantification of costs of third-party damages. Insurance payout and government cost as a function of required level of insurance: third-party damage (insurance payout + launch provider cost + government cost = constant).

The probability distribution within this damage range is not known. However, if it is assumed that the maximum value of damage (i.e., required level of insurance plus $1.5 billion) occurs with a probability equal to the threshold probability, an estimate may be made of the maximum expected value or benefit of the Amendments to the licensees per launch. This, it should be noted, is a slight overstatement of benefits because of the above approximation.[2]

The quantitative results of this analysis are shown in Table 1.

Finally, it is often helpful to combine an effort to bound potential impacts with a careful analysis of the structure of the impacts—essentially setting up an equation that would specify impacts if complete data were available. The example in Fig. 8, drawn from the analysis of the proposed financial responsibility requirement, represents such an approach. It identifies the mathematical structure of the impacts of the provisions pertaining to third-party liability, and specifies the points at which values are known.

Conclusion

This chapter has addressed the steps in analyzing the economic impacts of space transportation-related regulations. This type of analysis will become increasingly necessary in other areas as space activities become primarily business activities rather than government programs. There have been few decisions as yet even about who will regulate different types of activities. NASA continues to play a key role in providing technical expertise, but is not as it is currently structured equipped (or mandated) to act as a regulatory agency. New agencies, such as DOT (or a restructured NASA), will continue to enter into the regulation of space activities.

The principles discussed here apply to a broad range of such analyses. In summary, it can be said that a high-quality economic impact assessment is not one that quantifies every effect and purports to show with certainty what those effects will be. It is misleading to suggest certainty and it is time consuming and often useless to attempt to quantify impacts with inadequate data. A high-quality economic impact assessment is one that provides a well-structured framework for examining the effects of a proposed regulation and specifies the methodology used to quantify key impacts and identifies the weaknesses in the methodology and data. Stating that uncertainty exists should not be regarded as an admission of failure, but as the provision of a useful piece of information. The job of the analyst is not to serve as a black box out of which emerges the one true answer; it is to aid those to whom the analysis is provided (government decision makers, affected parties, and the public) in understanding the problem, the proposed solution, and the implications of both so that they can formulate informed opinions.

References

[1]"Economic Evaluation of User Fees Under Consideration by U.S. Department of Transportation," Prepared for Office of Commercial Space Transportation, U.S. Dept. of Transportation, Princeton Synergetics, Princeton, NJ, April 1991.

[2]"Initial Economic Impact Analysis: Proposed Financial Responsibility Requirements," Prepared for Office of Commercial Space Transportation, U.S. Dept. of Transportation, Princeton Synergetics, Princeton, NJ, April 1991.

Chapter 4. Economics of the Marketplace

Remote Sensing

David Moore*
Congressional Budget Office, Washington, DC 20515

UNTIL the mid-1980s, the use of satellites to collect information about objects on the ground was exclusively a public activity in the United States. The federal government developed, produced, and operated satellites that collected the raw data necessary for informational products that were mostly used for governmental purposes. A departure from this policy came in 1984, when the Congress decided to privatize the Landsat system. That system includes the satellites, ground stations, and processing equipment developed by the federal government to gather data about the land. The Land Remote Sensing Commercialization Act of 1984 began a 10-year experiment with the objective of transforming the Landsat system from a research project into the productive base for a new private industry.

Seven years into this experiment in space commercialization, its premise—that a new private industry can be created by the transfer of federal assets and with limited subsidies—is under challenge. While some progress has been made in privatizing Landsat, the prospects for a fully private system by the end of this decade are dim, an outcome anticipated by some observers as early as 1983.[1] The direction of current policy is toward returning Landsat to the public sector.

Remote Sensing Data: Current Outlook and Policy Options

The Earth Observation Satellite Company (EOSAT), a private firm owned by Hughes Aircraft and General Electric, has undertaken the commercialization of the Landsat system. It operates the government-owned satellites Landsats 4 and 5, processes and markets the data they produce,

---

This chapter is drawn from *Encouraging Private Investment in Space Activities*, prepared by the Congressional Budget Office, February 1991, and is declared a work of the U.S. Government and is not subject to copyright protection in the United States.
*National Resources and Commerce Division, United States Congress.

and invests in ground facilities and, to a very limited extent, in a new satellite now being developed. In return, EOSAT receives the revenues from the sales of data and also receives federal funds to operate the satellites and to cover more than 95% of the cost of building and launching a new satellite, Landsat 6. After the launch of the Landsat 6 in 1992, no additional support for EOSAT is planned. Appendix A provides an overview of remote sensing technology and products.

Available evidence indicates that the large-scale systems currently used for remote sensing, such as the U.S. Landsat and its predominantly French European counterpart, Systeme Probatorie d'Observation de la Terra (SPOT), cannot generate sufficient revenues from data sales to cover their cost. The cost of building, launching, and operating two Landsat-type satellites during the 1990s could range from $0.75 billion to $1 billion, exclusive of the cost of capital and insurance that a private firm would have to bear. This amount translates into an average annual cost of $85 million to $110 million between 1991 and 1999. EOSAT revenues stood at $25 million in 1989, up from $19 million in 1986, as shown in Table 1. One recent analysis of the industry estimated that even a 20% annual rate of growth would not be enough to cover the full private-sector cost of remote sensing by the end of this decade.[2] SPOT's situation is roughly comparable.

This pessimistic outlook for the prospects of land remote sensing as a private industry of the future—at least through the turn of the century— reflects a consensus of many studies. In 1983 the Department of Commerce, in preparation for transferring the Landsat System to the private sector, conducted several studies. One study, summarized in Appendix B, addressed the overall business including the market, costs, technology, competition, and other factors and concluded that the market was, and would remain for the foreseeable future, insufficient to support a viable business venture. This was later borne out by market studies performed for RCA/ Hughes prior to forming EOSAT. In 1988, the Department of Commerce as part of a Landsat policy review, released a set of contractor studies. (This review produced three contractor studies to support the Department of Commerce's process of deciding what to do about Landsat. The three

Table 1 Revenues of major land remote sensing companies (in millions of dollars)

Year	EOSAT	SPOT Image	Total
1986	19	5	24
1987	21	10	31
1988	23	16	39
1989	25	22	47

SOURCE: Department of Commerce, *Space Business Indicators*, June 1990, pp. 16 and 35.
NOTE: EOSAT is the U.S. privately owned Earth Observation Satellite Company. SPOT Image is the company that markets the data of the French satellite, Systeme Probatorie d'Observation de la Terra (SPOT)

studies are: The Analytic Sciences Corporation (TASC) study, the Kodak Remote Sensing study, and The Egan Group study. EOSAT produced an independent study on the same issue. See Refs. 2-4.)

The TASC study concluded: "Projected market revenues will not support a fully viable commercial civil Earth remote sensing system during the 1990s."[3] The Kodak study reached a similar conclusion. "The space segment cannot be commercialized during the present century, even under the most optimistic market projections."[2] EOSAT, in an independent study of its own presented to the Congress, was less direct: "This study finds that the special risk/return characteristics of an advanced satellite remote sensing system serving a wide variety of markets over the next eight to ten years precludes conventional private or multinational financing of the investment required for an advanced Landsat 7 system."[4] What is left of the vision of a fully commercial system of the Landsat or SPOT type according to the financial analysis included in the TASC study, is "a marginally profitable and financable system" if the government is willing to be cooperative and to implement "aggressive private sector risk reduction strategies."[3]

Currently, the total cost to society of producing land remote sensing data exceeds what the public and private buyers of the data are willing to pay for it. The market may fail to capture the full value of the data to society, for several reasons. First, the data produced by Landsat can be widely distributed at minimal cost beyond the immediate public or private purchaser and put to valuable use by others. Current prices fail to capture this value.

Second, a part of the social value of Landsat data may be inappropriately reflected in the revenues of private firms that buy remote sensing data and resell it. These firms, value-added producers as they are called, have sales two to four times greater than the sales of Landsat and SPOT data.[2,3] Regulations limiting the pricing and distribution policies of EOSAT or potential new suppliers of data may prevent the producers of data from charging prices and imposing terms of sale that would capture their share of the value of the data they produce. A similar case can be made about the international network of ground stations. Current law limits the annual fee EOSAT can charge these stations and prevents EOSAT from charging different stations different fees for the data they receive and subsequently resell.

Third, the value of Landsat data sold to public buyers may also not be fully reflected in the prices they pay for it. To begin with, the public goods to which Landsat data contributes are not well priced in the market—for example, better maps to support U.S. troops in the field, or the identification of the breeding areas of disease-carrying mosquitos. In addition, at the federal level, the agencies that find Landsat data most useful may not be paying its full value, hoping eventually to share the cost with other agencies. The unpaid-for value of Landsat data is covered by federal subsidies for current operations and in the larger appropriations necessary to build new satellites.

Policy toward privatizing remote sensing hinges around whether the federal government will fund new satellites. Appendix C shows the uncertainty and shifts in circumstances characterized policy and funding for Landsat during the 1980s. Ending federal funding would mean that as Landsats 4 and 5 reached the end of their lives, EOSAT would be forced to stand or fall in the mid-1990s as the market dictated. The option of ending federal support for the privatization of remote sensing is consistent with the belief that the value of this activity can be fully reflected in the market. Alternatively, a commitment to provide new public funding for remote sensing would imply a belief that the market currently undervalues the social worth of remote sensing data.

An option to encourage private investment in remote sensing activity if no additional funding for Landsat is provided would be to deregulate the market for remote sensing data by permitting a producer to charge different prices to different customers and to grant exclusive data rights to customers who are willing to pay. If private investors were to test the market under these circumstances, they would be likely to do so with satellite systems that are smaller, less expensive, and less capable than Landsat. The prospects of these "lightsats," however, have yet to be tested in the market. In addition, it is likely that deregulation would lead to commercializing the more lucrative pieces of the total "Landsat" market. The result is likely to be that the U.S. government would still have to provide support for space and ground systems to produce those data products that are required but for which no pricing mechanism exists.

In any renewed public commitment to finance remote sensing, the issues would be how to maximize the social benefits of the data obtained from remote sensing, and how to distribute these benefits. One option would be to publicly underwrite some or all of the capital costs while allowing private investors to operate the system and distribute the products. This option could be pursued in more than one way: for example, by continuing the current relationship with EOSAT and negotiating a federal contribution to a new satellite, or by auctioning off distribution rights to investors willing to share the cost of new satellites. A second option would be to return remote sensing to the public sector. A variant of this option would be to maintain a private marketing organization to distribute the data produced by a publicly financed and operated satellite system. A third option would be to conclude an agreement with other countries to create an international public institution that would coordinate the design, launching, and operation of remote sensing satellites. This approach might include a role for the private operation of satellites and distribution of data.

Conditions of Supply

Several factors determine the conditions under which remote sensing data are currently supplied. One is that the average cost of producing data declines as more data are produced. Declining costs are typical of public utilities, where a number of regulatory mechanisms have been developed to protect the public interest.[5] Appendix D provides a discussion of the cost characteristics of a declining-cost industry and its implications for pricing and economic efficiency.

A second factor is the way in which the market for raw data is regulated. The Land Remote Sensing Commercialization Act limits the flexibility of EOSAT and other potential U.S. entrants in setting prices and granting access to data. These regulations force providers of raw data to make data available to all customers on the same terms, under what is called the nondiscrimination principle. The benefit to the United States of maintaining the principle of nondiscrimination is the goodwill of nations over which remote sensing satellites fly. The cost is the limitation placed on the possibility of successful private investment.

Public investment in remote sensing systems is a last major supply-side factor to be taken account of. Competition in the current market for land remote sensing data is essentially between two mixed enterprises, this country's EOSAT and Europe's SPOT, both of them receiving public subsidies. In the United States, public policy and investments also define the domain of remote sensing activity that is treated as public and that which is considered private. For example, under current policy land remote sensing is to become a private activity, while remote sensing activity to gather weather data is to remain public. Planned increases in public investment in remote sensing systems for the atmosphere and oceans will make available a large volume of new remote sensing data for processing and distribution toward the end of the decade. Significant savings may be possible if the private sector proves to be a more cost-effective distributor of this information than the public sector.

Structure, Scale, and Cost

The structure of the industry producing land remote sensing information includes two levels—that of raw data production and that of value-added production. Each of the two main producers of raw data—EOSAT and SPOT Image—has a primary ground station and processing facility. In addition, each has an affiliated network of international ground stations that are publicly owned by the countries in which they are located, and pay annual access fees or royalties or both on sales of data to the raw data producer in exchange for receiving data. On the other level, that of value-added production, are many firms that buy data from EOSAT and SPOT, improve its information quality, and resell it to final users.

Raw Data Producers

The cost of current remote sensing activities is dominated by the cost of the "space segment" necessary to produce raw data—that is, satellites and their launch and ground support facilities. Landsat and SPOT are both large-scale systems that rely on large satellites carrying several different types of sensors. Landsat satellites are designed to last five years, but SPOT satellites for only three. Landsat 6 will cost $240 million to build and an additional $40 million to launch. [EOSAT provided the Congress with a broad range of options for new remote sensing satellites, from $160 million (1988 dollars) to $800 million (1988 dollars).[4]] Current estimates of the cost of SPOT satellites are comparable. The total cost of ground facilities and four SPOT satellites, two that are already built and two planned for the 1990s, is $1.2 billion exclusive of launch costs.[6]

Unlike Landsat, SPOT had a private and commercial aspect from its inception. SPOT Image, a mixed enterprise of which the French government owns almost half, was established to operate the system and to market and distribute its product on a commercial basis. Roughly a third of SPOT Image is owned by aerospace firms that build the SPOT satellites and related systems. While initially it was hoped that data sales from the first SPOT satellite would build a market sufficiently strong to pay for subsequent satellites, the $22 million of sales in 1989 only approached the level necessary to cover operating costs. By tailoring its product differently from that of Landsat (SPOT images cover less area in fewer spectral bands, but in more detail than EOSAT), SPOT has acquired customers that find the unique aspects of its data useful. In addition, SPOT has tapped the market for remote sensing data among those customers who will buy virtually any product on the market—for example, military and national security buyers. Some users have been able to enhance the value of Landsat data by merging it with SPOT data. These relationships among products and customers, and the youth of the industry in general, have led some observers to suggest that SPOT and Landsat products are more complementary than competitive.

Currently, operating systems on the scale of Landsat and SPOT cannot provide data at a cost that the market can cover. Market demand is so weak, relative to the scale of current systems, that even a producer with a world monopoly would have difficulty charging a price sufficiently high to cover the cost of production. The Kodak study illustrates this point in a projection showing that even if a single producer of land remote sensing data captured the entire market and grew at 20% annually during the 1990s, it could not produce sufficient revenues to cover the cost of the space segment until 2000.[2]

This characteristic of the current market suggests that a single land remote sensing system, as envisioned in the international consortium option, could be more efficient than two competing private systems by producing roughly the same data currently available with one system. The argument holds only so long as market growth and technology allow it to do so. Lower-cost technology could permit competing systems to cover their cost in the market. Dramatic growth in demand could have the same effect. Some analysts have argued that exactly such circumstances have rendered the regulated monopoly in international satellite communications, Intelsat, obsolete.[7] Public institutions that in the short run may appear efficient could in the long run become obstacles to technical change and the provision of services desired by consumers.

An alternative to the current approach to producing raw data would be to allow public and private sensors to share the same satellite "bus"—that is, the generic satellite including maneuvering engines, onboard power supply, and communication equipment. By allowing such sharing, it would permit large-scale operation at a lower cost. Private sensors could be placed in rented space aboard a public satellite, or a public sensor could rent space on a private satellite. Either arrangement could provide private remote sensing with economies of scope—in this case, the savings gained from sharing the satellite with sensors producing other types of data. Such arrangements would, however, require compromises in spectral coverage and in the time it takes to acquire a desired image.[3]

Small single-sensor satellites would also reduce costs of producing data, although the studies commissioned by the Commerce Department did not show that such a venture would be profitable.[2,3] The cost of such a system, variously referred to as "lightsat" or "cheapsat," has been estimated at $15 million to $50 million plus launch costs of $5 million to $15 million.[3] The lower costs would be associated with lesser capabilities, and hence with a system having less revenue-generating potential. Smaller firms are currently developing systems of this type, but their technology is unlikely to be tested in the marketplace as long as publicly supported competitors remain in operation. Ending public support for Landsat would thus provide a more open field for smaller satellites.

Operating Costs

Once a satellite is in orbit, the annual additional cost of producing remote sensing raw data with a Landsat or SPOT system is currently around $20 million. These costs include operating the satellite and programming its sensors to capture specific images, processing the data into digital tape or image form, and marketing and distributing these products. For most products, these costs are probably constant or increase as more data are demanded. For all policy options, these costs provide a benchmark for efficient pricing. For policy options that involve a private operator, these are the minimum costs that must be covered in the marketplace.

The claim that a private investor is better able to lower the cost of ground processing than a public operator is a key point in the case for continued private involvement in land remote sensing, even if the market cannot support a fully private system. The claim is based largely on the conceptual argument that the profit motive ensures better control of costs than do administrative directives.

EOSAT has probably lowered the ground operations cost of producing land remote sensing information, although the evidence is inconclusive and difficult to evaluate. EOSAT claims that the federal budgetary resources necessary to operate Landsat have declined as a consequence of commercialization. By EOSAT's estimate, federal spending to operate the Landsat systems has fallen from $36 million in 1985—the last year of federal operation—to below $20 million in 1990. (Statement by Pete Norris, Executive Vice President, Earth Observation Satellite Company, before the Subcommittee on International Scientific Cooperation and the Subcommittee on Natural Resources, Agriculture, Research and Environment of the House Committee on Science, Space and Technology.[8]) During 1985, however, data sales and international ground station fees provided offsetting revenues of over $15 million. Thus, the net position of the government remains virtually unchanged under private operation.

It is likely that EOSAT has achieved a lower unit cost for the ground operations of producing data, however. The volume of digital products, which are more difficult to produce, almost doubled between 1985 and 1989, more than offsetting a large drop in photographic images, which are less difficult to produce.[9] Over the same period of time, revenues have increased from about $15 million to $25 million, with the price of products held roughly constant. EOSAT claims that costs will be further reduced in the future as the 1970s-vintage processing equipment, inherited by EOSAT from the government, is replaced with new equipment.[10]

If indeed the private sector can operate satellites and distribute their product more cheaply than the public sector, the difference may be accounted for by other factors than inefficiency in the public sector. Cost control in the public sector may take second place to other objectives—for example, basic research and development. In the case of Landsat, the public agencies operating the system did so under mandates unlikely to allow for cost reduction. The National Aeronautics and Space Administration was running a research program. The National Oceanic and Atmospheric Administration was the caretaker of a system in transition. Consequently, the comparison of EOSAT with its predecessor does not show conclusively whether some future government-operated system with the explicit mandate to control costs could perform as well as EOSAT.

Value-Added Production

The cost of the value-added activities performed by firms that buy data from EOSAT and SPOT in order to resell it to final users depends on the amount of processing they do. The Kodak study of the remote sensing industry examined the computer equipment necessary for many value-added activities; it identified a range of systems from a personal computer-based system costing $25,000 to one based on a mainframe computer costing as much as $1 million.[2] These costs, unlike those of the space segment, do not represent a barrier to entering the industry, which has seen a proliferation of value-added firms during the 1980s. Value-added producers are expected to benefit from technical change in the computer industry that will lower the cost of current applications and create new applications. The increase in value-added sales in recent years implies a growing demand for raw data in the future.

Not enough information about value-added producers is available to determine whether allowing raw data producers to pursue different pricing strategies could allow them to capture a part of the revenues currently held by value-added producers. The TASC study suggested this possibility. The competitive structure of the value-added sector, with its many firms, suggests that cost savings in raw data production will ultimately be passed along to final consumers. Some value-added activities, however, involve special expertise or possessing mathematical formulas that permit extracting specific information from a set of digital data. Firms with these advantages might be able to charge relatively high fees for their products, thus capturing high profits. A raw data producer allowed to charge this type of customer higher prices could share in these returns.

Regulation and Competitive Behavior: EOSAT and SPOT Image

By law the Landsat system operates under the condition of "open skies" and nondiscriminatory access to remote sensing data. (President Eisenhower put forward the earliest formulation of the open skies policy. For discussion, see Ref. 11. The policy is formalized in the Treaty on Principles Governing the Activities of States in the Exploration and Use of Outer Space, Including the Moon and Other Celestial Bodies. See Ref. 12.) The open skies principle declares outer space to be an international zone open for all to use. The nondiscrimination principle holds that all parties should have access to raw data on the same terms. The Land Remote Sensing Commercialization Act acquires EOSAT and any other U.S. entrant to adhere to this principle, effectively prohibiting price discrimination. Underlying the U.S. adherence to the nondiscrimination principle are a desire to be fair to countries over whose territories U.S. satellites gather information, and the perception that there are foreign policy benefits in allowing other countries equal access to U.S.-developed systems. The issue of fairness in land remote sensing arises because only technically advanced countries are able to build and launch satellites, while the information they gather about all countries could be of commercial advantage to anyone possessing it, particularly as concerns the natural resources sector.

EOSAT might have a better chance of closing the gap between revenues and cost if it was permitted to charge different customers different prices for the same information or service. For example, clients willing to pay higher prices could be provided data on an exclusive basis. EOSAT is also restricted from practicing price discrimination among its international ground stations. In this case, while EOSAT's income increases with greater sales of data because it receives a royalty on sales to those ground stations, it has no way to capture a larger share of the returns from those stations' sales of raw data to others.

While the nondiscrimination principle does not explicitly prohibit EOSAT from moving downstream into value-added services and products, that principle is an element of the environment shaping EOSAT's strategy. Success in value-added markets could evoke suspicion that EOSAT's value-added operation was being given discriminatory access to raw data. Instead, EOSAT relies on firms in the value-added sector to expand the market for raw data through their effort to sell value-added products and services. In this context, EOSAT contends that a move into the value-added sector would represent competition with its own customers. The TASC study of the remote sensing market in the 1990s specifically addressed the question of whether or not a move into the value-added market could allow a large-scale remote sensing data producer to become commercially successful.[3] Its conclusion was that a Landsat-scale producer would still be unable to increase revenues enough to cover the full cost of raw data production.

SPOT Image's choice of strategy reflects a different view. Its sales of value-added products—five types of maps—could by one estimate generate as much as 30% of its expected $25 million revenues in 1990.[3] In addition, SPOT will reprogram its satellite images for customers who are willing to pay. At one time, the U.S. government offered similar Landsat services and applied special acquisition charges. EOSAT, however, while willing to acquire images requested by its customers, has discontinued charging the additional fees. SPOT Image drew as much as 10% of its 1989 revenues from charges for priority, according to one account.[6] Once SPOT produces an image, however, it becomes available to all buyers on standard terms, consistent with SPOT's adherence to a policy of open skies and equal access. Unlike EOSAT, SPOT also has flexibility in the prices it charges its international receiving stations. A minimum charge of $800,000 is required, but annual fees of as high as $2 million have been reported.

Public Investment in Remote Sensing

Most remote sensing activities under U.S. sponsorship are conducted by the public sector, including military, weather forecasting, and scientific research activities.[14] Proponents of commercialization argue that the present policy of restricting the scope of private activities to land remote sensing greatly decreases the prospects of successful private investment. From this point of view, the problem with remote sensing commercialization is that its market is artificially limited, and that a private system could succeed only if the government were to retire from the playing field. The Comsat corporation at one point proposed privatizing both the land remote and the weather satellites.[15] The law commercializing Landsat, however, explicitly prohibits commercializing the weather satellites.

In the future, the relation between public and private remote sensing is likely to grow more complicated. Foreign governments plan major new systems that will produce data overlapping that of EOSAT and SPOT. International concern with global climate change is leading to new investment in very large and capable remote sensing satellites that are likely to begin producing data in the late 1990s. These developments present opportunities and challenges to private investment in land remote sensing.

Japan, Canada, India, and Russia will, by the mid-1990s, all have remote sensing systems that produce data potentially competitive with Landsat and SPOT data in certain segments of the information market. However, just as SPOT and Landsat data are to some extent complementary, there is complementarity between the data produced by all other public systems and SPOT and EOSAT. The availability of more data generally creates more interest in its applications. The increased data allow different streams of information to be combined in value-added products. For example, some maps produced with satellite data combine data from SPOT and Landsat. International research into global climate change will potentially increase the market for remote sensing data, create more opportunities to combine land sensors with other types of sensors, and generate large flows of new data that private investors could help to distribute. The commercialization policy of the mid-1980s sought to open only one type of data to the private sector. The proliferation of data types in the future enhances the prospect of commercializing a part of remote sensing activity—satellite operation and/or data processing and distribution—among all types of data.

Conditions of Demand

Experience to date with the demand for remote sensing data suggests that full commercialization will be difficult, but perhaps not impossible. It also suggests that policy options in which the government would provide additional funds for satellite construction would be a remedy for the declining cost problem. The historical evidence cannot, however, fully answer the fundamental question of whether or not Landsat's social value is enough to justify its cost.

Market Size and Comparison

The estimated total revenues of the providers of raw data, SPOT Image and EOSAT, allow an approximation of the value of sales of worldwide remote sensing raw data since 1973 (see Table 2). The increase in the value of Landsat data sales in the late 1970s and early 1980s, from an annual level of $1.5 million in 1977 to more than $6.5 million in 1982, was a consequence of the introduction of digital products and also reflected the growing number of international ground stations—which increased from only three in 1975 to 10 in 1981.[16] In 1982, an increase in annual access fees for international ground stations from $200,000 to $600,000 also contributed to increased revenues. The large jump in revenues occurring in 1983 was caused by a major price increase for all products. Beginning in 1986, SPOT sales added to the total sales of raw data providers. The Persian Gulf War reportedly worked to increase the sales of both EOSAT and SPOT.[17]

In 1987, sales to public and private buyers in the United States accounted for half of the combined data sales of the two companies; United States and European sales accounted for over 80%. These figures, however, exclude the sales of international ground stations and accordingly overstate the U.S. and European shares. The Commerce Department reports that foreign sales, particularly in Asia, are the most rapidly growing part of Landsat's market, increasing by almost 50% in 1989.[18]

The Kodak study broke down the 1988 U.S. market for raw data by major government agency and private-sector buyer, as shown in Table 3. The major customers for civil remote sensing products in 1988 were the extractive industries and the Defense Department, each accounting for about a quarter of the domestic market. The study indicated that the distribution of the private-sector purchasers in the world market is similar to that in the U.S. market. The similarity in the distribution by activities of domestic and international value-added producers buying raw data from EOSAT supports the Kodak conclusion (see Table 4).

During the 1990s, EOSAT's revenues are projected to range between $30 million and $60 million annually, according to projections from EOSAT and the TASC and Kodak studies.[2-4] The forces driving the future remote

Table 2 Revenues of raw data providers, including access fees for foreign ground stations and sales of raw data (In millions of dollars)

Years	Revenues
1973	0.2
1974	0.5
1975	0.9
1976	2.2
1977	2.3
1978	2.8
1979	3.4
1980	3.8
1981	4.5
1982	7.6
1983	12.3
1984	14.5
1985	16.9
1986	21.0
1987	31.0
1988	39.0
1989	47.0

SOURCES: Congressional Budget Office; Office of Technology Assessment; The Analytic Sciences Corporation; National Oceanic and Atmospheric Adminstration; Department of Commerce.

NOTE: For 1973–1982, total revenues are the sum of direct data sales as reported in Office of Technology Assessment, *International Cooperation and Competition in Civilian Space Activity*, July 1985, pp. 298–299, and foreign ground station access fees from various sources. Before 1976 no access fees were charged. For 1976–1978 the Congressional Budget Office estimated access fees based on the number of commissioned foreign ground stations and the annual access fee of $200,000. For 1979–1982, foreign ground station access fees are as presented in The Analytic Sciences Corporation (TASC), *A Study of an Advanced Civil Earth Remote Sensing System*, April 1988, p. 2–40. For 1983–1985, total revenues are provided by the National Oceanic and Atmospheric Administration. For 1986–1990, total revenues are the sum of EOSAT and SPOT Image revenues as presented in Department of Commerce, *Space Business Indicators*, June 1990, pp. 16 and 35.

sensing market are improvements in sensor coverage and in products, with the value-added sector accounting for an influx of new customers. A continued demand for EOSAT products may depend on an assurance that Landsat will continue, since customers may not be ready to make the investment necessary to use data unless they are reasonably certain of its future supply.

Two recent developments indicate that the government is likely to remain a prominent customer for land remote sensing data. First, in 1990 the U.S. Global Change Research Program was initiated. During the 1990s, it will require considerable amounts of Landsat and SPOT data in order to address questions about ecological systems and their dynamics.[19] A second source of public demand not fully reflected in the market projections above comes from recent U.S. military involvement in the Middle East. While the reported record sales for EOSAT in August 1990 probably represent only a

Table 3 Sales of raw data from land remote sensing in U.S. markets, 1988

Market	Sales Millions of dollars	Percent
Private sector		
Extraction	2.7	26
Mapping	0.4	4
Forestry	0.4	4
Agriculture	0.2	2
Engineering	0.2	2
Other	0.4	4
Public Sector		
Department of Defense	2.6	25
Department of Agriculture	1.7	17
Agency for International Development	0.4	4
Department of the Interior	0.5	5
State and local governments	0.4	3
Academic	0.4	4
Nonprofit Organizations	0.1	1
Total	10.2	100

SOURCE: Kodak Remote Sensing, *Study for an Advanced Civil Remote Sensing System*, Department of Commerce, August 1988, Vol. 2, pp. 1–11.

Table 4 Services provided by domestic and international value-added firms purchasing data from EOSAT (in percentages of all firms purchasing data)

Service	Domestic[a]	International[b]
Petroleum and mineral exploration	56	52
Civil engineering and land use	73	65
Agricultural assessment	59	54
Forestry	64	53
Coastal studies	51	46
Oceanography	37	33
Hydrology	59	51
Engineering	33	40
Cartography	53	55
Media	19	11

SOURCE: Congressional Budget Office from EOSAT.
[a]Based on a total of 70 firms.
[b]Based on a total of 209 foreign firms.

temporary increase in demand, they raise the question of whether the full social value of Landsat data is reflected in the prices paid for the data. It is difficult to assign a value to data that enable more accurate maps to be made, possibly saving lives in time of war.[17]

Market Characteristics: Historical Evidence

The Landsat experience demonstrates that the market for raw data resembles the markets for other goods in that the quantity of data demanded will increase if prices are reduced and decrease if prices are increased. For example, the announcement in 1981, 1983, and 1985 that the prices of digital products were going to be increased led to roughly a doubling in the quantity demanded of these products in the months preceding the increases, and to a drop in sales to very low levels immediately following the increases.[20]

These aggregate sales figures conceal differences in the responses of buyers in different segments of the market. In general, buyers in the public sector were less responsive to price increases than buyers in the private sector. For example, when the price of Landsat products tripled in 1983, the value of government purchases increased more than fivefold, while the value of private purchases fell by a third.[14] That the private sector is more sensitive to price changes means that price increases will not serve to increase revenues sufficiently to cover costs. Also, the greater willingness of the public sector to pay for data indicates that price discrimination offers a solution to the problem of declining costs. Under that remedy, the producer covers costs by charging more willing buyers a higher price and less willing buyers a lower price—as low as the additional cost of producing the last unit of output. When all consumers pay the same price, some will benefit because they would have been willing to pay more. In a declining-cost industry, price discrimination allows the producer to cover costs by capturing this difference. While EOSAT is not allowed to discriminate in its price setting, the federal subsidies it receives, when added to the prices it receives for data, approximate price discrimination, with the federal customer paying a higher price and the private customer a lower price.

The experience with Landsat and SPOT indicates that the introduction of new or improved products can increase market demand. For example, from 1979 through 1984 the sales of new digital products increased each year, climbing from $750 thousand to $3.3 million. The point is reinforced by the jump in digital product sales following the introduction in 1984 of the new thematic mapper data.[20] The ability of SPOT to penetrate the U.S. market so rapidly also illustrates the appetite of some current buyers for any and all data and the expansion of the market likely to follow the introduction of new data in any form.

Finally, the history of the market for remote sensing data, with the recent entry of SPOT, demonstrates that supply creates demand—at least in the public sector where the government is both producer and consumer. This point is illustrated by the increase in European purchases of raw data that accompanied the entry of SPOT into the market. In 1984, the European sales of the Landsat ground station were just over $1 million, while by

1987, after SPOT's first full year of operation, the data sales of SPOT by itself were almost five times greater.[16,21] Europe having a system of its own, supported by direct capital subsidies, must explain a large part of this increase. The presence of government on both the supply and demand sides of the market suggests that when new public systems enter the arena sales will increase, but that these sales will be proportionately larger in new countries entering the market as governments seek to support their own national systems.

Policy Options

What roles should government and private investors play in the area of land remote sensing? Before addressing this question, it is necessary to decide whether the government ought to fund new Landsat satellites. That decision in turn depends on the answer to another question: since the data obtained by Landsat cost more to produce than EOSAT can sell them for in the market, do the data have sufficient social value to justify a subsidy of Landsat? If the answer to this question is negative, and no additional funding is provided for Landsat satellites, then the issue becomes whether the government has any role at all other than as a purchaser of raw data. If not, an option would be to deregulate the industry—by permitting the private-sector producers of raw data to provide their services on an exclusive, discriminatory basis.

If, alternatively, the benefits of Landsat are seen to exceed its cost, and the Congress decides to continue public support, the issue of private investment hinges on whether a private provider is best able to deliver to society the benefits of land remote sensing. If so, a private role is possible under a variety of circumstances: the current arrangement in which EOSAT negotiates capital contributions with the government, an open auction of distribution rights among private firms, or participation by private investors in an international consortium.

End Federal Support for Landsat?

The major uncertainty in ending federal support for Landsat is whether EOSAT, left to itself, could go forward with a new satellite in the mid-1990s. The consensus is that it could not. Nevertheless, steps could be taken to make investing in a new satellite easier for EOSAT or for new entrants from the private sector.

The government could encourage producers of raw data in several ways. It could choose to abandon the principle of nondiscrimination by allowing producers to offer exclusive rights to data and to charge their customers higher prices for those rights. Exclusivity could be limited in time, so as to permit eventual full scientific use of all data. Equal access could be maintained by requiring an open auction procedure for those parties interested in exclusive rights. EOSAT's revenues could also be increased by permitting it to adopt SPOT Image's policy of negotiated annual access fees for international receiving stations.

A second step would be to include weather satellites in the area of commercialization, thus allowing private producers to offer a broader array of revenue-generating products. Economies of scope and scale would result. The costs of operating satellites could also be lowered if common platforms could be used for some weather and land remote sensing operations. The scope of private activities could also be broadened by permitting private producers to bid for the commercial rights to distribute the raw data expected to be produced by the Earth Observation System sometime in the late 1990s.

The weight of available evidence suggests, however, that commercial providers operating on the Landsat scale are unlikely to be successful, at least in the next decade. Without additional federal funds to build satellites beyond Landsat 6, EOSAT could fail, leaving the federal government the task of establishing U.S. policy without an operating Landsat system.

The major advantage of ending federal support for Landsat would be the savings to the federal budget from not buying new satellites—ranging from $0.5 billion to $1.4 billion through the 1990s. The cost to the public sector of ending federal support would be the loss of the data currently purchased and used for various purposes, among them national security. A part of this gap could be closed by substitutes—aerial photography, SPOT data, and perhaps eventually data from new private sources. Private buyers would have the same alternative sources of data open to them, but could suffer financially. Value-added firms that have developed products requiring Landsat would suffer the most. However, since the decision not to support new satellites would be made far in advance of the actual loss of data, all buyers—both public and private—would have adequate time to adjust.

Maintaining an environment conducive to new entrants—likely of the lightsat scale—would be easier if Landsat were out of the picture, since parts of the market it formerly occupied would be open. Currently, the U.S. government is supporting the development of lightsat technology that may have commercial applications in remote sensing. More aggressive support—free launches, for example—would not be consistent, however, with a policy of ending federal subsidies. The most important posture of the federal government toward new entrants, with or without Landsat, would be as a customer.

Commit to Long-Term Public Support

The federal government could choose to support civilian land remote sensing in a variety of ways. Among the options most discussed are:

1) Emphasizing private investment in the marketing and distribution of Landsat data, with the objective of providing currently available data at the lowest cost to the government. EOSAT, or a different firm, would conduct these activities on a profit-making basis. The capital cost of the system could be shared between the public sector and the private sector, with the private contribution determined by negotiation with EOSAT, as occurred in 1987, or by an auction open to all potential system operators. A federal agency would oversee the process.

2) A return to the pre-1984 arrangement, with a remote sensing system financed and operated by the government.

3) Creation of an international consortium. This option would require SPOT, Landsat, and other land remote sensing systems to merge their physical assets and data streams. Private investors could operate the combined system and distribute its product.

Each of these options would require new spending to support the space segment of civilian remote sensing systems during the 1990s.

Costs of Options

The costs of the options through the rest of the decade would differ in several dimensions. The option to maintain a private operator of the Landsat system would reduce satellite costs for two reasons: first, because the private operator of the system would presumably carry as much as a third of the cost of new satellites; and second, because the new satellites would be duplicates of current satellites, reflecting the orientation of this option toward the current market for raw data as opposed to a broader and more expensive focus that would include developing and operating research and development sensors. The cost of returning to the public sector would be higher, since an effort would be made to raise the state of the art rather than merely to serve the current market. The cost of the international consortium option could be half that of returning to the public sector, because of international cost sharing.

Estimates of the federal spending required under each option are shown in Table 5. The estimates do not include offsetting revenues from data sales and international ground station access fees, or federal spending on data purchases and research and development. Part of the offsetting rev-

Table 5 Estimated spending for remote sensing options, 1992–2000 (in millions of 1990 dollars)

Option	Space segments[a]	Operations	Total
Commercialize operation and distribution	530–800	0	530–800
Return to the public sector	1230	150[b]	1380
Create an international consortium	615	100[b]	715

SOURCE: Congressional Budget Office.

[a]Space segment costs include those for two additional Landsat-scale spacecraft, sensors, launches, and ground investment. Estimates are taken from a range of configurations in Earth Observation Satellite Company, *Landsat 7: A Comprehensive Study*, January 1989, p. 15. The estimate for the option to commercialize operations and distribution is EOSAT's preferred configuration. The higher estimate for a return to the public sector is based on a configuration that includes research sensors. The international consortium estimate is set at half of the public-sector option estimate to illustrate the potential savings to the U.S. government of international cost sharing.

[b]Assumes continued support for Landsats 4 and 5, and $20 million annual operating costs beginning in 1996. Excludes revenues from private data purchases and international access fees that could offset a substantial part of all of these costs.

enues from data sales would in effect be a transfer from one part of the federal government to another, since the government would probably maintain its position as the most significant customer for remote sensing data.

Other Issues

Each of the three options would carry a number of advantages and disadvantages aside from its effect on federal outlays. They would all share the advantage of providing the data currently available through Landsat to its many users. They would have the common disadvantage of deterring new private investment in remote sensing by providing federal support to a specific competing system.

Emphasizing Private Investment

Maintaining the private sector as an operator of Landsat would probably permit the current assortment of land remote sensing data to be produced at the least cost to society. EOSAT's track record supports the belief that private enterprise, with its freedom to make new cost-saving investment, can lower production cost. Moreover, if satellite costs were shared between the federal government and the private operator, and the federal contribution was limited to a fixed amount, the government could avoid the risk of overruns in satellite construction—as it has in the current arrangement to build Landsat 6. A private contribution to capital cost would have to be reflected in higher prices paid by users for data, which would doubtless limit the demand for data. In the long run, however, keeping private investment in remote sensing alive would retain the possibility of commercializing the entire system, or of broadening the scope of private investment to include the distribution of data produced by other systems.

The drawbacks of this option arise from the same elements as its strengths. New research and development sensors would not be emphasized. A private operator might have more difficulty than a public operator in reaching an accommodation with the public operators of future systems like EOS, thus establishing a potential conflict between the public and private sectors. Finally, privatizing the operation of future Landsats would not solve some of the current problems that the system is perceived to have. For example, data would not be provided to researchers without charge. The solution to this problem would be to treat data as a cost of research and include this cost in the funding of research projects.

Should EOSAT be retained as an operator of the system under this option? Any change from the current organization of production would introduce new stress in a system that has already been transferred from one federal operator to another and then to EOSAT. A new operator would essentially start from point zero yet again. Moreover, EOSAT maintains that the vision of commercialization embodied in its 1985 agreement with the government—a vision including two satellites, Landsat 7 and Landsat 6—is still viable.

One drawback to maintaining EOSAT as an operator, particularly if the arrangement was worked out through negotiation rather than through an open solicitation or auction, would be that operating a Landsat system with the federal government carrying part of the capital cost might prove very profitable. An open auction for the rights to distribute data would be one way to limit profits, although EOSAT would certainly enjoy an advantage in such a competition. Another alternative would be direct negotiations between the government and EOSAT, although this approach would be sensitive to charges that the government had been a weak bargainer. EOSAT's status as a wholly owned subsidiary of the likely manufacturers of a new satellite—Hughes and General Electric—would add a further complication. One solution might be to apply some of the methods used in public utility regulation to limit prices and profits.

Returning Landsat to the Public Sector

Administration and Congressional proposals are currently calling for returning Landsat to the public sector. Specifically a successor to Landsat 6 would be funded by a joint NASA/Department of Defense program office. A return to public operation would be consistent with the belief that the net benefits to society are greatest when available data are as widely distributed as possible, and when new sensors and new types of data are being aggressively developed. The aim would be to maximize social welfare by maximizing benefits, including the unpriced public benefits of new applications that might or might not have potential in the private market. Satellites would carry not only sensors for which there was an established or anticipated market, but also experimental instruments.

Public operation would also be consistent with the declining-cost characteristics of the current industry, and with the legal prohibition on price discrimination. Data could be priced at the marginal or additive cost of production, an efficient solution from society's point of view. The gap between revenues and costs would be closed by subsidies, a practice that is justified when the total value of benefits exceeds their cost of production. A complication in this solution is the apparent ability of the private sector to lower the cost of processing and distribution, leaving it an open question whether public additive cost would be greater or less than the price charged for data by a private operator, even if this price reflected a contribution to the fixed costs of the system. The strategy of returning Landsat to the public sector would emphasize the benefits to be gained from a wide distribution of currently produced data, as well as from research and development and the production of new data, and would rely on the federal operator to adopt additive costs as a basis for pricing. Pressure within the government would possibly tend to push up prices in order to recover costs. Such pressure occurred in setting prices for the private use of NASA's Space Shuttle, when the price to private users was driven above the level of marginal costs.[22]

A fully public provider could choose sensors less suited to commercial market demand, and thus limit the growth of the private value-added sector. But consumers of raw data—among them value-added firms—would receive the full value of the subsidy. To the extent that the value-added market is competitive, the subsidy would be passed through to the ultimate consumer of information. Certain equity problems would remain, however, because among the principal private-sector beneficiaries would be large, profitable private firms—for example, petroleum companies. The obvious solution of raising prices to these users would not be permissible for a public provider.

A variant of this option, discussed in Appendix B, would be for the government to own and operate the space and ground systems, but to set up a private sector marketing and sales organization. This organization would be required to make minimum annual guaranteed payments to the government, plus a royalty on sales. Government would share in any upside potential, and because private capital was at risk the organization would have an incentive to market remote sensing data aggressively. This type of arrangement is also consistent with the international consortia option discussed in the following section. As with the first option, because private-sector participation is maintained, this variant of returning Landsat to the public sector would keep open the possibility of full commercialization at a later date.

Creating an International Consortium

This option would require pooling and distributing data from the existing independent systems of the United States, Europe, Japan, Canada, Russia, and other interested countries. By the mid-1990s, as existing spacecraft reached the end of their lives, new systems would have to be launched. A consortium would define, procure, and be responsible for the launch of these systems. Membership would almost certainly extend beyond countries with spacecraft in orbit to those with ground stations, and even beyond them to countries without any investment in remote sensing. Forming this type of consortium would require international agreement on national contributions, pricing policy, geographic marketing rights, revenue sharing, contracting procedures, and the attributes of future satellites.

On the assumption that a set of terms was available that could be accepted by all parties, creating an international consortium would also require a complex transition from the current system to the new arrangement. Revenues in excess of costs would probably not be a lubricant smoothing the transition, since the creation of a consortium would not in itself create demand sufficient to cover the full cost of a remote sensing system on the scale of Landsat or SPOT (unlike the experience of the Intelsat international communications organization). Private investment in general, and EOSAT in particular, could play a role in a consortium.

A consortium would offer two principal advantages. Costs could be shared among the international partners, avoiding the duplication of facilities that is likely to occur without an international agreement. However, if satellite procurement and design reflected political rather than technical requirements, these savings would be decreased. An international consortium could also deal with the problem of decreasing costs by adopting a policy of price discrimination, without running afoul of the access problem. The broad international membership of a consortium would mitigate the suspicion that technically advanced rich nations and their corporations might take advantage of less advanced poor nations. To maximize social benefits, a consortium could sell data to poorer countries at prices as low as the additive cost of producing data, and compensate through higher revenues from sales to the private sector and to governments willing to pay higher prices.

A disadvantage of an international consortium for the United States would be the loss of control over data important for national security. Another disadvantage is the possibility that such a consortium might impose limits on competition and private investment at some time in the future.

Appendix A: Remote Sensing Products and Technology

In its essentials the information provided by satellite remote sensing is simple. A photograph from space allows the object on the ground to be identified because the observer knows what it looks like. Similar to a visible light photograph, an infrared image of an agricultural area allows an observer to estimate yield because the reflective properties of a good crop or a bad crop are known. Remote sensing data can be converted into useful information in a variety of areas. In agriculture, satellite data can be used to identify crop types, acreage planted, and the health of specific crops, and to monitor soil moisture. The timber industry uses remote sensing data to discriminate among the types of timber, to assess timber quality, and to evaluate the damage caused by forest fires. The mineral and extractive industries have found satellite data useful in surveying large areas of land for topological signs of mineral and petroleum deposits. Land use and water resource applications of remote sensing information include mapping, regional and urban planning, commercial location, and monitoring of coastal areas and flood plains. Satellite data can also be used to monitor both air and water pollution. Finally, although commercial remote sensing is usually described as a civilian enterprise, the large sales of remote sensing data to the military indicate significant military applications.

Raw Data

In its earliest use, remote sensing was synonymous with satellite photography. Aerial photography remains a substitute for some of these applications of satellite remote sensing data. Satellite data are sold to customers in their most basic form either as photographic images or in a digital computer-readable format on floppy disks or tapes. Data that are processed to this extent only, are referred to as unenhanced or raw data. The quantity unit of measurement for raw data is a scene that may be purchased as a whole or in parts. Scenes that provide greater detail are more expensive than less detailed scenes. Digital-format products are more expensive than photographic images but are potentially more valuable. The geographic size of a scene depends on the characteristics of the remote sensing system. Landsat covers an area of about 110 miles wide in a scene and SPOT considerably less. SPOT is capable of providing more detailed scenes, however.

The sensors a satellite carries determine the information it gathers. A broad range of options are open across the electromagnetic spectrum from ultraviolet through visible light up to infrared frequencies. The choices made in this dimension, referred to as "special resolution," determine which applications and markets will find the data gathered useful. For example, construction, engineering, and mapping applications require only the visible light portions of the spectrum. Agricultural forecasters require infrared readings as well as visible light. Energy and mineral prospectors prefer even broader infrared coverage. The Landsat series of satellites has consistently expanded its spectral range. Currently, the thematic mapper sensor aboard Landsat 5, and to be carried on Landsat 6, offers the broadest spectral resolution commercially available with its capability to read seven different spectral bands stretching from visible light into the infrared frequencies.

The level of detail discernible in a remote sensing image defines a system's "spatial resolution." As with spectral resolution, different applications require different levels of spatial resolution. Engineering, construction, and intelligence community users prefer fine resolution, 10 m by the current industry standard. For other applications, such as forest yield or crop forecasting, less fine resolution in the range of 30 to 80 m is desired. While the most advanced military satellites reportedly can produce a visual image of an object only inches in diameter, SPOT's 10-m resolution is the finest commercially available.[24] Coarser resolution of 250 to 500 m is necessary for large-scale environmental surveys. Finer spectral resolution sensors capture a narrower area of territory, or swath, as a satellite passes over the ground. Thus, there is a tradeoff between the size of the scene and spatial resolution.

Remote sensing products can also be defined in two time dimensions: the time it takes to acquire, process, and deliver new data to a customer, and the length of the historical record of usable data a system provides. The Landsat system orbiting the Earth's North and South Poles revisits the same location on the globe each 16 days. SPOT, while in a similar orbit, can provide limited coverage as frequently as every seven days because its sensors are able to take off-angle readings. (Off-angle sensor movements also permit stereoscopic images to be gathered, allowing a better reading of topology.) Some potential markets for remote sensing data, for example, the media, require a turnaround time of hours or days. Currently, once an image is acquired, EOSAT's turnaround time is upward of a week. SPOT claims a slightly quicker turnaround. Public and private planners alike can use time series images of the same area. Landsat's time series goes back to the mid-1970s for its most basic images and to the late 1970s for its more sophisticated thematic mapper data.

Once an image has been captured, it is transmitted to a ground receiving station. Both the Landsat and SPOT systems have a primary ground station and an international network of receiving stations. All ground stations receive data obtained while the satellite is overhead, limited by the ability of the rapidly moving satellite to "see" the ground station as it passes from horizon to horizon. International receiving stations pay EOSAT and SPOT Image for the right to these data and resell them to foreign and domestic buyers. Primary ground stations receive additional data from a wider geographic area. SPOT Image employs an onboard storage recorder that captures data from around the world and transmits it to the primary station when the satellite is overhead. EOSAT uses a relay satellite to transmit data from areas beyond the sight of the primary ground station. Both the SPOT and Landsat systems collect data tapes captured by their international networks and maintain an archive of images from around the world.

Value-Added Services

The data produced by a remote sensing satellite are in many cases an input for subsequent processing and interpretation. The value-added sector improves the information quality of the raw data. At the simplest level, a value-added service is expert interpretation of a photographic image. As computational power has become readily available and less costly, value-added services increasingly involve developing and using software to refine and increase the value of satellite data. Value-added producers, of which there are a large number, have been able to charge their customers a price sufficient to cover the cost of acquiring and transforming the data, and to make an acceptable rate of return. The marketing activities of these firms could be a factor expanding the market for remote sensing data.

Appendix B: Landsat Commercialization as Viewed in 1983*

Various studies made over the past decade have concluded that the potential benefits from a Landsat system might run as high as a billion dollars annually. Why then has commercialization not taken place? Simply put, benefits could be seen but not *profits*.

For a business to achieve a profit requires revenue and a pricing mechanism. Unfortunately, many of the potential benefits from land remote sensing have not been linked by a pricing mechanism; in other words, they are a public good. The value of Landsat data has basically been equated with the cost of collecting data by other means on such things as mineral deposits and crop yield. Thus new or increased markets will only develop as an appreciation grows for the added usefulness and value of Landsat data. If the private sector cannot see net profit in such an operation, whatever its benefits, government ownership and operation or subsidization may be necessary to gain agreed-on societal benefits. The Reagan administration has been trying to get out from under this conclusion. What are its chances?

Landsat-D provides a class of observations for a finite life—possibly three years. When D fails, it will be replaced with D' (not a certainty since launch reliability is not 100%). D' will likely last two to three years. If action is not taken swiftly, it is likely that both D and D' will have failed before a replacement is possible. This series of events, coupled with France's expected competitive SPOT system, and other potential national remote-sensing systems, force the issue for continuing Landsat service in some manner.

---

*This is a report of an article entitled "Civil Remote Sensing Will Likely Need Federal Government Support," authored by Joel S. Greenberg (currently President, Princeton Synergetics, Inc.) which appeared in *Astronautics & Aeronautics*, June 1983, Vol. 21, No. 6, pp. 23–24.

Moreover, certain users, both U.S. and foreign, have come to rely on Landsat information. Service interruption may cause a reverting to pre-Landsat operations or switching to other sources, notably SPOT.

The known market (assuming uninterrupted service) for Landsat products appears to many to be insufficient for commercialization (expected returns do not offset the perceived risk). It is therefore likely that government support will be required to achieve a *goal* of commercialization. Loss of any piece of the market (due to interrupted service) presumably would later have to be made up by additional government support.

The anticipated sequence of events dictates rapid decisions on commercialization or public retention. Yet officials have lacked sufficient information for decisions to be made with respect to commercialization or retention. And the investment and timing involved make it likely that a bad decision could not be reversed for a decade or more. Delays, moreover, increase the likelihood of service interruption. Therefore, informed decisions imply the need for an additional spacecraft to back up Landsat-D'. Work must be started immediately to sustain service.

The inability to be specific about launch-vehicle cost likewise adds uncertainty to any private-sector investment decision. NASA has not established definitive plans for support of the Delta launch vehicle beyond FY86. Commercialization of Delta and other expendable launch vehicles is uncertain. And Shuttle pricing policy for Western Test Range (WTR) launches has not yet been established. Ariane is a possible alternative.

To understand the financial implications of commercialization, it is necessary to consider specific business ventures. At Econ, under Department of Commerce sponsorship, we recently developed a "business system" for providing uninterrupted service using Landsat-D and D' and then, slightly before the demise of D', phasing in new satellites with 80-m and later 30-m sensors producing stereoscopic images. It is emphasized that *many business systems are possible—we examined one* and this did not include value-added businesses. Metsats were also not considered. We evaluated the following options:

• Continued ownership and operation by the federal government (planned phase-out)
• Continued ownership and operation by the federal government (establishment of the necessary budgetary line-items)
• Private ownership and operation of an entity competitively selected
• Phased private ownership (government ownership and operation with private-sector marketing guaranteeing minimum annual payments plus royalty on sales to the government)
• Legislatively chartered but privately owned corporation (government-equity position)

Each option used the same demand forecast and the same schedule of events. Government net cash flow was developed for each. For the private-sector scenarios, financially viable business ventures were developed based upon achieving return on capital and other financial measures deemed necessary to achieve financing. The required rates of return were obtained by use of government subsidies.

As explained, great uncertainty attends this market or demand forecast (due to insufficient understanding of the relationship of demand to information product attributes such as price, resolution, and number and location of spectral bands). Government actions obviously influence demand. This is extremely important, for we find (for the cases considered) *insufficient demand for commercialization to take place without major government involvement through ownership and operation, subsidization, or a combination of both.*

Can the federal government withdraw from the scene? Withdrawal implies a continuing cost either through the acquisition of information products or through benefits foregone. *Withdrawal decreases the likelihood of commercialization.* That would likely increase prices charged by SPOT or other systems (less competition).

Government withdrawal (phase-out) can be the lowest-cost action *if* potential benefits are not significantly larger than costs and the cost of information products does not rise significantly.

Figure B1 summarizes the present value of government cash flow and average annual government cost associated with each of the considered

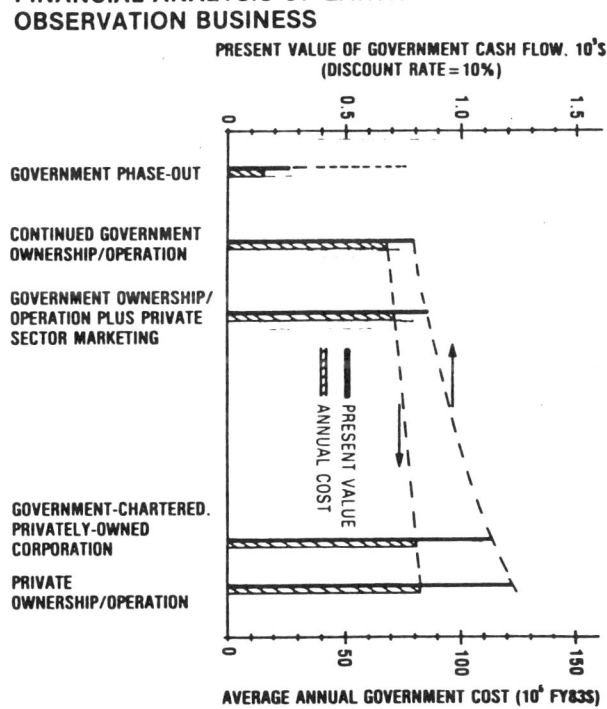

Fig. B1 Financial analysis of earth-observation business.

commercialization or retention scenarios. Cash flow and costs take into account government expenditures such as payments for information products, operations, research and development, subsidy payments, and equity purchases. These may be offset by receipts (from the private sector) in the form of lease payments, asset recoupment payments, TDRSS fees, profit sharing or royalty on sales, generated tax revenue and dividends.

For the business scenario considered, government phase-out can be the lowest cost (from a budgetary point of view, not necessarily from a benefits point of view). But for continuation of Landsat, continuing government ownership and operation produces the lowest cost. *Government costs increase as private-sector involvement increases and government operational involvement decreases.*

If private-sector operations can be conducted more efficiently than similar government operations, this situation might turn around. But I would note that as the private-sector involvement increases so does the significance and complexity of nonfinancial issues. Usually that does not spell efficiency in operations.

Government costs associated with continuation of land remote sensing should vary only slightly with the commercialization or retention scenarios. It is anticipated that government annual costs will average on the order of $70–80 million (FY83 dollars) through, and possibly beyond, 1994. The effect of asset transfer and recoupment policy appears to be minimal. In the private ownership and operation scenario, for example, average annual government cost (constant FY83 dollars) comes to $82 million with recoupment vs $81 million without recoupment. Return on equity and cost of borrowing do not become major factors. Reducing required return on equity and the cost of borrowing by some 20% (from return on equity of 17% and cost of borrowing of 11% to 14% and 9%, respectively) reduces average annual government cost from $82 million to $75 million (constant FY83 dollars) for the private ownership and operation scenario.

These financial results say that a full transfer of a Landsat-type system to the private sector with the expectation of a viable, self-sustaining enterprise would be *premature by a number of years.* They argue either justification for an unusual degree of government support (recognizing the high value of land remote sensing information, the possible lack of pricing mechanisms that reflect this value, and the resulting high risk of transfer to the private sector), *or* need for continuing government leadership in this area, with a limited but possibly gradually increasing private-sector role.

Besides examining government costs, we studied (for the basic business system considered) alternatives to government retention from the perspective of a potential private-sector investor, and draw the following conclusions. *To attain an acceptable rate of return, the alternatives to government retention involve either major subsidy or, if the private sector takes on the marketing function, continued government ownership and operation of the space and ground segments.*

Of private-sector options, government ownerhship and operation appears to be the most attractive choice from the standpoint of financing (size of investment).

In short, should the government choose to phase out Landsat, we do not see financial gain from having the private sector in the U.S. step in *without some form of subsidy* yet provide the same services furnished by the federal government. We do see that requirements for commercial success may include politically fragile government guarantees and policy actions.

Major nonfinancial issues affect international relations, foreign policy, and national security. Earth observation is a particularly sensitive activity internationally because nations take sovereignty seriously. Thus any framework selected for commercialization should help prevent international conflicts and should not compromise U.S. Security or foreign policy. In that light, *any* private system will need fairly close government oversight.

Action in this area should retain means to revise the public/private relationship.

Appendix C: A History of Remote Sensing Policy

During the 1970s, the U.S. public sector developed technology for land remote sensing, operated the satellite systems, explored new uses for the data, and was the primary purchaser of data once uses were developed. The U.S. government pioneered the use of satellite images in agriculture, geology, forestry, planning land use, and mapping, among other applications. (Morain and Thome provide a review of NASA's efforts to promote the use of satellite data during the 1970s and first half of the 1980s.[23]) As a part of its foreign policy the government encouraged the construction of foreign ground stations to receive satellite data, and through foreign aid programs, built ground receiving stations and promoted use of remote sensing data in the cause of economic development. Private petroleum and mineral producers were also among the earliest customers for remote sensing data. These companies paid fees for the data that covered no more than a modest fraction of the cost of operating remote sensing systems and none of its capital costs. The cost were substantial. Before 1984, the government spent $1.5 billion on Landsat, the U.S. civilian remote sensing system.[24]

The situation changed in 1984. The government decided to commercialize the Landsat system. The Land Remote Sensing Commercialization Act legislated this change and outlined a transition plan that included both operating and capital subsidies from the government to the future private owner/operator of the system. The law also opened the door for new private entry into satellite remote sensing and established the Department of Commerce as the agency responsible for developing a procedure to license new entry.[14]

The move to commercialization in the United States was not abrupt, but has proved to be difficult. During the Carter Administration, the government weighed options to commercialize or to privatize the Landsat system. (Reference 25 presents the results of the consideration of Landsat options initiated during the Carter Administration.) By the end of the Carter Administration, the favored option was privatization. A private entity would operate the system and distribute its products, but the federal government would maintain a financial role into the indefinite future. By 1983, responsibility for the Landsat system was transferred from the National Aeronautics and Space Administration to the National Oceanic and Atmospheric Administration, reflecting a change in the public policy perception of the system from a research and experimental system to an operating system.

By 1985, EOSAT had won the competition to commercialize Landsat. EOSAT was to assume responsibility for marketing, pricing, and distributing Landsat data, operating the government's two satellites, Landsats 4 and 5, and building two new Landsat satellites to be launched in 1988 and 1992. Revenues from data sales beginning in 1985 would flow to EOSAT rather than the government. The government was to pay for operating Landsats 4 and 5 (which were expected to fail in one or two years), for two new satellites, and for their launch on the Space Shuttle. A resource commitment of $295 million was anticipated.[26] This commitment was less than the $500 million figure understood to be available by most of the bidders at the time of their submissions and led Eastman Kodak, EOSAT's final competitor, to withdraw from the bidding.[15]

A conspiracy of events has resulted in changes in the major aspects of the original agreement. Disagreements among the line agencies of the federal government, between the Office of Management and Budget (OMB) and federal agencies and between the Administration and Congress led to an uncertain budget outlook for the construction funding of the new satellites. In 1987, this situation became acute, and EOSAT halted production of Landsat 6. The Challenger accident resulted in a change in launch vehicles from the shuttle to an expendable launch vehicle, and a change in the design of the satellite. The corporate ownership of EOSAT changed when General Motors acquired Hughes and General Electric acquired RCA. Finally, both the Landsat 4 and 5 satellites exceeded their expected lives, shortening the gap between the end of their useful lives and the operation of Landsat 6, but creating the question of whether EOSAT or the federal government will pay for their continued operation.

Currently, a 1988 agreement between EOSAT and the federal government is in place. One new satellite will be built instead of two with federal support valued at $250 million, $210 million for satellite construction and a government-provided launch valued at $40 million. In addition, continuing federal support has been provided to EOSAT for the operation of Landsats 4 and 5, roughly $20 million during 1990. For its part, EOSAT continues to invest in ground processing equipment and marketing, and most notably will invest $30 million in the construction of Landsat 6 to cover its current estimated cost of $240 million. (Reference 26 provides a summary of the changes in the Landsat commercialization agreement and the federal support necessary to implement the agreement.)

Appendix D: The Decreasing-Cost Industry

In most industries, producers operate with constant or increasing costs. Market forces establish a price that leads the individual producer to supply just enough of a product so that the additional, or marginal, cost of the last unit produced is equal to the market price. Typically, the cost of producing additional units rises as more are produced, eventually causing the average cost of all units to rise. In an ideal competitive industry, an individual firm produces at the level where the cost of the last unit produced is equal not only to the market price but also to the average cost of all the units produced. This ensures that all costs are covered by revenues and that society's resources are efficiently used.

In a decreasing-cost industry, however, high fixed costs mean that the average cost of production keeps on declining as more units are produced, for all quantities that consumers are likely to demand. Consequently, a single producer can supply the entire market at the lowest cost to society. For this reason, decreasing-cost industries are sometimes referred to as natural monopolies. In the decreasing-cost industry, market prices, marginal costs, and average costs are never equal.

Decreasing-cost industries are usually regulated or subsidized, thus allowing for a variety of solutions to their pricing dilemma. Waterways and dams often charge a low or even a zero price to assure wide use of the product, with taxes making up the difference between costs and revenues. Electric utilities price their power under a variety of regulatory strategies, often charging different rates to different customers for the same product. Through such price discrimination, a second-best solution to the decreasing-cost industry is achieved. The full cost of production is recovered from those customers willing to pay the highest prices. Wide social benefits are secured by charging other customers less, even as little as the marginal cost of production. The logic of economics thus suggests that price discrimination is an effective solution in the production of raw data by remote sensing. The logic of international relations, however, argues against price discrimination. Thus, the Land Remote Sensing Act requires that all parties have access to data on equal terms.

References

[1]Greenberg, J. S., "Commercialization of the Civil Land Remote Sensing System," Statement to the Committee on Science and Technology, U.S. House of Representatives, June 1983.

[2]Kodak Remote Sensing, *Study for an Advanced Civil Earth Remote Sensing System*, Dept. of Commerce, Aug. 1988.

[3]The Analytical Sciences Corporation (TASC), *A Study of an Advanced Civil Earth Remote Sensing System*, Dept. of Commerce, Aug. 1988.

[4]Earth Observation Satellite Company (EOSAT), *Landsat 7: A Comprehensive Study*, Jan. 1989.

[5]Kahn, A. E., *The Economics of Regulation: Principles and Institutions*, Vol 1, John Wiley, New York, 1970.

[6]Chenard, S., "SPOT's Subsidized Success Story, "*Space Markets*, Feb. 1990, p. 102.

[7]Hahn, R. W., and Kroszner, R. S., "Lost in Space: U.S. International Communications Policy," *Regulation*, Summer 1990, pp. 57–65.

[8]Earth Observation Satellite Company, *Images of the Earth from Space: Innovative Applications, Advanced Technologies, and New Markets*, 101:2, Feb. 15, 1990.

[9]U.S. Geological Survey, EROS Data Center, "Summary of Worldwide Landsat Data Sales 1989," prepared for Landsat Data Distribution and Marketing Working Group, Canberra, Australia, May 21, 1990, pp. 1, 3.

[10]EOSAT, Personal communication, Oct. 26, 1990.

[11]McDougall, W. A., *The Heavens and the Earth: A Political History of the Space Age*, Basic Books, New York, 1985, pp. 127–128.

[12]Office of Technology Assessment, *Commercial Newsgathering from Space*, May 1987, p. 49.

[13]Saunders, R., "Spot Draws New Customers with Maps," *Space News*, June 18–24, 1990, p. 18.

[14]Office of Technology Assessment, *International Cooperation and Competition in Civilian Space Activities*, July 1985.

[15]Taranik, J. V., "Landsat Privatization, Commercialization, and the Public Good," *Space Commerce*, Vol. 1, 1990, p. 75.

[16]U.S. Geological Survey, EROS Data Center, "A Summary of Worldwide Landsat Data Sales," Sept. 1985.

[17]"EOSAT Sees High Demand for Gulf Images," *Space News*, Sept. 24–30, 1990, p. 3.

[18]Department of Commerce, *Space Business Indicators*, June 1990, p. iv.

[19]National Research Council, *The U.S. Global Change Research Program: An Assessment of FY 1991 Plans*, National Academy Press, Washington, DC, 1990, p. 78.

[20]U. S. Geological Survey, EROS Data Center, *Annual Report of Landsat Sales for Fiscal Year 1986*, 1987.

[21]Euroconsult, *World Space Industry Survey: Ten Year Outlook*, 1988 ed., Paris, 1988, p. 315.

[22]Congressional Budget Office, *Pricing Options for the Space Shuttle*, May 1985, Chap. iv.

[23]Morain, S. A., and Thome, P. G., *Situation Analysis: A NASA Perspective on Commercial Earth Observation*, Technology Applications Center, Albuquerque, and Center for Space and Advanced Technology, Arlington, Jan. 1989, pp. 16–22.

[24]"Sensing the Earth from Space," Harvard Business School, N9–389, 154, 1989.

[25]Department of Commerce, *Planning for a Civil Operational Land Remote Sensing Satellite System: A Discussion of Issues and Options*, June 20, 1980.

[26]Roher, K. A., and Smith, M. S., "The Future of the Land Remote Sensing Satellite System (Landsat)," CRS Rept. for Congress, April 1989.

Product and Service Pricing: Launch Vehicles

Eric Gabler*
Silver Spring, Maryland 20901

Introduction

OVER the last 12 years, a growing number of domestic and foreign enterprises have entered or expressed interest in entering the international market to provide commercial launch products and services. The supply of launch products and services now exceeds projected demand, and competition among providers is intense. This heightened competition has led to private and governmental concerns about the conduct of international launch competitions, particularly with regard to pricing practices.

This chapter utilizes economic theory and publicly available information to provide an overview of the economics of the commercial space launch industry and its pricing practices. This is a difficult task, particularly because the commercial launch industry reveals little about its economic decision making to the public. Nonetheless, it is hoped that a sufficiently accurate picture of the industry emerges to assist persons interested in understanding and addressing the complex problems facing this industry.

An outline of this chapter will be beneficial to the reader. The first section provides a brief history of the commercial launch industry. The second section describes the competitive environment of the international launch market and the entities that participate in this market. The third section covers the cost structure of the commercial launch industry, highlighting the extraordinary importance of government space launch programs to the competitiveness of commercial launch providers. The fourth section presents a pricing behavior model for the commercial launch in-

---

Copyright © 1992 by Eric Gabler. Published by the American Institute of Aeronautics and Astronautics, Inc. with permission.

*Eric Gabler is a former employee of the Office of Commercial Space Transportation of the U.S. Department of Transportation. All views expressed in this chapter are those of Mr. Gabler, and do not necessarily reflect those of the Office of Commercial Space Transportation.

dustry that is similar to one often used by economists to interpret industries with oligopolistic market structures. This pricing behavior model frames the discussions in the fifth through eighth sections, which explain the various terms and conditions that affect the price and value of a launch service. The ninth section is a note on the role of governments in maintaining free and fair markets for launch products and services. The chapter concludes with a short summary of its major points.

Although launch products (e.g., launch vehicles) and the specific services required to conduct a launch can be treated as separate components, they are most often bundled together for sale as a complete package known as a launch service. It is this service that a commercial customer normally purchases for transportation of satellites into space. Consequently, the emphasis of the analysis in this chapter is on the production and marketing of launch services rather than the separate components that make up such services.

Development of the Industry

The current international market for commercial launch products and services must be understood in terms of the political and economic developments that have fostered the growth of the industry. Although space policy matters are described in more detail elsewhere in this book, the following summary captures the principal events influencing the launch sector.

Government Monopoly

Prior to the 1980s, the U.S. government was essentially the only provider of space launch services to the western world. These services were provided through a fleet of expendable launch vehicles developed from military missiles. With the advent of the Space Shuttle, the official U.S. policy was to use the Shuttle to provide launch services for all U.S. government missions, as well as all interested commercial and foreign customers.

The monopoly position held by the U.S. government ended in December 1979, when the European governments successfully launched their own launch vehicle, the Ariane. Immediately thereafter, Europe initiated an aggressive program to provide commercial launch services to the international market through Arianespace, a French company jointly owned by private and government entities. At the same time, some U.S. companies also became interested in providing commercial space transportation, either by marketing launch services on expendable launch vehicles that were to be phased out of production by the U.S. government in its transition to the Shuttle, or by developing and marketing new vehicles.

The U.S. government sought to accommodate domestic efforts to privatize the government-dominated launch sector. It introduced supportive space policy in 1983, and granted permission to commercialize the existing expendable launch vehicles. In 1984, Congress passed the Commercial Space Launch Act and the President established an office within the Department of Transportation dedicated to commercial space transportation. These policies were successful in encouraging several U.S. entrepreneurial firms to enter the market for providing small-lift and suborbital launch services.

However, these policy initiatives were not sufficient to induce the large U.S. aerospace companies to enter the market for commercial launch services, largely because the U.S. government continued to make the Space Shuttle available to all authorized commercial and governmental, domestic and foreign users. In fact, the U.S. government actively competed for commercial customers. By the mid-1980s, competition between the U.S. government and the European-backed Arianespace to launch commercial and foreign government satellites had driven prices down to levels where private U.S. launch companies could not compete.

Entry of Large U.S. Firms

The Challenger accident in January 1986 forced the U.S. government to reconsider its policy of using the Shuttle to launch commercial payloads. In August 1986, the United States announced that the Shuttle would be dedicated to payloads important to national security and foreign policy, explorations, pioneering, and developing new technologies and uses of space. Commercial payloads then on the Shuttle's manifest were largely removed. These displaced satellites created a major new source of demand for commercial launch providers.

The withdrawal by the U.S. government from the commercial market left Arianespace as the only operational launch service available to commercial customers in the western world. Prices for commercial launch services subsequently increased by a reported 20%.[1] These higher prices created a strong incentive for U.S. companies to enter the commercial launch market.

Other actions by the U.S. government served to encourage market entry by U.S. firms. In January 1987, the U.S. Air Force ordered 20 Delta II vehicles under the Medium Launch Vehicle (MLV) contract from McDonnell Douglas Space Systems. With this order, and a previous order by NASA, McDonnell Douglas was able to restart its Delta production line. General Dynamics Space Systems Division announced in January 1987 that it would build 18 commercial Atlas Centaurs (rechristened the Atlas I) on speculation, and in May 1988, won the Medium Launch Vehicle II (MLV II) contract to provide 10 Atlas II vehicles to the Air Force. Both the Delta II and Atlas vehicle types are well suited to the commercial launch market.

Earlier, in February 1985, the Air Force had ordered 10 Titan IV vehicles from Martin Marietta Astronautics Group under the Complementary Expendable Launch Vehicle (CELV) contract. The Titan IV, which is a much-enhanced version of the Titan 34D (Titan III), was meant to supplement the Shuttle for Air Force mission needs. Although the CELV contract is a conventional military procurement, and the Titan IV has not been marketed as a commercial vehicle, this contract enabled Martin Marietta to keep its Titan production lines open.

By 1987, in response to the incentives of supportive U.S. space policy, higher commercial launch prices, and government orders, the three major U.S. expendable launch vehicle manufacturers—General Dynamics,

McDonnell Douglas, and Martin Marietta—entered the commercial launch market in earnest. The commercial arm of General Dynamics became General Dynamics Commercial Launch Services. The commercial launch operations of McDonnell Douglas eventually became known as McDonnell Douglas Commercial Delta. Martin Marietta adopted the name of Martin Marietta Commercial Titan. The first commercial U.S. launch of a communications satellite was successfully completed by McDonnell Douglas in August 1989.

Entry of Other Launch Firms

Interest in providing launch services to the international market was not limited to U.S. and European launch firms. Beginning in the mid-1980s, the Soviet Union and the People's Republic of China (PRC) attempted to market their own space capabilities. China, through the state-owned trade company China Great Wall Industry Corporation (CGWIC), became particularly active in this area, and has successfully sold launch services to western customers. The Soviet Union, represented principally by the state technical and management organization Glavcosmos, was not as successful, in part due to western technology transfer restrictions. It is expected that organizations of the former Soviet Union will continue to market launch services to the international market.

Japan has invested heavily in developing an all-Japanese-produced launch vehicle, known as the H-II. A recently formed consortium of Japanese companies called the Rocket System Corporation will build the vehicle and is reported to be bidding for commercial launches that will occur in the middle of this decade. Other countries, such as India and Brazil, are not as far along in their launch vehicle development programs, but are said to be interested in commercial programs as well.

Market Environment

Projections of future supply of and demand for commercial launch services are subject to much uncertainty. Future supply of launch services could vary greatly depending on whether or not launch providers from the former Soviet Union enter the commercial market. Projections of demand that are based on current space applications could prove conservative, as the advent of new commercial uses of space-based systems could generate unanticipated customers. The projections below reflect widely reported estimates developed using conservative estimating techniques.

Outlook for Supply and Demand

The existing U.S. and European launch capacity dedicated to the commercial market could launch as many as 25 to 30 medium (Delta-class) and large (Atlas-class) satellites per year. The planned Japanese supply of launch services to the commercial market is uncertain, but could be sufficient to launch four or more commercial satellites per year by 1995. The Chinese launch service, CGWIC, may also have the capacity to launch four or more commercial satellites per year in the 1990s, although China has agreed not to permit its launch providers to launch more than three communications satellites for international customers in any 1 year through 1994, and no more than a total of nine during the 6-year period ending in 1994. Although the status of launch providers of the former Soviet Union is unclear at this time, launch providers in that region could double world commercial launch capacity should they enter the market with their capabilities intact.

The projected number of payloads for launch is less than the supply of launch services. Medium and large geostationary telecommunications satellites, which will probably continue to be the primary source of commercial payloads, will not average much more than 20 per year through 1995, and may decline to fewer than 20 satellites per year in the later years of this decade. In addition, approximately five to seven medium to large scientific and civil remote sensing satellites will be launched by western nations each year on commercial vehicles, although these will be primarily restricted to the national launch providers of the nation or national region funding the payloads. The number of civilian payloads that will be launched by the former Soviet Union is uncertain.

A promising but less predictable payload market is the one for small, low Earth-orbiting, mobile communications satellites. Deployment of these satellites may begin by the middle to late 1990s, and some forecasters predict that as many as 100 or more of these satellites will be launched in the latter half of the decade. However, other estimates are significantly lower, and there is no certainty that this market will materialize. Moreover, many of these satellites would be launched in clusters of multiple units per launch. If so, available launch capacity could readily accommodate this new demand. Thus, it is likely that competition to provide commercial launch services will remain acute.

Military satellites comprise a separate, distinct category of launch demand. In the United States alone, they account for approximately 15 launches of medium and large launch vehicles each year. Military satellites, largely launched by the United States, the former Soviet Union, and China, are usually restricted to militarily-procured launch vehicles. As such, they are not counted as commercial payloads in this analysis.

Launch Providers

Three principal classes of participants characterize the commercial launch services market. These are the launch providers, launch customers, and national governments.

As noted above, the major launch providers currently in the commercial launch products and services market are the following: General Dynamics Commercial Launch Services, McDonnell Douglas Commercial Delta, and Martin Marietta Commercial Titan in the United States; Arianespace for Europe; CGWIC in China; Glavcosmos (or its successors) in the former Soviet Union; and Rocket System Corporation in Japan. Each of these entities offers launch services on vehicles capable of lifting medium and large geosynchronous communications satellites into orbit. Table 1 lists the particular types of vehicles and/or vehicle configurations offered by each

Table 1 Launch companies and their vehicles

Vehicle	Company	Country	Launch site	Lb to LEO[a]	Lb to GTO Dedicated[b]	Lb to GTO Dual[c]	Reliability
Delta II (7925)	McDonnell Douglas	U.S.	Cape Canaveral	11,100	4,000		99%[d]
Atlas I	General Dynamics	U.S.	Cape Canaveral	13,000	5,200		90%[e]
Atlas II	General Dynamics	U.S.	Cape Canaveral	15,000	6,100		One flight
Atlas IIA	General Dynamics	U.S.	Cape Canaveral	15,700	6,400		untested
Atlas IIAS	General Dynamics	U.S.	Cape Canaveral	18,000	8,000		untested
Titan III	Martin Marietta	U.S.	Cape Canaveral	28,000	N/A		84%[f]
Titan III	Martin Marietta	U.S.	Cape Canaveral	N/A	10,200	9,500	—[g]
Ariane 40	Arianespace	Europe	Kourou, French Guiana	10,100	4,200	3,700	—[h]
Ariane 42P	Arianespace	Europe	Kourou, French Guiana	13,200	5,700	5,300	—[h]
Ariane 44P	Arianespace	Europe	Kourou, French Guiana	14,300	6,600	5,700	—[h]
Ariane 42L	Arianespace	Europe	Kourou, French Guiana	15,400	7,100	6,200	—[h]
Ariane 44LP	Arianespace	Europe	Kourou, French Guiana	15,400	8,200	7,300	—[h]
Ariane 44L	Arianespace	Europe	Kourou, French Guiana	15,400	9,500	8,600	92%[i]
Ariane 5	Arianespace	Europe	Kourou, French Guiana	42,000	15,000	13,000	untested[j]
Long March 3	Great Wall Industry Corp.	P.R.C.	Xichang, P.R.C.	11,000	3,100		app. 80–85%
Long March 2e	Great Wall Industry Corp.	P.R.C.	Xichang, P.R.C.	19,400	7,000		1 flight[k]
Long March 3a	Great Wall Industry Corp.	P.R.C.	Xichang, P.R.C.	18,700	5,500		untested[l]
Proton (SL 12/13)	Glavcosmos	Soviet Union	Tyuratam (Baikonur)	44,100	10,100		88%[m]
Zenit	Glavcosmos	Soviet Union	Tyuratam (Baikonur)	33,100	9,900		92%[n]
H-2	Rocket System Corp.	Japan	Tanegashima	20,700	8,800		untested[o]

Sources: Refs. 2 and 3. [a]Capacity to low Earth orbit (LEO). [b]Capacity to geostationary transfer orbit (GTO), dedicated launch. [c]Capacity to GTO, two satellites on one vehicle. [d]71 successes in 72 launches since 1977, various versions of Delta vehicle. [e]18 successes in last 20 attempts, Atlas Centaur versions. [f]Test launch scheduled for 1995. [g]Test partially successful. [h]Scheduled for flight test in 1992. [i]36 successes in 39 commercial flights, all Ariane versions. [j]Scheduled for test flight in 1993. [k]Titan III rate applies. [l]Last 28 launches successful as of October 1991. [m]Past 25 launches as of end of 1988. [n]As of end of 1988. [o]See reliability rate for Ariane 44L. [p]21 successes in last 25 Titan III launches. Intelsat VI mission failure excluded, in that vehicle functioned properly.

firm. Organizations in China and the former Soviet Union may offer additional vehicles to the commercial market in the future.

Dedicated launch services for small payloads (those payloads of less than 2000 lb to low Earth orbit) are offered primarily through entrepreneurial U.S. companies. Of these, Orbital Sciences Corporation is the best known, and, in October 1991, sold a commercial launch on its Pegasus vehicle to Brazil. However, Arianespace can transport small satellites to low Earth orbit on some of its missions, and CGWIC is offering light launch services on a dedicated vehicle. (The need to deploy small mobile communications satellites may lead to future demand for light launch services. However, as of yet, a significant commercial demand for small launch vehicles has not emerged.)

As of September/October 1991, Arianespace was the largest single provider of commercial launch services, with approximately 38 payloads awaiting launch, or 59% of the market. General Dynamics was the second largest provider with 15 payloads on its commercial manifest, followed by McDonnell Douglas with 7 payloads. Martin Marietta had one remaining commercial mission to carry a civil payload and has been largely inactive in the commercial market since 1990. CGWIC had two firm launches scheduled, and Glavcosmos had one commercial mission to launch an Indian civil payload. Actual counts of payloads on commercial launch manifests are subject to wide variations, depending on whether a satellite is classified as commercial or military. The counts listed here refer to nonmilitary payloads, and thus include civilian payloads (e.g., weather satellites) as well as commercial communications satellites.

Many firms sell space transportation products, such as engines or vehicle components, and certain organizations in the former Soviet Union have marketed whole space launch systems such as the Zenit for export. However, most customers for space transportation hardware (whether parts or whole systems) will be launch services firms and not satellite owners and operators. With the exception of purchasing upper stages for satellites, firms owning or operating satellites are not generally interested in obtaining launch hardware and conducting launch activities.

Launch Services Customers

Customers for launch services range from commercial telecommunications firms, as is the case in the United States where telecommunication activities are carried out by firms like Hughes Communications, GTE Contel, and GE Americom; to international not-for-profit cooperatives such as the International Telecommunications Satellite Organization (Intelsat) and the International Maritime Satellite Organization (Inmarsat); to governmental telecommunications organizations such as national postal ministries (PTTs). PTTs may, in turn, behave as commercial buyers on the world market, or as buyers captive to their domestic launch providers. In all, there are currently no more than 50 direct customers of medium to large commercial launch services worldwide, although this number will grow over time.

Owners and operators of satellites may or may not be the launch customers. Traditionally, satellite owners purchased satellites from manufacturers and then arranged for launch services. However, some satellite purchasers now require that satellites be turned over to them by the manufacturer as fully operational, on-orbit satellites. In such cases, it is the satellite manufacturer who must arrange to have the satellites launched.

National Governments

A fundamental reality of the commercial launch market is the role played by national governments in developing and fostering launch industries. The vehicles currently operated in the commercial market were developed by governments, are produced on government-owned and/or -financed production lines, and are launched from government-built and -owned facilities.

Governments develop launch systems principally to provide for national defense and to guarantee national access to space. National governments support commercial space launch industries for any or all of the following reasons: reduction of the cost of civil and/or military access to space through larger, more economic vehicle production runs; generation of more robust national space transportation systems; creation of national prestige; generation of hard currency; improvement of balance of trade; creation of jobs; and improvement of the national technology base. The complex relationship of national governments to their respective launch services is discussed throughout this chapter.

Costs of Production

Were it not for the role of governments in developing launch systems and purchasing launch services, entry by commercial firms into the medium and large-lift launch markets would be impossible. The multibillion dollar investments required to establish launch systems cannot be supported by existing or projected commercial demand.

Development Costs

The cost of designing, developing, and testing a new medium or large lift launch system, including the tooling and equipment needed to build the vehicle and the infrastructure needed to launch it, is the largest cost component associated with a launch system. It is reported in the industry press that it will cost participating European governments approximately $5 billion to produce and test an operational Ariane 5 launch system by 1996. If amortized at an 8% discount rate over a production run of two to four vehicles per year from 1996 through 2000, and eight vehicles per year thereafter for the next 15 years, development costs would amount to approximately $100 million per vehicle launched. The cost of developing a new U.S. launch system may be $10 billion or more. Only in the case of some recently developed small-lift orbital and suborbital vehicles has private capital played a major role in vehicle development.

Governments develop launch systems principally for national space and security needs, not commercial ones. Thus, governments do not attempt to recover the development costs of launch systems from commercial launch companies or their customers. Moreover, any attempt by governments to recover a significant share of development costs through *pro rata* charges to commercial customers could have highly negative consequences to the launch companies and their customers. Were a government to attempt to recover its development costs in isolation from other governments, its national launch companies would be at a severe competitive disadvantage in the commercial market. Alternatively, were all governments to agree to recover development costs, the resulting higher charges to customers of national launch providers could significantly reduce commercial demand for space launch services.

Although governments bear the large fixed costs of developing launch systems, some companies may incur significant costs for the following capital expenditures: expansion of production capabilities beyond those established by the government; replacement of worn-out tooling, equipment and buildings; and improvements to existing vehicles and launch infrastructure. For instance, General Dynamics reports that it has invested more than $400 million in upgrading its commercial and military Atlas programs. However, even these costs remain a relatively small share of the cost of developing a new launch system.

Operating Economies Built into New Vehicles

Launch systems vary with regard to the quality and efficiency engineered into them. In general, higher development costs should yield lower long-run operating costs. Thus, even if an international trade regime were to be established whereby operating subsidies were forbidden and companies were expected to recover fully their recurring costs of production, one company could enjoy significant cost advantages because its government invested more money up front in the development of a more advanced launch system.

It is reported that the objective of the Ariane 5 program is to produce a launch vehicle that will cost no more than the Ariane 44L to build and operate, but will lift approximately 60% more payload to orbit.[4] Similarly, if built, the proposed U.S. national launch system would greatly lower costs per pound to orbit. However, given its later start, the new U.S. system would not become available to the government before the end of this decade. Future availability of this system to commercial customers is uncertain, but would probably not occur before 2005—approximately one decade after the introduction of the Ariane 5. Thus, if the Ariane 5 meets its design objectives, Arianespace would be in a much stronger competitive position relative to U.S. launch companies.

Disadvantages of Dependency on Government Vehicle Development

Although government participation is essential to the development of medium and large launch systems, some disadvantages result to commercial operators from the dominant role of government. The greatest disadvantage is that government-developed vehicles are optimized to meet government mission needs, which do not necessarily coincide with commercial needs.

Governmental policies associated with vehicle development can also serve to decrease competitiveness of national launch companies. The European Space Agency is a multinational agency that must award contracts among suppliers from different countries on the basis of a "fair return" for each, which is proportional to funding provided.[2] Arianespace pursues a similar policy towards its subcontractors and is thus unable to minimize costs by easily changing suppliers.

Economies Attributable to Larger Production Runs

The capacities of individual U.S. and European launch companies are constrained by production and/or launch pad constraints to a dozen or fewer vehicles per year. Nonetheless, the ability to make the fullest use of this capacity is critical to the economics of the launch enterprise. Tables 2 and 3 show that production costs are highly dependent on the size of production runs. Production costs decrease with greater utilization of fixed production facilities due to the more efficient use of variable inputs, such as labor.

The lower costs associated with fuller utilization of fixed production facilities are not "economies of scale" as understood in economics. Economies of scale are the lower average production costs that result when the fixed capacity of the production facility itself is increased. Cost advantages attributable to economies of scale may characterize the launch industry of

Table 2   Government Atlas G/Centaur D-1A cost per flight estimate (1984 costs in millions of 1990 dollars)[a]

	Flights per year			
	2	4	6	8[b]
Booster hardware				
Airframe	$27	$24	$22	$22
Guidance/avionics	$10	$9	$8	$7
Liquid engines				
Atlas	$17	$13	$11	$10
Centaur	$10	$8	$7	$6
Subtotal	$65	$53	$48	$45
Upper stage hardware[c]	—	—	—	—
Payload fairing	$1	$1	$1	$1
Launch operations				
Launch support	$15	$8	$7	$6
Government services	$5	$2	$1	$1
Subtotal	$20	$10	$8	$7
Other government costs	$16	$9	$8	$7
Range operations	$2	$2	$2	$2
Cost per flight	$104	$76	$67	$62

[a]Predecessor to the Atlas I. Dollars adjusted to 1990 using Producer Price Index.
[b]Realistic demand/production/launch rate with military orders included.
[c]Included in booster hardware.

Table 3 Government Delta 3920/PAM cost per flight estimate (1984 costs in millions of 1990 dollars)[a]

	Flights per year			
	2	4	6	8[b]
Booster hardware	$29	$25	$24	$23
Upper stage hardware	$4	$4	$4	$4
Payload fairing	$1	$1	$1	$1
Launch operations				
Launch support	$15	$9	$7	$6
Government services	$5	$2	$1	$1
Subtotal	$20	$11	$8	$7
Other government costs	$3	$3	$3	$3
Range operations	$2	$2	$2	$2
Cost per flight	$60	$48	$43	$41

Source: Ref. 6.
[a]Predecessor to Delta II. Dollars adjusted to 1990 using Producer Price Index.
[b]Realistic demand/production/launch rate with military orders included.

the former Soviet Union (if the industry remains intact). Economies of scale could also be reached if the three major launch systems in the United States were consolidated into one launch system. However, the cost of this consolidation would be a loss of vehicle diversity.[2]

Lower average production costs associated with larger production runs have important implications for company production and marketing strategies. Companies are under pressure to sell enough launches so that they can produce at efficient levels. Moreover, to lower average production costs as much as possible, the launch company has an incentive to commit to large production runs in advance of actual sales. This strategy has been pursued by Arianespace, which ordered 50 vehicles in a single lot from its subcontractors in 1989. Associated with this order, Arianespace has undertaken reforms to its procurement strategies to lower costs, reportedly by 20%.[5] General Dynamics initiated its commercial program with a production run of 18 Atlas vehicles even without firm commitments for launches. Subsequently, General Dynamics expanded this commitment to 60 vehicles.

Role of Government in Attaining Larger Production Runs

Even when commercial demand cannot support large production runs, government demand may enable firms to attain economic levels of production and to compete commercially. It is the practice and/or policy of the governments of all nations with launch capabilities to utilize national firms to the maximum extent feasible for the launch of government payloads. This practice is fully consistent with the national objectives that led to the development of the launch systems used by the national firms—the creation and maintenance of an industry critical to national space and security objectives. Of course, government customers also benefit from larger production runs made possible by commercial sales.

In recent years, the U.S. government has further encouraged the development of the commercial launch industry by purchasing civilian and some military launches on a launch service, rather than vehicle, basis. Through this method, the government makes its requirements analogous to those of the commercial market, permitting the launch companies to standardize their production and marketing procedures.[2]

The presence of government vehicles on production lines benefits launch firms selling to the commercial market in another important way. In cases where a commercial vehicle encounters hardware problems while on the pad, parts from government vehicles under production or in inventory can be substituted. This advantage is frequently associated with U.S. vehicle manufacturers, but applies to any company that provides launch products and services to its national government. Of course, governments may also benefit if parts from a commercial vehicle under production are used to replace defective hardware on a government vehicle on the pad.

Pricing of Commercial Launch Products and Services

For competitive reasons, the launch companies are secretive about their marketing strategies and the prices and terms of launch transactions. Those data that do exist are often anecdotal, incomplete, or incorrect. Consequently, any discussion of pricing practices of the industry is handicapped by a lack of information.

However, economic theory offers important insights into the behavior of firms in other industries that operate in markets similar to that of the commercial launch industry. By using this theory as a framework to interpret the observable actions of commercial launch firms, a potentially useful model of the industry emerges. The economic model proposed here for the commercial launch industry is one that applies to a noncollusive oligopoly where the motivation of the principal participants is profit maximization, but where price offers are constrained by concerns about initiating destructive price competition.

The following paragraphs develop the assumptions and logic of the proposed model. However, before proceeding, it should be noted that other possible economic models of launch industry behavior can be formulated, and that no claim is made that the proposed model is definitive.

Market Characteristics

The international market for launch services most closely fits the economic concept of an oligopoly. As defined in economic theory, an oligopoly is a market situation in which the number of sellers is small enough for the activities of one to affect the activities of others, and for the activities of any or all of the others to affect the first. This contrasts with the situation of firms in a market characterized by a large number of sellers, where no one firm is large enough to significantly affect the market through its pricing activities.

Oligopolistic firms face downward-sloping demand curves for their products. That is, all else being equal, the firm is able to sell more of its product at a lower price than a higher one. However, because each launch firm offers a product (i.e., launch services) that is somewhat different from its competitors with regard to quality, customer service, and other factors, the demand curve facing each firm will be different from the demand curves of other firms. Thus the quantity of the first firm's product that will be demanded by customers at a given price will be more or less than the quantity demanded of the second firm's product at the same price.

Profit Maximization

In classical economic theory, firms are assumed to be motivated by the desire to maximize their profits, or more precisely, maximize the net present values of their cash flows at acceptable levels of risk. Of course, profit-maximizing behavior will not always yield profits. In particular, firms may err in their projections of future costs and/or sales used in pricing their products and services. (This situation may have occurred in the case of General Dynamics, which has written off $400 million in space launch investments over the last 2 years. However, this amount includes investments associated with the MLV II contract with the Air Force. Moreover, the write-offs were taken in conjunction with a wider corporate strategy that reflects pressures facing the entire aerospace industry. The actual profitability of General Dynamics' current commercial launch operations is unknown.) Even if losses occur in the short run, a profit-maximizing firm will continue producing as long as it can at least cover its recurring costs of producing additional units, even if fixed costs are not covered.

It is clearly the goal of privately-owned commercial launch companies such as those in the United States to maximize profits. Although this maximizing behavior is constrained by the cost, production, and performance parameters inherited from government vehicle and infrastructure programs, commercial marketing decisions will remain free of noneconomic considerations. The assumption of profit maximization should also be valid in the case of a firm such as Arianespace, which is jointly-owned by private and government entities. Firms meeting the criteria of private or private/government ownership dominate the current international launch market.

However, the motivation for profit maximization becomes murky in the case of wholly government-owned launch entities, such as CGWIC. In such instances, noneconomic factors associated with government programs and objectives (e.g., hard currency generation) could enter directly into commercial marketing decisions. Without international agreements to constrain possible noneconomic behavior, these situations could lead to market disruption.

Noncollusion

There is not now, nor is there ever likely to be, a general economic theory of oligopoly. To the extent theory exists, it tends to be tailored to the degree of collusion among the oligopolistic firms. Oligopolistic firms in some industries openly collude, as in the case of formal cartels such as OPEC. Monopolistic pricing theory would apply in this latter case. Some oligopolies are characterized by imperfect collusion, in which one firm assumes the responsibility for price leadership of the entire industry. Finally, some industries have noncollusive structures characterized by independent pricing actions by participating firms.

The commercial launch industry is treated here as a noncollusive one, i.e., it is assumed that firms act independently. This assumption is made because there is no documented evidence of formal or tacit arrangements between firms to set prices or market shares in the launch industry, nor have any charges of collusion been made. Rather, launch companies frequently and openly engage in disputes with each other concerning specific competitive practices.

Various pricing theories have been proposed that attempt to capture the effects of individual firm pricing decisions in a noncollusive oligopoly. The principal theory of this type postulates that a firm can expect one type of reaction from its competitors if it lowers the price of its product, and another type of reaction if it raises its price. In particular, if it lowers its price to attract more customers, other firms will lower their prices as well in order to protect their market shares. This response would be particularly likely in a market such as the commercial launch market where competition is intense and the loss of even a few sales can be damaging to production costs and schedules. Thus, the price reduction by one firm can potentially trigger an industry-wide erosion in prices with little gain in sales for any given provider.

On the other hand, if the firm increases its price in an attempt to earn more revenues per unit sold, other providers will not follow suit. By holding their prices at existing levels, the other firms will gain market share at the expense of the firm initiating the price increase. (This pricing constraint would not apply in situations where only one firm is capable of meeting a particular customer's needs. In this case, the firm could charge a premium price without concern for losing the sale.

In effect, a firm in this situation finds itself faced with what is known as a "kinked" demand curve (shown in Fig. 1). The kink occurs at an established, going market price for the product in question (e.g., a Delta-class launch service). As long as firms in the market sell at or near this price, there is comparative stability. However, the flatter part of the curve above the kink describes the large reduction in sales to a firm that would result if it were to increase its price above the going price, whereas the steeper part of the curve below the kink indicates that the firm would realize only a minor gain in sales were it to price below the going level. In either case, it can be shown that the overall effect of price changes will be to lower overall net revenues, either through greatly reduced sales or lower profit margins. (A technical, economic explanation of the kinked demand curve theory is beyond the scope of this discussion, particularly with regard to marginal cost pricing concepts. However, the reader may wish to consult any intermediate economic textbook for a more detailed development of this theory.)

Conformity to prevailing prices can be explicit or implicit. That is, firms may or may not specifically address the issue of possible price retaliation in their daily marketing activities. However, at some level, knowledge of accepted or prevailing price ranges will set the bounds for pricing decisions.

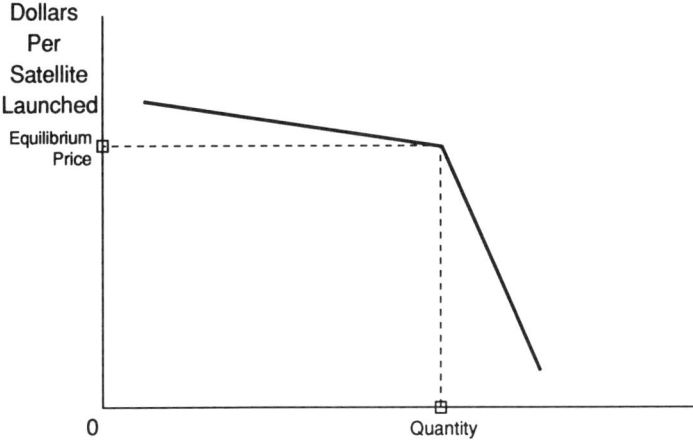

Fig. 1 Kinked demand curve facing firm.

Nonprice Competition

If firms behave in the manner described by the kinked demand curve pricing model, it would be expected that market participants would attempt to avoid competing on the basis of price, and would focus instead on less sensitive "nonprice" competitive issues. Through such measures, a firm will attempt to shift its demand curve to the right, so that more of a product can be sold at the going price. Examples of nonprice competition in the launch industry would include marketing strategies designed to advertise qualitative advantages (e.g., reliability, accuracy) of a particular launch service, the offer of special financing and payment terms, accommodation of special customer needs and scheduling requirements, and provision of other services. Examples of all these practices are discussed in the following sections.

It should not be assumed that nonprice marketing strategies are without cost to launch providers. However, by nature of being nonprice, they are less subject to retaliation by competitors, in part because their values may be difficult to quantify.

Price Stickiness with Changing Production Costs

Firms facing a kinked demand curve will be reluctant to change their prices, even if they experience minor shifts in their costs of production. This tendency for prices to remain steady even through production costs change is referred to as *price stickiness*.

Price stickiness could account for the apparent lack of volatility in the publicly quoted price ranges for commercial launch services since 1988. Despite keen competition in the market and apparent reductions in production costs by some companies due to large block buys and improved production techniques, the reported representative prices for different launch vehicle classes have generally held constant at $45 to $50 million for dedicated Delta-class launches, approximately $60 million for dedicated Atlas I-class launches, approximately $70 to $75 million for Atlas IIA-class launches, and approximately $85 to $90 million for dedicated Atlas IIAS-class launches. Representative prices for dedicated commercial Titan III/Ariane 44L-class launches are not often cited in that very few satellites require these dedicated services. Rather, these vehicles will generally be used to launch two Delta/Atlas class payloads at one time (see below).

Despite periodic press reports of price cutting, reported prices for actual launch competitions seem to support the stability of the above representative price ranges. Reported prices may or may not include the price of reflight insurance, usually 10 to 12% of the base price of the launch. However, only a comparatively small number of actual launch prices are reported, particularly for Atlas-class competitions. Thus, it is impossible to make any definite conclusions as to how well the industry conforms to the pricing behavior predicted by the model. It is possible that the fear of triggering price competition contributes to a reluctance to report any information on winning bids, particularly when these bids are from the lower levels of prevailing price ranges.

Price Competition

This is not to say that price competition does not occur on occasion in the international launch market. Price bids below accepted price ranges are reported to have occurred from time to time without serious repercussions on prevailing market prices, although such deviations may be noted pointedly by competitors. Launch companies apparently make allowances for particular competitive situations or tolerate isolated deviations from established price ranges. In addition, as will be discussed, lower prices will be offered for multiple-launch buys or special launch arrangements.

The most spectacular example of price competition in the international launch market was the widely reported ARABSAT 1C launch competition involving launch companies from the United States, Europe, and China. The ARABSAT 1C is a Delta-class satellite for which launch services normally would have been expected to sell at or near $45 million. However, according to press reports citing European and Chinese officials, the European and Chinese launch companies ultimately bid the price down to $35 and $25 million, respectively. This acrimonious and drawn-out battle eventually went to the Europeans and appears to have been settled on issues other than price.

The motivations of the European and Chinese participants in the ARABSAT 1C competition are unknown. In other industries, aggressive price reactions are often triggered by perceived encroachment on a competitor's established customer base or market share. In any case, the vehemence of the price competition (particularly by the European firm) may have served to warn other companies of the potential consequences of price competition that falls below prevailing ranges.

Collapse of Market Equilibrium

The market equilibrium price will tend to hold (or change slowly) given industry-wide adjustments to production costs, demand, and competition.

However, at some point, pressures may build that cause the price structure to collapse.

The kinked demand curve model offers little insight into the consequences of major disruptions of the market price equilibrium. However, it is likely that the resultant price erosion would cause some of the currently established commercial providers to incur major losses and leave the commercial market (unless extensive government aid was forthcoming). Assuming that more than one competitor remained after the shakeout, it is likely that a new going price would be established, although not necessarily under a noncollusive market framework. However, where the new price would be established is uncertain.

Such a collapse could be triggered by the emergence of a major new, low-price launch provider, or by a sharp reduction in production costs due to a technological breakthrough. It is also possible that a launch provider, with support from its government, might attempt to initiate a large-scale price war leading to an industry shakeout. Following the shakeout, the supported company could re-establish higher prices around a larger or dominant market share.

Establishment of Going Prices

One frequent economic criticism of the kinked demand curve theory is that it does not explain how going market prices become established in the first place. However, Tables 2 and 3 offer a possible explanation for the origin of the prices generally cited for the launch industry. Data in these tables show costs (actual and estimated, inflated to 1990 U.S. dollars) of the major classes of launch vehicles to the U.S. government during the period prior to their commercialization. Although the government cost categories are not directly comparable to those for commercial operations, the similarity of these cost data for medium- to large-volume production runs to cited prices for Delta- and Atlas-class launch services is notable.

This similarity raises the possibility that current going prices for the market have been influenced by known, historic U.S. government costs for its expendable launch vehicle services. U.S. government expendable launch vehicles were used for selected launches of commercial and foreign-government satellites as late as 1987. These historic cost structures could reflect current cost structures of the commercial companies, but probably overstate costs. However, prices tied to these costs may represent levels that were initially tolerable to all launch providers from the point of view of profits.

Behavior of Customers

Given the modest numbers of buyers of launch services, it is possible that imperfect competition may exist on the buying side of the market as well. That is, some organizations may buy so many launches that their purchasing decisions may drive up the price of launches for all buyers. However, given the present and projected oversupply of launch services, it is unlikely that this effect, if present, would be pronounced.

Special Considerations Affecting Price

Under special circumstances, launch providers will offer prices that are at or occasionally below the lower ends of standard price ranges. These circumstances appear to be understood and accepted by other firms. Situations where lower pricing is encountered are described immediately following.

Multiple-Launch Contracts

Launch companies prefer multiple-launch contracts over single-launch contracts. The average costs to the company for providing launch services to a series of satellites are less than for an individual satellite due to reduced marketing expenses, the spreading of vehicle–satellite mating and mission planning costs, efficiencies from experience with the satellite, and other factors tied to volume economies. In addition, multiple-launch contracts reduce risk to the launch company by facilitating planning and guaranteeing revenues of future missions. Consequently, launch companies will offer prices from the lower portions of their standard price ranges for multiple-launch contracts. Similar considerations will cause them to offer better prices to loyal customers.

Dual Launching

Launch firms are able to lower costs by using larger, more efficient vehicles to lift two or more satellites at a time. (Larger launch vehicles, while costing more per launch, are more cost efficient than small vehicles when measured on a per-pound-to-orbit basis. Whereas price-per-pound to low Earth orbit may equal $9000 for a small launch vehicle such as the Pegasus, it is only $4500 for a Delta II.) However, over the range of medium- to large-lift vehicles, some or most of the efficiency advantages of "dual" launching are lost due to the weight of the special launch structure used to mount the satellites to the vehicle. A greater incentive to dual launch probably stems from the ability to launch more satellites from a fixed amount of pad capacity. Arianespace is the principal company associated with this practice, although Martin Marietta also makes use of dual launching.

However, sharing a ride with another satellite imposes certain restrictions on launch customers that can mitigate the attractiveness of a dual launch at normal prices. Special orbital requirements cannot be accommodated in a dual launch. More significantly, the satellite owner is subject to launch delays due to problems with the other payload. Consequently, the launch provider may offer prices somewhat below those that would be offered for a dedicated launch service, particularly for a satellite that is needed to fill out a dual mission.

"Introductory" Pricing

A company offering a new launch product or service may believe that a low "introductory" or promotional price is necessary to attract customers who might otherwise be reluctant to rely on unproven technology. The practice of introductory pricing is specifically addressed in a special launch agreement signed by the United States and China (see below). However, with regard to this agreement, use of introductory pricing is viewed to be appropriate only for the first, and in extraordinary cases, the second, successful commercial launch by a new vehicle.

Terms Affecting Prices of Launch Services

The price quoted for a launch service must be considered in light of the payment and financing terms attached to it. Terms can be used to modify the reported price of a launch by several million dollars. By such means, a launch provider can attempt to gain market share through lower effective prices without actually quoting lower prices.

Payment Schedules

Customers pay for launch products and services according to schedules and terms that vary among launch providers. In general, U.S. and European launch firms prefer that customers begin payments for launch services at the time of contract signing, followed by periodic installment payments to be completed by the time of launch. Payment periods usually range from 30 to 36 months—which is the preferred time that it takes to prepare and launch a mission.

If necessary, launch companies will accommodate shorter payment periods or special financing needs of their customers. Some missions have been undertaken and completed in little more than a year. In cases where special payment schedules must be accommodated, the U.S. and European launch companies report that they require larger initial payments and larger installments to compensate for the shorter time periods. In effect, the company will structure a nonstandard payment schedule so that the present value of payments that it receives is equal to the present value it would have received under a normal payment schedule. Were such adjustments not made (and in some cases they may not be), the effective price to the customer would be reduced significantly.

Also in the way of nonstandard payment schedules, the industry press reports that at least one company (Arianespace) is willing to consider receiving some payments from commercial customers after the successful completion of a launch.[1] This policy represents a form of incentive payment for reliable launch performance. Given competitive pressures facing launch companies, it is likely that other companies will be pressured to match such terms.

Form of Payment

Payments for launch goods and services are almost always made in cash, usually in U.S. dollars. Currency fluctuations can work for or against launch companies. Arianespace maintains that the strength of the dollar and lower European interest rates enabled it to compete against the Shuttle.[4] The weaker dollar would now seem to favor U.S. launch companies, although Arianespace continues to enjoy more marketing success.

In exceptional cases, launch companies are receptive to receiving some percentage of payment in the form of barter or "offset" goods. For instance, lumber has been a proposed form of payment in each of the last two competitions for the Indonesian PALAPA satellites, although the contracts seem to have ultimately been awarded based on cash payments. Valuation of the bartered goods must consider the costs of transportation, handling, marketing, and other administration necessary to liquidate them. Failure to do so could constitute a form of hidden price concession. These transaction costs are especially high for a privately-owned launch company, particularly if the goods in question are outside of its normal business enterprises. Launch providers with close links to government-owned or -controlled enterprises have an advantage in this area, as state enterprises may be more willing to cooperate in selling the received goods.

Customers may also require that the selected launch provider offer products and services not related to launch services. Offers of this type, if uncompensated by the customer, effectively reduce the price of a launch.

Equity Shares

In order to gain a launch contract, a launch provider may assume a limited equity share in the satellite company. Such participation clearly helps to defray the cost of the launch to the customer. In turn, the launch company may benefit in the future from the proceeds of its investment. This situation occurred recently in a widely reported case involving General Dynamics, in which it won a contract for two launches from a company in which it purchased an 8% ownership position. Arianespace is reported to be prepared to enter into similar arrangements up to a 10% equity share.[1]

Financing

Financing for launches is usually arranged through commercial lending channels. However, export–import loan guarantees can be arranged through national governments for qualifying customers. Because such guarantees eliminate much of the risk to lending institutions, they can result in significantly lower commercial interest rates for the borrower. Limited direct financing may be offered by launch companies in some cases. Arianespace reportedly is prepared to provide such financing, not to exceed 15% of the contract.[1]

Vehicle Performance and Reliability

The products and/or services offered by different launch providers are not homogenous in quality or capability. Differences exist with regard to lift capability, reliability, accuracy, customer services, scheduling flexibility, and other factors. All of these factors are important to customers, although their relative importances will vary for each customer. In any case, launch providers aggressively advertise those attributes of their services that they regard as superior to other services in the market. Conversely, they will attempt to mitigate their perceived disadvantages through special provisions to customers.

The range of differentiating characteristics among launch services is so broad that it is broken here into two major categories. The first category, vehicle performance and reliability, is covered in this section. The second category, schedules and launch range characteristics, is the subject of the following section.

Lift Capability

The lift capability of the vehicle is, of course, an essential consideration to the customer. Current telecommunications satellites are often too heavy to be carried by Delta-class vehicles. The ability of the vehicle to transport extra stationkeeping fuel for the satellite may also be a consideration (see below).

Incremental improvements of vehicle lift capability will not necessarily be reflected in higher prices for a launch service. For instance, the Delta launch vehicle has improved significantly in lift capability in the last several years, whereas the price charged for a Delta-class launch has remained largely unchanged. This ability to hold to existing price levels probably stems from improved production methods, but the improved lift itself represents a form of nonprice incentive to customers.

Reliability

Reliability is usually measured as the percentage of previous launches in which a vehicle successfully delivered its payload to orbit. All of the world's major launch vehicle systems have reliability rates of 80% or higher, with the Delta launch vehicle standing out with a reliability factor of almost 99% over its last 71 launches (see Table 1).

There is a significant amount of art in the generation of reliability statistics, and it is no surprise that launch companies employ the method that yields the best statistical accuracy for their vehicles. Failures of test vehicles are usually (and understandably) excluded, and different companies may select different time frames for calculating their reliability statistics. In situations where a company believes a launch failure was not the fault of its vehicle, this may also be excluded. Moreover, launch vehicles have evolved significantly over time, and the performances of several generations of a vehicle are frequently averaged into one reliability statistic.

Particularly in the case of new vehicles offered by new market entrants, but also in the case of new vehicles offered by established market participants, there may be little historical data with which to estimate reliability. In these cases, customers and their insurers must consider the degree to which a vehicle has been built from proven components, the demonstrated record of the vehicle under its government development program, and/or the proven competence of the launch company.

Launch companies can help alleviate customer concerns about reliability by allowing the customer to inspect their facilities and the facilities of their major associates or subcontractors. Arianespace has such an arrangement with at least one customer.[7]

Insurance

The degree to which a customer will discount the value of a launch service due to the perceived risk of launch failure is dependent on a variety of factors. Perhaps foremost is the ability to obtain reasonably priced insurance that will compensate the customer for the value of the satellite and the launch in the event of a failure. This insurance typically covers both the safe launch and successful startup of the satellite.

Insurance rates have fluctuated significantly over the last several years in response to the fortunes of the launch industry. Following the failure of several different launch systems in the 1985 to 1987 time frame, rates hit levels exceeding 25% of the insured value. Following a period of successful launches extending into 1990, rates fell to levels of 16 to 18%. Currently rates are reported to be between 15 and 20%.

Insurance for the replacement of the satellite and launch expenses cannot eliminate all risk to the satellite owner. This is particularly true when a satellite must be deployed at a specific time to meet contracted customer requirements and/or a ceiling exists on the amount of insurance that can be purchased. In such cases, the satellite owner may seek the highest launch reliability possible. Alternatively, if the payload is to be deployed into a satellite system possessing sufficient capacity to meet demand for the amount of time it would take to launch a replacement satellite should the first launch fail, then the risk to the customer should be easily manageable in a normal insurance market.

Accuracy of Orbital Insertion

The accuracy with which the launch service can deliver a satellite to its specified orbit is another important factor affecting customer perception of the launch service. This is because accurate orbital insertion by the launch vehicle minimizes the need to use the satellite's own fuel reserves to correct the orbit. Satellites require stationkeeping fuel to maintain position on orbit. Because fuel reserves are frequently the limiting factor on satellite life, use of fuel reserves to reach geostationary orbit can shorten the satellite's operational life by as much as several years.

One launch service may also be able to lift more weight to orbit than alternative launch services. In such cases, the option of using the extra lift capacity to carry more stationkeeping fuel may be an important selling point for the launch service.

Much has been made of the life-on-orbit issue in recent launch competitions. Some naive assessments simply estimate the number of years that a satellite's fuel reserves may be reduced using a particular launch service relative to another, and add to the "price" of that launch service the undiscounted value of foregone transponder revenues that may occur 10 or more years in the future. However, consideration of the effects of accuracy/life-on-orbit in the selection of a launch service is considerably more complex than the above approach would suggest.[8]

In particular, the economic consequences of satellite fuel exhaustion must be evaluated relative to the discounted value of future net revenue streams from satellite operations, the risk of failure of the satellite due to a nonfuel–related component (e.g., the electrical system), and the opportunity costs of extended satellite operation. Continued operation of an inefficient satellite in a valuable geostationary orbital slot could entail substantial opportunity costs, although the early replacement of the satellite with a more efficient one carries its own costs and risks. (A computerized model that could perform such an assessment is currently under development for the Office of Commercial Space Transportation of the U.S. Department of Transportation.)

In some instances, the value of extended lift-on-orbit is more easily determined. Operators of satellites used primarily to relay signals between large Earth stations can initiate special satellite stationkeeping maneuvers at any time while the satellite is on orbit to extend the useful life of dwindling fuel stores, if added life is desired. One such technique is the patented "Comsat maneuver," in which a satellite is allowed to drift north and south of the equator within certain parameters as it is tracked by the Earth stations. This allows fuel sufficient for 1 year of operation under normal stationkeeping conditions to last 5 or more years. The discounted cost of employing this maneuver to extend a satellite's life by 5 years may not exceed a net present value of $2 million. It was reported in 1989 that the Comsat maneuver was to be implemented to extend the lives of the ARABSAT 1A and 1B satellites.[9]

Vehicle Environment

The price of a launch service must also be evaluated with regard to the acceleration and acoustical environment of the vehicle's payload bay. Most commercial vehicles offer safe environments. However, in some cases, the environment of the vehicle could damage the satellite unless special modifications are made to the vehicle, satellite, and/or flight plan. The costs of these modifications and who pays for them will enter into the customer's decision on whether to use the launch service.

Vehicle–Payload Interface Requirements

A satellite must be built to conform to the interface requirements of the type of vehicle that will be used to launch it. If the satellite manufacturer knows sufficiently far in advance which launch service will be used, it can build the satellite to fit that vehicle at a cost that should not vary greatly among vehicle types. However, once the satellite is constructed, a decision by its owner to use a different launch vehicle than the one the satellie was designed for will involve a satellite modification cost of up to several million dollars. This cost must be considered along with price by the customer when selecting a launch service provider.

Schedules and Launch Range Characteristics

Launch facilities and range services offered by commercial launch providers are critical elements affecting a customer's appraisal of the value of launch services. The quality and breadth of these facilities have important ramifications for the preparation and launch of the satellite and the accommodation of customer launch date requirements.

Launch Sites

There are only a few locations in the world that can currently support launches of medium and large satellites into geostationary orbits. In fact, there is essentially only one such location for each nation or political region with geostationary launch capabilities. In the United States, it is Cape Canaveral; for Europe, it is Kourou in French Guiana; in China, it is Xichang; in the former Soviet Union, it is Baikonur/Tyuratum; and in Japan, it is Tanegashima. These and other sites can also support launches to low Earth orbit, although some sites are not well situated for polar launches.

Western launch sites offer sophisticated facilities and services that can support virtually any technical requirement of the customer. Facilities in nonmarket economy countries are not as sophisticated and may require that the customer make special arrangements. These arrangements could entail expensive customer-provided payload preparation and clean room equipment and services.

Customer personnel involved in the launch also require hotel and restaurant accommodations and medical, transportation, communications, and entertainment services. The cost and quality of these services, or, in their absence, the cost of independently providing them, vary significantly among launch sites. In addition, customers will prefer ready access to the launch site for company personnel and visiting customer clients. Strict security measures or other restrictions that block or impede access to the launch site will cause the customer to discount the value of a launch service.

Accommodation of Customer Scheduling Requirements

A customer generally has special requirements concerning when a satellite must be launched. These requirements factor into the choice of a launch service in several ways and, in some cases, may be determinant.

The customer may require a launch date, or slot, that a launch provider cannot provide. Launch providers can be constrained in their ability to provide launch slots requested by customers because of manufacturing/inventory limitations, or more likely, because of constraints at the launch facility or range from which the vehicle would be launched. Existing medium- and large-lift launch vehicles generally require expensive pad facilities specifically designed for them. This capacity is essentially fixed unless the government chooses to expand it. Western launch facilities maintain at most two pads for each vehicle type, and can support no more than 8 to 12 launches per year for any given vehicle type.

Because missions generally require 2 to 3 years to plan, and because launch companies must maximize the use of facilities within this planning horizon, availability is less flexible for nearer-term requests, particularly those within 2 years of the present. Once all pad capacity has been committed for a given time period, a new request for a launch in that time period can be accommodated only if scheduled downtime (if any) is utilized, or if an existing customer can be shifted to a later launch date. More about scheduling flexibility is said below.

Ability to Meet Launch Schedules

The customer may also believe that one launch provider would be better able to meet an agreed upon launch schedule than would be another. A launch provider may fail to meet a scheduled launch date for several reasons, including: voluntary suspension of service following a launch accident; involuntary suspension of service due to damage to the launch pad or other launch facilities caused by a launch accident; delays due to technical problems with a launch vehicle while on the pad; resource conflicts with other users of a shared-use facility; or overly optimistic scheduling.

A special case of potential resource conflicts at a shared-use facility arises at U.S. launch facilities, where commercial launch slots are subject to preemption by the military in times of national emergency. This potential for preemption or other delays caused by the military is frequently cited to customers by competitors of the U.S. companies. However, in reality it is very unlikely that preemption would ever occur, and the Air Force has been accommodating of commercial schedules. For instance, no delays or preemptions due to military activities occurred during the Persian Gulf conflict.

Delays in launching are expensive to customers. A 1-month launch delay is reported to cost the customer the equivalent of 1% of the value of the satellite.[1] A customer who believes that a launch has a high probability of being delayed will either discount the value of the launch service, or will seek to assure the receipt of financial compensation from the launch provider should such a delay occur. Compensation for delayed launches is a relatively new practice, first offered by General Dynamics, followed thereafter by Arianespace.[1] This represents an additional type of nonprice competition.

Launch providers can also alleviate customer concerns about maintaining launch schedules by the demonstrated ability to adjust for delays through rescheduling. For instance, Arianespace has shown an ability to reorder its manifest to meet time-critical missions that would otherwise be delayed. This ability was particularly evident in the periods just before and after the failure of Flight 36 in February 1990. Shifting of manifests is accomplished through negotiations with commercial customers and/or through cooperation by the national government regarding the rescheduling of its own payloads—another means by which governments can assist their commercial companies. It is generally believed that European governmental organizations cooperate with Arianespace in accommodating commercial customer schedules.

Customer Penalties

Finally, in some cases, it will be the customer who delays the launch. Typically, the launch provider will charge its customer a significant penalty if the customer fails to meet expected mission dates. This penalty becomes more severe the closer the delay is to the scheduled launch date. A customer seeking a fixed launch slot, but who anticipates a possible delay in acquiring or preparing a satellite, must factor the different penalties charged by the launch services into its decision. Launch companies charging lower penalties will gain a nonprice advantage with regard to these customers.

Risk Management for Possible Launch Delays

Even with some degree of financial compensation for delays, the customer remains at risk in the event that the launch company might suspend operations for a prolonged period of time. To reduce this risk, the customer may pursue a strategy of double sourcing, in which two launch providers are contracted to share in the launch of a series of satellites. In association with a strategy of double sourcing, the customer will request launch options from the launch providers. Launch options are commitments by a launch company to provide launch services at agreed terms and conditions should a customer choose to exercise an option in the future (as it might following the suspension of launch services by its other source). Through this risk-reducing strategy, a customer can maintain access to space even if a major accident should cause one of its providers to cease launch activity indefinitely.

Reflight Provisions

A special need for schedule accommodation may arise in the event of a launch failure in which a customer's payload is destroyed. The customer can make arrangements with the launch company to assure in advance that, should a failure occur, a new launch can be obtained quickly and at reasonable terms. Launch companies offer customers priority for a relaunch within a set time period from their request, and reflight insurance to cover the cost of a new launch. The terms of such services can vary among providers.

Government Policy Factors Influencing Market Transactions

Beyond the essential role of governments in establishing launch systems and purchasing launch services, they influence launch companies through policies that affect the workings of the international market. These policies range from encouragement of company marketing endeavors, to efforts to establish a fair international trade environment, to control of technology transfer and other national security issues.

Government Support of Marketing Efforts of Commercial Launch Providers

Some forms of government support to the marketing efforts of their national firms, such as the previously described provision of export–import loan guarantees, are accepted practices in the international market. In addition, governments are expected to provide official representation of national launch provider interests to foreign governments and to create supportive and consistent policy environments for their launch providers.

However, it is generally considered undesirable for governments to become too closely involved with the marketing efforts of their national companies, as this participation becomes dangerous to market stability. For instance, government intervention by one country in the form of direct operating subsidies or price supports could lead to similar supports by one or more other countries. Competing subsidies would result in great and unnecessary expenses to all participating governments and the rapid termination of commercial services by those companies not receiving equivalent subsidies from their governments.

Other generally unacceptable forms of government support fall under the heading of government inducements offered to potential customers of national launch providers. Inducements are defined to include the following types of government intervention: unreasonable political pressure; provision by governments of resources of commercial value unrelated to the launch service competition; and offers of favorable treatment under or access to government policies and programs related to defense and national security, development assistance, and general economic issues (e.g., trade, investment, debt, and foreign exchange policies).

The governments of all countries with space launch capabilities deny that they engage in market-distorting practices. However, two-tier pricing, in which the national government customer pays a higher launch price for an equivalent launch service than does the commercial customer, may exist. This practice would represent a form of direct subsidy.

Establishment of Free and Fair Trade Rules

Because market-distorting practices such as direct subsidies would prove highly destabilizing to the international market, formal means to preclude their introduction are being explored within the context of "Rules of the Road" negotiations between the governments of the United States and Europe. Eventually, these talks will be expanded to include other nations, possibly leading to the creation of a multilateral framework for free and fair trade in the international launch market.

It is not known how closely the prices offered by launch providers of current or formerly nonmarket countries approximate their costs of production. The lack of free markets in these countries would make the valuation of resources used in production difficult, even if production accounts were available for analysis. Largely because of the concern that nonmarket economy pricing practices could lead to inadvertent subsidies to national launch providers, the "Memorandum of Agreement Regarding International Trade in Commercial Launch Services" was signed by the United States and the People's Republic of China in 1989. A key provision of this agreement states that, "The PRC shall require that its providers of commercial launch services offer and conclude any contracts to provide commercial launch services to international customers at prices, terms, and conditions which are on a par with those prices, terms, and conditions prevailing in the international market for comparable commercial launch services." The agreement further requires that China's insurance providers conform with international rates and practices, and contains specific clauses to reduce misunderstandings in the areas of inducements and introductory pricing.

Surplus Military Assets

The recent and welcome easing of the arms race, as well as the normal obsolescence and replacement of existing missile systems, have led to a surplus of military missiles. The governments of the United States and the former Soviet Union must dispose of these assets in a safe and economical manner. One option would be to sell them for use in the commercial launch market.

The sale of certain larger surplus missiles could infringe on existing markets (government and commercial) for medium launch vehicles. However, the greatest share of these surplus assets would impact the small launch vehicle market. As this market is the least developed at this point, the governments must proceed cautiously.

Technology Transfer Controls

Because space technologies and space launch systems have potential military uses, governments cannot be indifferent to market transactions involving these items. Consequently, western governments have been reluctant to allow western space technologies, especially satellites, to be exported to certain countries.

To assuage U.S. concerns about the potential transfer of sensitive U.S. technology to China while U.S.-built satellites are in China awaiting launch, China signed the "Memorandum of Agreement on Satellite Technology Safeguards," one of three special space agreements it concluded with the United States in 1989. Exports of U.S. satellites to the former Soviet Union were still restricted as of January 1992, although these restrictions will be reconsidered in light of recent developments there.

Governments may also impose restrictions on trade to accomplish other foreign policy objectives. In June 1989, the U.S. government banned the export of satellites for launches from China and other items on its munitions list to China to protest the Tiananmen Square incident. This ban was lifted for satellites in December 1989. In May 1991, the U.S. government announced a new ban on the issuance of export licenses for U.S. satellites and satellite technology for launch from China. This ban was in response to ballistic missile sales by China to Pakistan and Syria. This ban remained in effect as of January 1992, but could be lifted at any time following satisfactory assurances that ballistic missile sales will cease. These bans necessarily create uncertainties for potential customers of Chinese launch services who would require U.S. export licenses prior to shipping their satellites to China. However, the imposition of such bans is not arbitrary and can be predicted based on certain provocative actions.

International Organizations

Whereas some commercial customers will seek only the best deal for launch services, international organizations such as Intelsat and Inmarsat are under pressure to spread launch awards among national launch providers of member countries. However, launch companies compete aggressively for these contracts, indicating that the organizations retain a significant amount of choice as to the launch providers they select.

Summary and Conclusions

The commercial launch industry emerged in the 1980s as an outgrowth of national space programs. In all cases, commercial providers have been established and/or encouraged by their respective national governments, and the role of governments in this industry remains large.

Costs of operation for commercial launch providers include the recurring costs of resources and labor needed to produce launch hardware and conduct launches, but may also include capital costs for equipment replacement, new or expanded facilities, and some vehicle development expenses. These costs do not include the very high costs of developing vehicles, production equipment, and launch infrastructure, which are paid for by governments to support national space and security objectives. Government procurements of launch services from national providers are critical to reaching economic levels of production.

Pricing strategies of launch companies are difficult to investigate due to the proprietary nature of most information in this industry. However, economic theory provides some insights into how firms in similar market situations may behave. The pricing model proposed in this analysis for the commercial launch industry is one that applies to a noncollusive oligopoly where the motivation of the principal participants is profit maximization and where pricing strategies are constrained by concerns about initiating destructive price competition.

Government intervention in commercial marketing activities is destabilizing and could lead to costly subsidy wars and/or damage to unsubsidized competitors. Thus, governments must necessarily set up the foundations of space launch services, but should maintain an arm's length relationship to commercial launch markets. Current efforts by the United States and Europe to establish "Rules of the Road" for government conduct in the international launch market will contribute to the further development of a free and fair international market for commercial launch services.

References

[1]Chenard, S., "Selling Ariane," *Space Markets*, Vol. 3, May/June 1990, pp. 121–128.

[2]Congressional Budget Office, "Large-Capacity Launch Vehicles," *Encouraging Private Investment in Space Activities*, U.S. Government Printing Office, Washington, DC, Feb. 1991, pp. 9–43.

[3]Jane's Information Group, *Interavia Space Directory, 1989–90*, edited by A. Wilson, Interavia, S.A., Geneva, Switzerland, 1990.

[4]Heydon, D. A., "Ariane Program Plans and Outlook for Commercial Launch Services," AIAA Paper 90-0890, presented at the 13th AIAA International Communication Satellite Systems Conference & Exhibit, Los Angeles, March 11–15, 1990, p. 4.

[5]Chenard, S., "Streamlining Ariane," *Space Markets*, Vol. 5, Sept./Oct. 1989, pp. 294–299.

[6]Blond, E., and Knittle, W., *Space Launch Vehicle Costs*, prepared by the Aerospace Corp. for the Dept. of Transportation, Office of Commercial Space Transportation, 1984.

[7]Bartlett, F., "Plan Successes, Not Failures," *Space News*, Nov. 4–10, 1991, pp. 15–16.

[8]Greenberg, J. S., and Christensen, C. B., "The Selection of a Launch Vehicle," Canaveral Council of Technical Societies, 26th Space Congress, April 1989.

[9]Bulloch, C., "Arabsat: A Neglected Asset," *Space Markets*, Vol. 1, Jan./Feb. 1989, pp. 36–44.

Product/Service Pricing: Support Facilities (Space Facilities)

Chester M. Lee*
Spacehab, Washington, DC 20024

Introduction

SPACEHAB, Inc. is a privately owned company whose objective is to generate income by developing and building, with private financing, a pressurized module that will augment the high-value pressurized middeck area and satisfy the demand for conducting experiments in a man-tended pressurized volume on the Space Shuttle Orbiter, as well as provide an additional habitat for longer duration missions.

SPACEHAB will, in essence, provide additional facilities on board the Orbiter for which it will charge a fee. The setting of the fee structure is the subject of this chapter. Since 1981 the Space Shuttle has successfully completed more than 40 missions. During this period, NASA obtained valuable experience in a wide range of manned spaceflight activities, including the successful deployment, retrieval, repair, and redeployment of satellites. One of the significant lessons learned was the demonstrated value of a pressurized volume of space in the Orbiter in which the crew could conduct "hands on" experiments, and which would provide additional living space for the crew in longer duration missions. This experience clearly showed that the middeck area of the crew compartment was a most valuable piece of real estate in the Orbiter for the conduct of scientific and commercial research and living. It also demonstrated that the Orbiter's pressurized cabin was limited in its capacity for accommodating longer duration missions.

The SPACEHAB module fits into one-quarter of the payload bay of any orbiter. The modules will enhance the capabilities of the Space Transportation System by doubling the existing orbiter habitation volume and

---

Copyright © 1992 by the American Institute of Aeronautics and Astronautics, Inc. All rights reserved.
*Vice President.

quadrupling the volume available for man-tended space experimentation while still leaving room for other major payloads in the rear three-quarters of the cargo bay.

The SPACEHAB module concept provides the simplest and most practical way possible to augment the high-volume pressurized middeck area and satisfy the demand for conducting experiments in a man-tended pressurized volume on the Space Shuttle Orbiter.

Historical Background

Concept

The founder's original concept of SPACEHAB was to provide a passenger carrier to be mounted in the cargo bay. However, based on discussions with NASA representatives and advice of consultants, the concept was changed to that of providing a small pressurized volume for laboratory-type work. NASA's experience with the Orbiter had clearly demonstrated the value of additional pressurized working space for the astronauts.

Financing

The early financing of the SPACEHAB project was provided by a loan and equity from SPACEHAB's founder, Robert Citron, and later from private investors. In addition, SPACEHAB management negotiated several cost-shared studies with aerospace companies. Through mid-1987, private placements raised approximately $2.2 M, but as a result of the Challenger accident, a fourth private placement in 1987 was not fully subscribed. This was a significant setback to SPACEHAB's efforts to obtain the financing support necessary to complete its phase B studies and to move on to phases C and D. (Phases A, B, C, and D are NASA identifiers reflecting: phase A—Studies; phase B—Preliminary design; phase C—Design; and phase D—Construction.) It was already recognized that flying to space and conducting activities in this new environment was expensive. The Challenger accident vividly verified that it was also risky. In as much as the SPACEHAB concept obviously needs the Shuttle for transportation to orbit, support on-orbit, and return from orbit to Earth, the difficulties to SPACEHAB's continuing viability were hardly insignificant. Confident that "man in space" was a national policy and a national commitment, and that the Shuttle would resume flight operations, it became increasingly important that SPACEHAB secure an agreement with NASA that would convince investors that NASA would agree to put SPACEHAB on the Shuttle manifest for a series of flights. NASA's requirement for evidence of SPACEHAB's technical and financial viability before it would commit to an agreement, plus the requirement of financial investors of the reality of a commitment of NASA to fly the SPACEHAB module, produced a "chicken and egg" scenario not uncommon in startup situations.

SPACEHAB's new management was able to raise an additional $1.8 M funding through a fifth private placement. Major participants in this private placement were two financial firms, Poly Ventures, L.P., a new high-tech venture capital fund, and BEA Associates, a large New York investment management firm. This new infusion of funds and the continuing support of SPACEHAB's prime contractor, McDonnell Douglas, and McDonnell Douglas's subcontractor, Aeritalia (now Alenia Spazio), by investment in SPACEHAB equity, as well as accepting subordinated debt conditions, allowed the design effort, the seeking of additional financial support, and the negotiations of a Space Systems Development Agreement (SSDA) with NASA to proceed in parallel. In March 1988, the official Shuttle manifest included, for the first time, SPACEHAB flights beginning in 1991; and in August of 1988, NASA and SPACEHAB reached an agreement on a SSDA. The NASA SSDA is essentially a NASA Launch Services Agreement with several special considerations added to assist new embryonic commercial programs getting started. In the case of SPACEHAB, the special consideration was to defer the progress payments to be paid to NASA for transportation to space on the Shuttle until 1 month after launch. SPACEHAB would, however, pay the escalation and interest charges on the deferred payments. The attainment of these milestone events were important in obtaining funding support through bridge loans from equity investors. These loans were necessary in order to carry the project forward as efforts to obtain additional equity and a bank loan continued.

In the meantime, the efforts to obtain additional equity had moved offshore. Early interest had been shown by both the government of Taiwan and separate entities of Mitsubishi Corporation, Japan, in participating in the SPACEHAB project. The Taiwan commitment was delayed for a time due to a change in government, while the Mitsubishi Corporation negotiations and discussions continued for an extended period of time due to questions concerning NASA's commitment to SPACEHAB and commercialization of space activities and of the overall market for SPACEHAB services. In 1990, SPACEHAB, Inc. successfully competed for a NASA contract to provide services of a Commercial Middeck Augmentation Module to fly crew-tended experiments developed in cooperation with NASA at the Centers for Commercial Development of Space (CCDS) and through Joint Endeavor Agreements (JEA). NASA Joint Endeavor Agreements are agreements with private enterprises and/or universities in which NASA provides standard transportation services at no charge and the private enterprise or university provides the hardware and principal investigator of the experiment and pays for any optional services required or requested. The signing of this contract was coincident with both Mitsubishi's and Taiwan's decision to commit finally to significant equity participation.

The final linchpin in completion of the required financial support to augment the almost $40 M of SPACEHAB equity was to obtain a bank line of credit. The principal obstacle to obtaining this credit from the commercial banking institutions was their concern about the U.S government's commitment to its pronounced policies and the commitment of the U.S. Congress to provide the appropriations to meet NASA's contractual commitments to service contracts. The breakthrough on overcoming this obstacle occurred when SPACEHAB, Inc. was able to obtain insurance policies to cover the bank debt in case of termination by the U.S. government of the Commercial Middeck Augmentation Module contract with

SPACEHAB for lack of appropriations. In March 1991, the Chase Manhattan Bank of New York and SPACEHAB, Inc. agreed to the terms and conditions for a revolving credit line of $64 M. Now, for the first time, SPACEHAB, Inc. had established a total financial structure sufficient to permit authorization of its prime contractor to proceed with completion of the final design, construction of the module structures, and the procurement installation and integration of the subsystems into the module. Until this financial structure was in place, and in order to have any hope of meeting a schedule, all of the design elements, construction of the module, attainment of agreements and contracts, and completion of all negotiations for the necessary financial support had to be worked in parallel. The risk of failing in any one of these requirements with the collapse of the entire program was always there and will always be there for any embryonic commercial space venture. However, the nature of the "chicken and egg" scenario makes it necessary to take the risk. Therefore, any similar venture must be prepared for this approach. It is mandatory to develop a convincing argument that there is a market for the space product or service that is of sufficient magnitude that an adequate return in investment is likely, before embarking on such a venture. In SPACEHAB's case, the confidence and commitment of the equity investors and the support of the prime contractor McDonnell Douglas and their principal subcontractor Alenia Spazio (formerly Aeritalia) made this parallel activity possible.

SPACEHAB Facility

The SPACEHAB Middeck Augmentation Module is designed to provide additional pressured working space to the orbiter middeck. The module, which is shaped like a cylinder with a truncated top, measures approximately 10 ft in length and 13.5 ft in diameter. The height from the keel of the cylinder to the truncated top is 11.2 ft. There is a 40-in. clearance between the module and payload bay doors at the centerline. The truncated top has two viewing ports for Earth or space observation. Optical viewports can be removed to permit penetrations for controls for experiments that could be mounted external to the module to be exposed to the vacuum. The aft bulkhead is bolted and will be removed between each mission to provide access for space station-type double-rack-size experiments.

The volume of the module permits mounting of up to 61 orbiter middeck-type lockers, or various combinations of lockers, single and/or double racks. A locker and its structure occupy 2.2 ft^3 with 2.0 ft^3, available for actual experiments. The locker mounting structure has the capability to carry a total of 80 lb. This includes the locker's inert weight of 18 lb. Each double rack provides an equivalent volume of 45 ft^3 and its mounting structure can carry up to 1200 lb including the double-rack weight of 150 lb. Single racks essentially provide one half of the double-rack capability.

Resources available to experiments include power, cooling, data, and communications. Although the SPACEHAB module occupies less than 20% of the Orbiter's cargo bay, it provides the necessary interface to accommodate two orbiter standard mixed cargo harness (SMCH) cables for 3.2 kW power continuous and 5.7 kW peak for 15 min/3 h. Both air

PRODUCT/SERVICE PRICING

Table 1 Payload dc power accommodation

	Power, kW				
	1 SMCH		2 SMCH		Time limit on
Mission phase	Max. cont.	Peak	Max. cont.	Peak	peak power
On-orbit	1.40	2.65	3.15	5.70	15 min/3 h
Ascent/descent	0.60	0.60	0.60	0.60	2 min mission phase

and water cooling are provided. Tables 1 and 2 summarize the power and cooling resources available. Table 3 summarizes the command and data capability.

For example, in acquisition channels there are total of 160 discrete low-voltage signals to be divided among the experiments that are flown on a mission. There are a total of 16 of these discrete signals available per rack. What each experiment can be allocated is mission dependent on what the total complement of experiments requires.

Developing Pricing Concept for SPACEHAB Services

As often occurs in making early pricing and market estimates, the initial estimates of cost to build the product, cost of capital, cost of transportation to space, unanticipated requirements, changes in the market, and so forth are underestimated or not considered adequately. This was also true in the case of SPACEHAB. The price sometimes used in early marketing efforts and assessments was unrealistic and underestimated by as much as 40–50%. In addition, some of the cost factors initially considered, such as transportation to space, insurance charges, and so on, changed considerably after the Challenger accident.

In order to develop a price, it was necessary to better understand the numbers and types of payloads that could be accommodated by available SPACEHAB resources. SPACEHAB, Inc. and its prime contractor McDonnell Douglas in 1986 conducted an analysis of the resources (i.e., power, cooling, weight, volume, data, astronaut time, etc.) that would be available from the Orbiter/SPACEHAB combination vs the average utilization by the individual experiment. The sources of information for the

Table 2 Payload heat rejection capability

	Type of heat rejection		
Arrangement	Air/water	Air	Water
Total experiment	4000		
All locker		1400	2600
Rack		2000	4000 airload

Table 3 Command and data resources

Accommodation	Total	Locker	Per rack	Remarks
Data Acquisition channels				Minimum rack allocation based on locker requirements. Two-rack allocation can be supplied to one rack on mission basis.
Discrete input low, 0–5 V	160	Mission dependent	16	
Discrete input high, 0–28 V	72	Mission dependent	12	
Analog input high, 0–5 V	160	Mission dependent	32	
Serial digital input, 8 bits	8	Mission dependent	1	
Telemetry downlink, kbps	16	Mission dependent	16	Includes Subsystem Data
Onboard display, RS-422	1 Microcomputer			Orbiter GRID 1530
Recording				Orbiter Payload Recorder Interface
Analog track, 1.9–125 kHz@ 7.5 ips	1	Mission dependent		
Digital track, 32–64 kbps @ 7.5 ips	1	Mission dependent	1	
Closed circuit television	1 Ch. for user provided video source	Mission dependent	1	Orbiter CCTV System Interface
Timing				Orbiter Payload Timing Buffer Interface
Orbiter GMT signal	As required	Mission dependent	As required	
Orbiter MET signal	As required	Mission dependent	As required	
Command				
Discrete outputs, 28 V pulsed	32	Mission dependent	8	More can be made available
Serial digital output, 16 bits	8	Mission dependent		Orbiter GRID 1530

analyses were the middeck experiments flown before the Challenger accident. The results of this analysis are shown in Tables 4 and 5.

In addition, SPACEHAB, Inc. procured the services of a consultant group to develop a cost structure for the space flight of SPACEHAB modules. The consultant group, TEG International, Inc., reviewed SPACEHAB's operating cost estimates, market demand, competition, and cost per pound comparison and developed a series of pricing scenarios that assumed a break-even number of lockers or equivalents sold and payment of debt within a number of flights that depended upon such parameters as

Table 4 Number of lockers capable of being supported
(based on average experiment requirements)

Mission configuration	Module capacity	Limiting parameters			
		Crew time, man-hours	Energy,[a] kWh		Weight, lb
		2 Crewmen (100)	1 SMCH (215)[b]	2 SMCH (485)[b]	3000
100% Locker	61 Lockers	61 Lockers	34 Lockers	61 Lockers	50 Lockers
2 Single rack	41 Lockers	41 Lockers	0 Lockers	38 Lockers	34 Lockers

cost contingencies, and year-end cash requirements. Prices developed from these scenarios are fairly close to the current prices in SPACEHAB's existing pricing policy.

SPACEHAB, Inc. also initiated a reassessment of the global market for SPACEHAB services. It had been recognized that until microgravity processing developed a viable, potential, profitable product, private industry would not be ready to undertake the development and production of such a product. The initial enthusiasm of many, but not all, in the early 1980s, for early successes was recognized as overly optimistic. Rather it became more universally recognized that commercialization of microgravity-processed products was a longer term project, and that a great deal of basic data needed to be obtained in order that private industry could more effectively and efficiently determine which of the processes to be pursued had the greater probability of developing a profit-making commodity. Until a profit-making commodity is developed, the primary markets for SPACEHAB services would be government and/or government-sponsored agencies both in the United States and offshore, such as NASA, ESA, NASDA, NARDA, AZI, CNES, ESTEC, and so forth, which will utilize SPACEHAB services. These agencies are and will continue to be the source of "seed" funds for microgravity-processing experiments until private industry does develop a potential profit commodity. Estimates as to when this will occur

Table 5 Resource requirements used in matrix calculations

Resource	Average experiment[a]	Single-rack experiment
Weight, lb	97	485.0
Crew time, h	2.6	11.5
Energy, kWh	10.5	121.4

[a]Seven-day mission assumed.
[b]Standard total energy available in kilowatt-hours and is mission dependent.
[c]Nominal 60 lb per locker.

range from pessimistic to optimistic. However, it does appear certain that the effort to develop commercial applications in this new and relatively unknown environment of space will continue for some time to come.

Development of Current Pricing Policy

The pricing structure of any product and/or service must include the cost of the development of the product and/or service, the continuing operational costs, and ultimately a return on investment (ROI). In the case of SPACEHAB, the elements of the pricing structure were defined as

1) *Amortization of Development and Construction Costs*—Prior to determination of any ROI considerations, the embryonic commerical venture must recover the costs of construction of the commercial product as well as general and administrative expenses (G&A), additional facilities lease or construction costs, insurance, bank financing, and so on. In SPACEHAB's case the construction part of the total cost includes the McDonnell Douglas construction contract whose basic cost estimate was verified by using an industrial cost-estimating model. The contract contains the usual profit and penalties negotiated between SPACEHAB, Inc. and McDonnell Douglas. The amortization of this element of the total price is defined as the lease charge.

2) *Integration and Operation Costs*—The SPACEHAB concept visualizes repeated flights and, as noted previously, the NASA/SPACEHAB, Inc. SSDA covers eight flights of the module. Based on early experience with microgravity-processing experimentation, SPACEHAB, Inc. management determined that provision of frequent flight opportunities would be a key advantage of SPACEHAB and provided two flight modules, each of which could be turned around approximately every 6 months, thus permitting up to four flights per year if the demand required. To carry out integration and operations (I & O), it was necessary to obtain an I & O contract. Under the resulting negotiated cost plus incentive fee I & O contract, McDonnell Douglas would be responsible for refurbishing the module between flights as well as having the responsibility for integrating the various experiments, representing the integrated experiment payload to NASA for safety, and so forth, collating and preparing all the documentation required by NASA, and training the NASA ground personnel and flight crews on the module subsystems. McDonnell Douglas, as the I & O contractor, has agreed to a contract that covers eight flights and includes the estimated costs for the functions described above. In the SPACEHAB price, this element is called the *integration charge*.

3) *Transportation Cost*—The third element is the cost of transporting the SPACEHAB module and its experiments into space, supporting the SPACEHAB module while in space, and then returning the module to the launch site. NASA provides this service to SPACEHAB via the Space Shuttle Transportation system and charges SPACEHAB for this service. SPACEHAB's charge to its customers for this element is based on what NASA charges SPACEHAB, Inc. for Shuttle services and is basically a pass-through of the NASA charge. However, to ensure that the transportation charge of NASA is covered, SPACEHAB's price to its customers

includes a 5/4 factor to the pro-rata shares of transportation cost for a single locker. This 5/4 factor is linearly reduced essentially to zero as increased numbers of payload services are contracted for by a customer up to a dedicated module. In essence, the reciprocal of this factor is the anticipated utilization rate.

Once having defined the elements of the total price as 1) lease, 2) integration, and 3) transportation, the next steps were to determine revenue needed to pay off the line of credit from the banks, the interest, insurance, subordinated debts, and the ongoing administrative expenses, as well as to provided a minimum acceptable rate of return to the investors. The NASA SSDA included eight flights and, based on the Shuttle Transportation Manifest, SPACEHAB could anticipate an average of two flights per year after startup with a potential of additional flights if demand necessitated additional flights. Based on two flight units and a 6-month turn-around between flights, four flights per year were possible. However, only a maximum of three per year was considered. Several scenarios in which flight rate, locker or equivalent utilization, revenue income, outlays for interest and debt payment requirements, delays in schedule, and so forth were analyzed. Based on these analyses, SPACEHAB management developed a revenue stream that over a period of several years would permit paying off the operating expenses, the senior debt (bank line of credit), and the subordinated debt. However, by the end of this period of time (approximately 5 years), the equity investors would not see any return on their investment except for the assets then owned by the company. The integration price was the basis of the negotiated cost plus the incentive fee contract negotiated with McDonnell Douglas. The transportation price, as stated previously, was a pass-through to NASA based on the price charged to SPACEHAB by NASA for the Shuttle services.

Having completed the first step of setting the total price for a complete module, the second step was to define how the total charge would be prorated per experiment and what should be the governing item upon which to base the pro-rata share price of the total overall price such as complexity, volume, power use, and so forth. Based on the analysis of previous experiments flown in the Orbiter middeck (refer to Tables 4 and 5), SPACEHAB, Inc. management decided that the nominal load of the SPACEHAB module should be 50 lockers (even though more volume is available). Having determined the nominal number of lockers or experiments per flight, the next step was to develop the fairest and most equitable method of pricing the individual experiments. The 50 lockers should normally accommodate 17 or 18 experiments, based on the analysis of middeck lockers previously flown. This would be the maximum number of experiments that could be used as the basis for the experiment price. SPACEHAB management and its prime contractor attempted to develop a formula using each SPACEHAB resource individually. Complexity of the experiment was eliminated because it was a judgment evaluation and subject to controversy. Cooling was eliminated because it was basically dependent on power utilized. Power was considered doubtful because of the extreme variations in power requirements anticipated as to peak power, continuous power, and times required of each. It was also felt that the power variation in requirements of each experiment could be accommodated by judicious time lining during the orbital operations. However, it was a consideration and pricing formulas were developed for volume and weight and energy. The following is an example of an option that considered both weight and volume:

$$\text{Price} = \text{total lease cost} \times \frac{0.5 \times \text{payload wt (lb)}}{3000 \text{ lb}}$$
$$+ \frac{0.5 \text{ payload volume (ft}^3)}{110 \text{ ft}^3}$$

where total lease represents the revenue required from a single SPACEHAB module flight, 3000 lb is the maximum total payload weight that SPACEHAB can allow in the forward position in the cargo bay, and 110 ft^3, is the equivalent volume occupied by 50 lockers.

This option was not considered acceptable because several example runs indicated that the aggregate mission price determined could be less than the target price per mission. Because payload weight is limited and is, therefore, a critical factor, the loss of revenue on the particular payload example could not be made up. In another example, the particular payload example used allowed more income for one locker or its equivalent than the regular price charged. Although this could be considered acceptable, this case along with the one described before it was discarded because of inconsistency. Other options that considered volume had similar disadvantages.

Another option was developed and analyzed for weight, power, and volume limited to ½ locker equivalent increments:

$$\text{Price} = \text{total lease cost} \times \left[\frac{0.5 \text{ payload wt (lb)}}{3000 \text{ lb}} + \frac{\text{average energy}}{\text{total energy available}} \right]$$

where total lease represents the revenue required from a single SPACEHAB module flight, and 3000 lb is the maximum total payload weight that SPACEHAB can allow in the forward position in the cargo bay. Example runs indicated that if a single locker experiment required maximum power commensurate to its heat rejection capability, the price would be as high as the price for a locker including transportation. Therefore it was also considered unacceptable.

In the final analysis, SPACEHAB management decided that since the weight limitation was the most critical factor, the pricing formula would be based on weight alone. A nominal weight per locker equivalent was determined by dividing the total weight of 3000 lb by the nominal number of lockers that would be accommodated (50). For leases for multiple experiments, the average weight will be acceptable. In the case of single locker experiments, provisions are made for weight less than the nominal weight. Thus a pricing policy was developed as follows:

$$\text{Price} = \text{total lease cost} \times \frac{\text{payload wt (lb)}}{3000 \text{ lb}}$$

Having determined the charge for the smallest accommodation provided by the module, the next step was to determine the price for larger accommodations such as space station-type double racks, single racks, and so forth. These prices were determined in a single, straightforward equivalency factor. For example, based on the analysis of average weights of double-rack experiments previously flown on the Shuttle, it was decided that the average nominal weight should be 800 lb for the experiment(s) plus 150 lb of rack inert weight for a total of 950 lb. This is the equivalent of 16 lockers so the lease and integration (L & I) price was made equal to 16 lockers. The L & I price of a single rack will be one half of a double rack. However, the pro-rata share of the transportation portion included a 15% markup rather than the 25% markup of the locker (or equivalent) pro-rata share of the NASA transportation charge.

Once SPACEHAB's management defined the cost structure and determined the overall price of the physical accommodations and resources offered in the module, the next step was to define clearly for the customers what services and resources were to be provided for the price. Many of those involved in negotiating the early NASA Launch Service Agreements will recall the difficulties of defining standard and optional services. SPACEHAB management did establish a definition of standard services to be included in its pricing policy and user's guide. These services defined what SPACEHAB would provide in the areas of 1) analytical integration, 2) physical integration, 3) mission design, 4) flight phase, and 5) postflight phase. Recognizing that other extra services not covered by the standard services offered would be required, SPACEHAB offered to provide or negotiate with NASA for optional services on a case-by-case basis and attempted to define these services based on past experience with NASA and Spacelab.

It should be noted at this time that although SPACEHAB management considered discounts for a quantity buy, the only consideration that was deemed reasonable at this time was the decrease of the 5/4 factor included in the transportation charge that must be paid to NASA for its services. This factor was eliminated for a lease of services for a dedicated module.

In retrospect, the early low estimates of costs (which ultimately determine price) to provide the facility and services defined were due to a lack of complete appreciation and understanding of the impact of terms and conditions imposed by financial institutions that ultimately provided the major portion of the financial support. The negotiations of the terms and conditions required by the financial institutions were further complicated by the fact that the U.S. government (NASA in this case) was playing two different roles in its agreements with SPACEHAB. One of these roles is as a provider of services (i.e., transportation into space) and the second role is as a customer of the services offered by the commercial enterprise.

In the case of SPACEHAB, the bank required in the terms and conditions of the credit agreement that prime and subcontractor accept significant changes to their contracts which in effect increased their (contractor's) risk exposure. Further, since the financial institutions, both in the United States and offshore are concerned about the U.S. government budgeting process, particularly the U.S. Congress's willingness to fulfill the U.S. government's (NASA) long-term commitments, they demanded protection against an abrogation of the government's commitment either through an acceptable termination provision in the services to NASA contract or insurance coverage as a backup to the U.S. government's failure to meet its commitment. Although, in the opinion of some knowledgeable lawyers, NASA had the authority to provide an acceptable termination clause [i.e., one that would provide coverage of costs for research and development (R & D) or production of the facility as well as costs incurred by SPACEHAB to carry out the provisions of the NASA/SPACEHAB services contract] NASA refused to provide anything other than its standard termination clause, which would limit its liability to funds appropriated and allotted to the services contract, and further limited to SPACEHAB costs directly spent in supporting the NASA services Commercial Middeck Augmentation Module (CMAM) contract as measured by the progress of SPACEHAB toward completion of the lease arrangement. NASA would have no liability for the SPACEHAB expenditure involved in development or production should, for whatever reason, NASA reduce the flight rate of SPACEHAB missions and effectively delay SPACEHAB income. The delay insurance covers additional expense incurred (including additional interest on the line of credit) up to the specified total amount of the policy. Should NASA, for whatever reason, delay a launch 24 months, then the constructive termination would become effective. As a result of NASA's decision, SPACEHAB management was forced to seek termination insurance.

Insurance brokers and their underwriters initially took the position that they would consider termination insurance for almost any cause of termination by NASA, except termination that was due to lack of appropriations by the U.S. Congress. They did agree to provide such insurance if the Shuttle program was cancelled because of a Challenger-type catastrophic accident, but not for a direct lack of appropriations. Early on, the underwriters committed to a $30 M policy that would cover losses incurred due to delay of the scheduled launch for causes other than SPACEHAB or its contractor's fault. This policy would cover a delay of up to 24 months and then cover constructive termination after 24 months up to the remaining funds of the $30 M policy amount. *Constructive termination* is an event that causes something to happen. In this instance, for insurance purposes, if an event (i.e., launch) does not occur within 24 months, termination automatically occurs and payment under the policy is made. SPACEHAB's management together with its insurance broker, Frank B. Hall & Co., continued their efforts to obtain insurance coverage for termination due to lack of appropriations. A breakthrough finally occurred primarily through convincing arguments that the Congressional 5-Year Deficit Reduction Plan indicated that the NASA budget could be expected to increase rather than be reduced and that the risk of Congress failing to appropriate sufficient funds was mitigated. Lloyds of London underwriters agreed to underwrite a $50 M policy covering termination due to lack of appropriation of funds.

Although the delay and termination insurance policies were crucial to the bank, it should be noted that any commercial venture such as SPACEHAB requires a number of insurance policies to satisfy financial institutions. Such policies include

1) *Module Preflight and Postflight*—covering loss or damage to the modules on the ground, including consequential losses occurring when the modules are SPACEHAB's risk

2) *Space Flight Insurance*—covering losses resulting from destruction or damage to the modules occurring from intention ignition through launch, flight, and termination

3) *Space Flight Liability Insurance*—covering all third-party liability of SPACEHAB arising from space flight on the STS. Including, if necessary, the orbiter return flight on the NASA 747 from Edwards AFB, California to Kennedy Space Center, Florida

4) *Comprehensive General Liability*—covering SPACEHAB's third-party liability exposures other than space flight

5) *Workers' Compensation and Employer's Liability Insurance*—covering employees of SPACEHAB in compliance with statutory requirements

6) *Umbrella Liability*—providing excess liability limits over primary general liability, auto and worker's compensation policy limits

7) *Directors and Officers Liability*—protecting SPACEHAB and its directors and officers against loss arising out of claims made against the directors and officers for alleged wrong doing

8) *Fidelity Bond*—covering losses from employees dishonesty

The cost of all these must also be considered in the development of the total cost to be recovered in establishing a pricing policy. As mentioned above, the modification of the Construction and I & O contract required by the bank not only increased the prime contractors responsibilities but also the risk taken by the contractor as to the schedule and performance. This resulted in a higher fee commitment to the prime contractor.

The cumulative effect of the added insurance and increased requirements resulted in an increase to the amount of funds to be provided by the bank. The increase of the total loan in turn added to the total amount of interest paid by SPACEHAB for the bank credit. Obviously, these increased costs increased the price for SPACEHAB services. However, it should not be concluded that the government can provide the SPACEHAB services cheaper than SPACEHAB, Inc. because it is self-insured and has a lower cost of capital. It should be noted that in actuality, government self-insurance of space activities may be more costly than taking insurance. This is discussed by C.B. Christensen and Joel S. Greenberg in "Issues in Space Insurance," IAA-91-652, 42nd Congress of the International Astronautical Federation, October 1991.) It should also be noted that the launch agreement offered by SPACEHAB, Inc. to all its customers, including the service contact with NASA, is a fixed-price contract. In the negotiations with NASA for the Services contract, SPACEHAB management successfully negotiated out most NASA requested changes that were primarily hardware modifications that would increase the capability of the SPACEHAB module to handle more complex experiments downstream. Such changes, although of the "nice-to-have" category, are not beneficial to SPACEHAB at this time and would have increased the Construction and I & O contracts considerably. However, SPACEHAB, Inc. did agree to all changes related to safety. Today, 1 year since the Services contract was signed and additional reviews, including safety (representing the greatest potential cost impact) have been completed, there have been no increased costs passed on to the customers, including NASA.

Space Insurance

Daniel E. Cassidy*
Arlington, Virginia 22201

Introduction

INSURANCE has provided important economic benefit to commercial space operations, and has been a major factor in the cost of space ventures. Essentially all commercial satellites and most satellites of non-U.S. governments planned for commercial purposes have been insured against physical loss or damage during launch. Many have also been insured during in-orbit operations. Some organizations build backup satellites to hedge against failure, yet still find it advantageous to buy payload insurance.

The U.S. government policy has been to self-insure space launches. NASA, Department of Defense, and other agencies retain the risk of potentially losing program capability, and of possibly having to reprogram funds or seek supplemental appropriations. Commercial entities, however, typically transfer as much of the risk as possible through the mechanism of insurance. The payment of insurance premiums represents the cost of risk and must be included as a cost of doing business in space.

The business of space insurance has been almost exclusively insuring communications satellites which operate in geostationary orbit. It is unlikely that satellites would have played as significant a role in expanding communications in the United States and countries throughout the world without the availability of insurance. International satellite communications, on the other hand, would probably have been less affected by the availability of insurance due to the financial structure of the International Telecommunications Satellite Organization (INTELSAT). INTELSAT elected to self-insure many of its launches, including the most recent INTELSAT VI series. The United States representative to INTELSAT,

---

Copyright © 1992 by Daniel E. Cassidy. Published by the American Institute of Aeronautics and Astronautics, Inc. with permission.
*Consultant, Technology Programmatics International.

the public stock company COMSAT, did insure its ownership share of the satellites.

The supply of payload launch insurance has in most cases met the demand of the satellite communications industry, although a notable exception occurred in the mid 1980s following a series of space losses. (In November 1985, RCA Satcom K2 was launched uninsured on the Space Shuttle since the only available coverage was considered inadequate; co-riders AUSSAT 2 and MORELOS B were insured under earlier insurance placements.) The supply of insurance capacity contracted and drove-up the cost of the limited available supply. It was made clear during this period that availability of insurance for space projects could be substantially affected by conditions in the insurance markets.

Elements of Space Insurance

Insurable interests in space-related activities can arise from a number of different sources creating risk of financial loss. Exposure to loss can include the consequences of physical loss or damage to space systems and ground property, the loss of related property and services upon which the business depends, and the legal liability resulting from damage to third persons.

Payload launch insurance has been the most critical type of insurance from the business perspective of demand, cost, and amount of insurance capacity available from the worldwide markets. This covers the risk of physical loss or damage to satellites and other payloads during space launch. Satellite in-orbit insurance has been closely related to launch insurance underwriting and market trends and covers risk of loss after the launch phase. Payload insurance for the launch phase is the main focus of the discussions on insurance markets, insurance capacity availability, cost, and risk analysis considerations.

Launch liability insurance availability and cost have generally not been a problem, because of the recognized range safety practices and the fact that there have not been third-party claims. Some difficulties have arisen for manufacturers with product liability coverage, but as is discussed later, cross-waivers between parties have mitigated much of the difficulty. Liability is the principal concern of the government under space treaty obligations, and the concern for its citizens and other third parties. Liability insurance, therefore, is discussed primarily in terms of United States government requirements and the indemnification available from the government. It must be recognized, however, that a major third-party claim would have substantial effect on the availability and cost of launch liability insurance.

The importance of payload insurance to space projects is that the potential cost impact of a single catastrophic event during space launch creates a commercial imperative. In general, the risk of loss cannot or will not be born by a single company or government organization since the amount of potential loss is too high. It is necessary to divide the potential loss and transfer the pieces to those willing and able to take space risks at smaller amounts, through the mechanism of insurance. By spreading the risk, the insured exchanges risk of large potential loss for an essentially fixed amount of insurance premium. The premium is the cost of risk which has to be paid by the insured in undertaking the enterprise.

Risk spreading is also important to insurers. They are able to pass off part, and in some cases most, of the risk through reinsurance. It is also important when they are able to insure many and diverse events. As is discussed later, the relatively few launches have not given much opportunity for underwriters to establish a balanced book of business over a large number of space risks.

Space Satellite Insurance Markets

Space satellite insurance is arranged through worldwide markets. In the aggregate these markets recently have had insurance capacity for space launch risks of over $300 million, depending upon the particular risk, underwriting assessments, and level of premium that is being charged. (Capacity estimates are as high as $350 million, although underwriters rarely put their entire line on a single risk.) Premium rates are currently about 16–18% for communication satellite launches to geostationary orbit, and about 1–2% for in-orbit insurance. The $300 million that could be available to insure a single launch or separate satellites seems adequate for current requirements, although larger capacity may be necessary in the future.

Market Experience

Capacity and premium rates are driven by market forces and depend on the profit and loss results of the space insurance industry. An historical perspective is presented in Fig. 1, which illustrates the premiums collected compared to losses that have resulted from the space launch and in-orbit insurance business. This comparison is closely tied to the changes that have occurred in the cost and availability of space insurance. Net premium is the amount that was received by underwriters after paying the expenses, primarily brokerage fees in acquiring the business.

Insurance capacity available for space risks is not a fixed amount, but a result of a number of factors, including need, risk assessments, premium rates, and alternative market opportunities. Over the last several years the amount of this capacity has varied substantially. During the time period prior to 1984, world capacity increased in response to demand and reached a maximum of about $260 million. Following the large number of launch and in-orbit failures from 1984 through 1986, the space insurance market contracted and capacity declined below $100 million. (The Challenger did not involve space insurance losses, although it is believed that the disaster had a substantial negative effect on insurers' confidence in space risks.) Since 1986, capacity has again continued to increase year by year to the recent levels.

The substantial reduction in market capacity is attributed in part to the withdrawal of a number of cash flow underwriters. A large number of insurers were lured into the space insurance market in the 1970s and early 1980s by prospects of large premiums, without adequate knowledge of the risks involved. In addition to the general attraction of the space business, cash flow underwriting became a way to obtain additional premiums for investments. Insurers could take advantage of the profits from investing

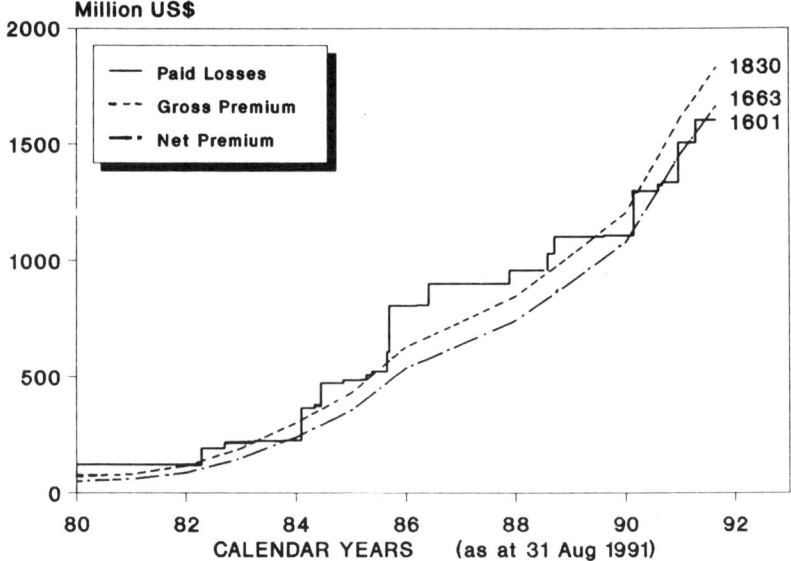

Fig. 1 All premiums and losses (cumulative). Presented by B. Pagnanelli, Assicurazioni Generali, at the Sixth International Conference on Space Activities, Rome, Italy, Sept. 16–17, 1991.

the funds at the high interest rates that were available in the financial markets at that time. (Cash flow underwriting also affected general property and casualty insurance with the consequence that many commercial lines such as medical, professional, product, and municipal were essentially left without coverage starting in the mid-1980s.)

Broad insurance market participation was welcomed by insurance buyers and contributed to increased market capacity and reduced premium rates, which in many respects were reduced to unjustifiably low levels. Capacity withdrawal, on the other hand, contributed to a near collapse in the space insurance market.

While rates were different depending on the particular space launch, in 1983 the launch rates for communications satellites (geostationary orbit) were about 5% for the Space Shuttle and 7–11% for expendable launch vehicles (ELVs). In-orbit rates were about 1% per year. Along with the substantial reduction in capacity in 1986, geostationary orbit launch rates increased to 25% and more in some cases. In-orbit rates increased to about 5% per year.

These higher rates resulted in a larger flow of premium to insurers, which had the concomitant effect of attracting the additional capacity to the market indicated previously, and the subsequent gradual reduction in rates to current levels. It is noted that some launches in the latter part of the 1980s were insured previously at the low rates agreed to prior to 1986.

Major Markets

The principal underwriting for space launch risks is conducted in a relatively small number of organizations. These organizations are located mainly in the United States and Europe. Recent estimates of satellite insurance capacity concentrations are in the following countries and corresponding major underwriting organizations. The percentages of space insurance capacity identified by country are based on world capacity of about $300 million.

United States	27%
International Technology Underwriters (INTEC)	
United States Aircraft Insurance Group (USAIG)	
Germany	17%
Deutscher Luftpool	
Munich Re	
France	17%
La Reunion Spaciale	
AGF/SCOR	
United Kingdom	14%
Ariel Syndicate	
Merrett Syndicate	
Rome Syndicate	
Marchant Syndicate	
Italy	14%
Assicurazioni Generali	
Others:	11%
Japan, Sweden, Australia, et al.	

Underwriting Approaches

The primary space insurance markets underwrite and manage risk retention by utilizing a number of different mechanisms. The objective of the underwriting process is to assure that the insurer's interests are served while meeting the needs of the insured. For a successful placement, ultimately the terms and conditions, total sum insured and premium rate will be agreed to between the insured and a sufficient number of insurers to meet the coverage requirements. For the most part, this process has been successful in achieving satellite coverage.

Some of the various market arrangements established for the purpose of underwriting space risks are identified in the following paragraphs.

Lloyds syndicates typically underwrite space risks by following recognized leaders within Lloyds who have specialized in the space line of business. The lead underwrites will negotiate with insureds, through their brokers, and commit to some fractional amount of the coverage required. (A Lloyds member broker is required for dealing at Lloyds.) Other interested syndicates will then sign onto the agreed coverage at various percentages, which are then aggregated and the coverage is issued on a Lloyds policy.

Other insurers, particularly in the United States, employ a Managing General Agent (MGA) to perform the underwriting function and commit the insurers to an authorized amount of capacity established for space risks. The MGA will negotiate with the broker acting on behalf of the client, and commit whatever fraction of his or her authorized capacity that he or she wishes.

One form of MGA is where a group of primary insurance companies are represented, and a single insurance policy is issued jointly by the several insurers for a specified amount of the coverage. Each of the insurers is primarily liable for its agreed amount of any space loss, and in turn may cede a percentage through its own reinsurance agreements. Another form is where an insurance policy of one primary insurance company is issued, which includes amounts ceded through reinsurance agreements. In either case the primary insurer is liable for any space losses and must collect any claims from his reinsurers.

Some insurance companies also underwrite space risks directly utilizing their own staff for this purpose. In this case, the insurer makes a substantial internal underwriting commitment to the space insurance business. The insurer is primarily liable for the amount of coverage on his or her insurance policy, although usually retains only a portion of the face amount of coverage and cedes the remainder through reinsurance agreements or treaties.

Reinsurers generally rely on the underwriting skills and decisions of the ceding insurance company. This is not always the case in the space insurance industry. Some space reinsurers are very active in assessing space risks and maintaining a leadership role in influencing the market reaction to specific risks, as well as establishing terms, conditions, and premiums charged to insureds.

Risk Analysis and Risk Rating

A major problem with determining the cost of risk is the uncertainty in predicting the probability of failure. There have been relatively few space events employing a number of different space technologies, with different experience levels, and therefore a small statistical base exists from which to quantify risk for any given launch or series of launches. On the other hand, a determination that all launch vehicles, or all launch vehicles and their predecessors within the same family, should be grouped together would ignore many important characteristics that distinguish relative risk.

Loss Probability Considerations

Sample size and the homogeneity of the samples are very important factors in establishing the actuarial bases for predicting the probability of failure. Unfortunately, space risk analysis has little of either to go by. The binomial distribution with outcomes of either success or failure, and a single-sided "confidence" concept are used here to illustrate the statistical limitations of the space data resource in making risk assessments. The single-sided confidence limit provides an approach to estimating the upper bound for the probability of failure, but not the lower bound.

For simplicity, the following examples assume no failure scenarios with five sample events in one case and 10 sample events in another case. The analysis can be extended to larger sample sizes and one or more failures in the sample events, with increased complexity of calculation. The objective here, however, is to describe an approach to demonstrating that due to the number of events and types of risks involved, space risk analysis has a weak statistical basis. In order to establish premium rates, reliance must be placed on overall market losses and the use of subjective underwriting to discriminate between risks.

To illustrate, if there were no losses in a sample of five homogeneous space events (e.g., identical combination of launch vehicles launching identical satellites), there is still a 50/50 chance that the single launch probability of failure (SLPF) for that combination could be as high s 13%. This is based on the result that if the SLPF were higher than 13%, there is about a 50% probability that there would have been one or more failures in the five sample events. Since the sample of five produced no failures, one can be 50% "confident" that the SLPF is no more than 13%. Accordingly, there would be less than 50% confidence that the SLPF is less than 13%.

Increasing the sample size to 10 events with no failures results in a 50% confidence that the SLPF is no more than 6.6%. The sample size has improved the result, but this is still a large estimate of failure rate, particularly when it is argued that there were no failures and 10 successes. Things get more divergent when higher confidence in predictions is required. For the 10 event/no failure case, a maximum 20% SLPF would have to be assumed to achieve a 90% confidence level. It would take 229 homogeneous sample events with no failures to have 90% confidence that the SLPF is no more than 1%.

Determining Probabilities from Space Data Base

If one looks at some 120 commercial satellites launched over the past 10 years, the composite failure rate (number of failures/number of events) has been about one in six, or .167, including launch and early in-orbit failures. If some sense of homogeneity in the events could be assumed, based on the above analysis, the data would provide a 50% confidence in an upper bound of 17% SLPF. At 90% confidence the upper bound would increase to 22% SLPF, which is over five percentage points more than the composite failure rate. Thus, on this statistical basis, insurers would have to charge 22% to cover the risk, plus expenses and desired profit, to have a 90% confidence that they would make at least the desired profit in the long run.

But the historical events have not involved the same launch vehicles and satellites, and many factors have intervened, including different technologies, technology changes and product improvements, redesigns, and upgrades. Many of these factors resulted in major changes within the same family of launch vehicles and satellites as well. Even when attempts are made to further divide past space events into groups, the sample sizes get smaller and the question still remains as to the homogeneity of the events within the groups.

The situation is that space risks are not actuarially based and requires that many other factors be taken into consideration in the underwriting process.

Market Losses vs Assessing Specific Systems

Notwithstanding the above situation, historical loss records of failures/events are at least the starting point for estimating the cost of risk. The data reflect a "technical" rate for different types of launch vehicles and spacecraft systems, but more often are used to represent the aggregate of launch failures or in-orbit failures. Losses in excess of premiums collected have been the main driving force in the space insurance markets. The aggregation of space failures are what produced the market losses and caused the large volatility in the capacity availability and cost of space insurance.

In the past, following major losses, little distinction was made between the various expendable launch and satellite systems seeking insurance. To a great extent this is still market practice. However, the underwriting approach is to seek discrimination and rate individual risks when it is justified from the insurer's point of view.

Figure 2 depicts the historical composite failure rate (number of failures/number of events) for the limited cases of communications satellites after separation from the launch vehicle. This figure represents the failures occurring during the latter part of the launch process, including apogee kick motor (AKM) firing, initial checkout, and early orbit operations. Failures attributable to the earlier part of launch are not included. These results are singled out to illustrate an overall composite failure rate of about 7½% resulting from the aggregation of insurance market losses attributable to

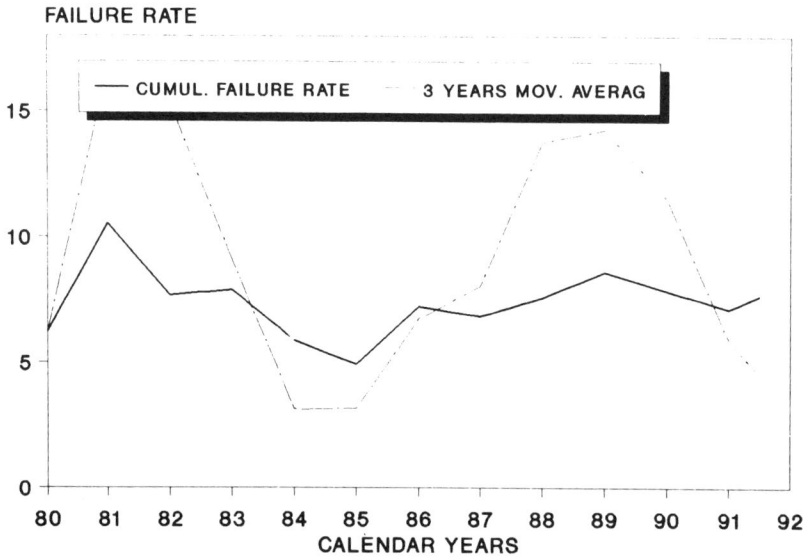

Fig. 2 Commercial satellites: AKM and early orbit failure rate. Presented by B. Pagnanelli, Assicurazioni Generali, at the Sixth International Conference on Space Activities, Rome, Italy, Sept. 16–17, 1991.

one type of risk. It is the challenge of those seeking insurance coverage, where this type of risk is involved, to distinguish their risks from the past results and obtain more favorable treatment.

Space Risk Underwriting

Case-by-case assessments of individual risks are necessary in order to discriminate effectively between risks. This involves a number of considerations including systems heritages, experience and demonstrations, design and testing, as well as manufacturing and integration oversight. The focus of underwriting will be on space systems reliability and performance, as reflected in the results. While it is unlikely that there will be significant increases in the statistical base for many years, lower inherent risk has the most potential for reducing the level and variability of insurance costs. Without substantial improvement in results, buyers will continue to be subject to market volatility due to the uncertainty and the level of risk.

As a final consideration, it is noted that market forces will continue to dominate, where the buyer (insured) and seller (insurer) negotiate price and conditions based on supply and demand. Presenting a number of insurable events to the markets, such as a series of launches which provide large premium volume, offers opportunity for more favorable rates than a single launch. Lower valued risks, with less sums insured, will have certain advantage in requiring the agreement and participation from a fewer number of underwriters to obtain the desired capacity.

Space Insurance Coverages

There are various spacecraft-related risks associated with transportation, storage and handling on the ground, space launch, operations in Earth orbit, return from orbit, and recovery. For the most part these risks can be insured under specified terms and conditions, and typically fall into the following categories:
1) Prelaunch, including transit and storage
2) Space launch
3) Relaunch guarantee
4) In-orbit life
5) Re-entry and recovery
6) Dependent property and services
7) Third-party liability

The amount of insurance coverage depends upon the insured's exposure to financial loss which can include direct and indirect damages, including business interruption. The bases for determining the amount of insurance, depending on insurance market capacity, includes considerations of the following:
1) Asset value
2) Replacement value
3) Launch services cost
4) Extra expenses
5) Operating expenses and debt service
6) Loss of revenues under contract
7) Liability amount stipulated by regulation

Prelaunch Insurance

Prelaunch insurance covers the physical loss or damage to the satellite, associated propulsion systems, and other space systems during periods after leaving the manufacturer's plant. This coverage can include transit and storage, satellite processing at the launch facility, and integration with the launch vehicle. The party having risk of loss during these periods depends upon the provisions of the particular contracts involved.

In addition to covering the direct physical loss or damage, prelaunch insurance also covers indirect or consequential loss. Coverage, for example, could include loss in revenues from the satellite, and extra expenses resulting from launch delays. Under the prelaunch insurance policy, the delays would have to be caused by the dirct physical loss or damage to the satellite or other dependent property.

Since prelaunch insurance does not cover the risk of space launch, it is important to assure that no gaps exist in coverage. Different insurers are involved and different insurance policy terms and conditions can apply. It is particularly important that certain periods are considered, e.g., when the risk of loss transfers from the supplier to the buyer and when the prelaunch coverage terminates and launch coverage attaches.

Space Launch Insurance

Launch insurance covers the physical loss or damage to the payload starting from some specified event during the initial launch sequence, such as intentional ignition, opening of the clamps, or liftoff. The specific event or time of attachment can be very critical, depending upon the supplier contract risk-of-loss provisions, prelaunch insurance termination, and the particular launch vehicle. Coverage continues through commissioning in orbit, or until some established time after the day of launch.

Coverage includes all risks of physical loss or damage to the insured satellite as a consequence of the launch, with certain exceptions such as war, riot, and nuclear exclusions. All upper stage firings, systems deployments and checkout, and the initial period of operations are covered under the launch policy. The period of coverage is typically 180 days and can be up to 12 months, although in the past it had been limited to 90 days. The insurance can provide for payment in the event of partial loss of satellite capability, or may pay only for total or constructive total loss.

Relaunch guarantee may also be offered by the launch provider as a form of launch insurance. The guarantee covers the risk of launch vehicle failure prior to payload separation, and the amount of coverage is usually the value of the launch services. An amount may also be available to cover part of the satellite value, although additional insurance is required for full satellite coverage. The guarantee is part of the launch services agreement (LSA) and any claim would be settled under the LSA.

The guarantee can offer the advantage of securing a substantial amount of initial launch phase coverage at the time of signing the LSA. It also effectively increases the insurance capacity available for a particular launch, if the launch provider does not obtain coverage in the commercial insurance markets. In addition to augmenting coverage during the initial launch phase, insurance is also required for post separation, including upper stage firings, satellite deployments, commissioning, and early orbit operations.

In-Orbit Insurance

In-orbit insurance covers the physical loss or damage to the satellite following termination of the launch coverage. The period of coverage, which previously had been 3 years, was limited to 1 year following the adverse market experience of the mid-1980s. It is now available up to 2 years with prospects of again increasing the period of coverage to 3 years.

The amount of coverage is the value to the insured, considering such basis as the asset value, replacement cost, or revenues. It is often changed over the time period of the coverage to reflect depreciated value, although there may be cases where the value is increased, such as when the cost to replace is higher or revenues increase. As in the case of launch insurance, partial loss or total loss coverage is available.

Liability Insurance

Liability insurance covers the legal liability arising out of claims from third parties for property damage and bodily injury or death. Launch liability insurance covers exposures during operations at the launch site prior to launch as well as during the launch, and includes the first year of operations in orbit. Extended in-orbit coverage is available after termination of the launch liability policy.

In the case of NASA launch services, the payload owner is responsible for obtaining liability insurance and including the United States government as a named insured. The amount of insurance obtained is usually $500 million in the case of one payload. Under the launch services agreement, NASA is authorized by law to provide indemnification to payload owners for claims that exceed the amount of insurance. (Users of NASA launch vehicles can be indemnified under provisions of the National Aeronautics and Space Administration Act, 42 U.S.C. 2451, as amended by the NASA Authorization Act of 1981, P.L. 96-48.)

Other parties such as manufacturers of the satellite and NASA contractors are not included in the insurance and government indemnification provisions. Although the payload owner can include its contractors as named insureds under the liability insurance policy, government indemnification is not available. NASA, however, does make determintions on a case-by-case basis whether to provide indemnification to its own contractors. (NASA Procurement Notice 83-3, Indemnification of NASA Contractors Involved in Space Activities, November 1983 was implemented based on P.L. 85-804 and Executive Order 10789.)

In the case of launches conducted by commercial companies under Department of Transportation (DOT) license, the launch provider is responsible for obtaining liability insurance covering actuaries of the U.S. government, the launch customer, and other participants. By law, the contractors of the launch provider and the customer, and government contractors are included. (The Commercial Space Launch Act Amendments of 1988, P.L. 100-657.)

The Commercial Space Launch Act Amendments of 1988 provides that the amount of liability insurance shall be determined by the DOT, but not in excess of $500 million. For the larger U.S. commercial launch vehicles, the DOT has established the amount of liability insurance required to be between $165 million and $215 million, depending upon the particular launch vehicle. In addition, the government provides indemnification covering all launch participants up to a maximum of $1.5 billion above the amount of insurance. This indemnification provision is due to expire in November 1993 unless reauthorized by Congress.

Manufacturers of launch vehicles, satellites, and other space systems are exposed to strict liability for damage that results from failure of their products. This exposure can include damage to third parties as well as damage to space systems. Although insurance coverage is provided for product liability, the coverage generally excludes space risks, in particular, damage to spacecraft.

A number of developments in the law have eliminated much of this exposure of manufacturers. In the case of DOT-licensed launches, all participating manufacturers are included under the government indemnification provisions of the Space Launch Act Amendments. In the case of NASA launches, subject to NASA's discretionary authority under P.L. 85-804, NASA contractors can also obtain government indemnification. In addition, cross waivers of liability and hold harmless provisions between the various participants in launch operations, whether NASA conducted or DOT licensed, have been used successfully to defend lawsuits against space product manufacturers. (See, for example, the California court case *Appalachian Insurance Company et al. v. McDonnell Douglas Corporation et al.*, 262 Cal. Rptr. 716.)

Financial Loss Insurance

Financial loss insurance is based essentially on *force majeure* risks, in particular, acts of government, which do not necessarily involve physical loss or damage to property. Small entrepreneurial companies that rely on government services for their business have an increasing need to demonstrate an acceptable degree of protection as a condition of financing. (The recent example of SPACEHAB demonstrates that lending institutions are not willing to take the risk of adverse government action when making space business loans.)

Coverage could include a specified level of protection from launch preemption, adverse change in the government's commercial launch policy, or nonappropriations of funds necessary to provide the government services. These risks are difficult to quantify and are subject to moral hazards, and, accordingly, the coverage is narrowly drawn and expensive.

Conclusion

Insurance has been a major factor in the use of communications satellites for commercial purposes. In the future, it is anticipated that new and innovative commercial space applications will be developed that will challenge approaches to insurance underwriting with different risks, scope of coverages, and insurance capacity requirements.

Space insurance is primarily market driven, although earlier in its history there was an element of space mystique in the spirit of the new frontier. With a fair amount of hindsight available to judge, the more pragmatic approach to underwriting space risks will be the vision of the future. Being market driven, however, means that supply and demand will play a major role.

As in the past, the investment value of premiums may override caution in the selection and pricing of risks. Likewise, the accumulation of large underwriting profits will encourage more capacity and increase competition between the underwriting markets, also driving down premium rates. On the other hand, large accumulation of losses can threaten the availability of insurance and the viability of future space projects.

The objective is a reasonable balance between the risks and the cost of their coverage. The demonstration of space systems performance and reliability as reflected in the insurance results will be the principal determining factor.

While space operations do not provide an adequate actuarial base, the expectation is that the space industry has the capability and expertise to present highly controlled and managed risks to the insurance markets. Nothing less than excellence in space technology and applications must be demanded of space systems to assure the stability and availability of space insurance for future projects.

Chapter 5. Relationship of Economics to Major Issues

Commercial Development of Space: Government/Industry Relationship

Joel S. Greenberg*
Princeton Synergetics, Inc., Princeton, New Jersey 08540

Introduction

BOTH government, the public sector, and industry, the private sector, undertake programs that are aimed at influencing private-sector investment decisions. These investment decisions encompass those of commercial organizations as well as those of individuals. For firms, these investment decisions range from corporate strategy decisions, such as whether or not to enter a new business area (for example, become a provider of satellite communication services), to capital investment decisions, such as which satellite launch service provider to utilize. For individuals, these capital investment decisions range from equipment replacement (replete with considerations of tradeoffs between higher nonrecurring cost and lower recurring costs and fuel choice) to the acquisition of new products and/or services made possible by the introduction of new/advanced technology.

The private sector undertakes research, development, demonstration, and advertising/promotion projects that lead to increased producer and consumer investments and to the formation of commercial ventures. To a large extent, these programs are undertaken with the expressed intent of maintaining or increasing the value of the firm. A vast array of government-sponsored programs are undertaken with the specific objective of developing technology and/or creating the environment that will lead to increased private sector investment and the formation of commercial ventures that are in the public interest. These encompass research, development,

---

Copyright © 1992 by Joel S. Greenberg. Published by the American Institute of Aeronautics and Astronautics, Inc. with permission.
*President.

and demonstration programs as well as incentive, regulatory, standardization, and other programs.

The Department of Energy's (DOE) programs of developing and demonstrating technology are directed at many sectors of the economy, including the nuclear power industry, the electric utilities, the appliance industry, and the construction industry. DOE's incentive programs, such as the energy tax credit program, are aimed at influencing consumer and producer decisions in such a manner that their optimal choices are altered. The setting of auto mileage standards influences both producer and consumer automobile production and consumption decisions, respectively. The result is that choices are made that are in the public interest, such as achieving increased oil and gas conservation. DOE also undertakes information dissemination programs (analogous to advertising) aimed at increasing consumer and producer awareness with respect to energy efficiency and other product attributes (i.e., the appliance labeling program), and thereby influencing consumer choices and producer product decisions. Finally, DOE also participates in many forms of regulatory programs ranging from the regulation of utility rates of return to the setting of appliance efficiency standards.

Other government agencies pursue similar types of programs with emphasis on different technologies and different industry segments. For example, NASA has an active program aimed at developing and demonstrating communication satellite technology, and aimed at both the communication carriers as well as hardware manufacturers (i.e., keep U.S. industry competitive).[1,2] In recent years NASA has instituted a joint endeavor program wherein NASA provides services at a cost below that of full cost recovery or other established pricing policy in return for a commitment by the firm to proceed at least with the initial steps that will lead toward a commercial venture. NASA has recently become an anchor tenant for Spacehab, a commercial venture aimed at providing and leasing additional locker space for performing experiments in space aboard the Space Shuttle, thereby providing the impetus for a commercial business venture that otherwise would likely not have been possible (Ref. 3, pp. 18-24). NASA has also entered into an agreement with GEOSTAR (Ref. 3, pp. 25-31), a commercial venture to provide position location through the use of satellites, such that GEOSTAR received space launch services to be repaid in the future in the form of a royalty on sales (thereby placing the government at risk while at the same time providing a loan at below market rate). Unfortunately, GEOSTAR has recently terminated its business operations.) Through this arrangement, which reduced GEOSTAR's initial capital requirements by more than $100 million, GEOSTAR was able to raise sufficient capital for the initiation of a commercial venture that would otherwise not have been possible.

NASA also supports research relating to crystal growth and materials processing in space which may lead to commercial ventures and/or improve the competitiveness of U.S. industry in the international environment. In addition, NASA has an active technology transfer program which includes the publication of *Tech Briefs* and *Spinoffs*. To help exploit the development of new high technology space ventures, NASA has established and provides financial support for the Centers for the Commercial Development of Space (CCDS) program. The CCDSs are not-for-profit consortia which represent joint undertakings involving U.S. companies, academia, and non-NASA government. Industry contributions provide leveraging of the government support. (It is interesting to note that in its support for the CCDSs, NASA is actually participating as a venture capitalist, but not acting as a venture capitalist. No formal financial goals and objectives have been established for assessing the need for and desirability of continuing NASA funding of the on-going CCDSs.) The CCDSs are expected to stimulate high technology space-related research which will lead to the development of new products and processes having either commercial potential or contributing to new commercial ventures. To date, NASA has established 16 CCDSs, of which eight are primarily engaged in materials processing in space research and/or development of biomedical, biotechnology, and agrichemical products and services.

The Department of Transportation is responsible for the overseeing of the commercial space transportation industry, including the licensing of commercial launches. The State Department and other government agencies are currently addressing fair trade issues relating to space transportation in order to ensure the competitive position of U.S. commercial space transportation providers in the international marketplace.

The Department of Transportation's Maritime Research and Development Agency (MARAD) developed and demonstrated ship- and dockside-related technology aimed primarily at the maritime industry.[4] The Housing and Urban Development (HUD) Agency has an active program to assist municipalities with the development of cogeneration energy facilities.[5] The Treasury offers tax credits aimed at encouraging industry investment in research and development (R&D) and allows for accelerated depreciation in order to influence industry capital expenditure policies.[6] In prior years, investment tax credits were allowed and were aimed at increasing capital investment. In addition, government agencies participate in the Small Business Innovative Research (SBIR) program wherein a percentage of government research funds are used to support R&D activities of small businesses that are likely to lead to commercial products and/or services. Through the Industrial Research and Development (IR&D) program, government provides funding (for FY1989 this funding was on the order of $4 billion) for R&D activities aimed at maintaining the productivity and competitiveness of U.S. high technology-oriented industry.

In recent years the commercial development of space has become an important subject. The Department of Commerce has for many years been concerned with the commercialization of the civil land remote sensing system and the weather satellites[7] as well as other space commercialization activities.[3] Expendable launch vehicles that had been developed for government use (notably the Delta, Atlas, and Titan vehicles) were "privatized" and a commercial launch service industry was encouraged via a series of government actions culminating in the Commercial Space Launch Act of 1984 and significant amendments to the Act in 1988.[8,9] The Office of Management and Budget (OMB) recommended that NASA commercialize (or at least achieve off-budget financing for) a number of infrastructure elements.[10,11] The National Space Council, under the guidance of Vice-President Quayle, has recently issued a national space policy on the subject of commercial space[12] and NASA is actively considering an agency-wide space commerce opportunities plan.[13] This plan delineates nine initiatives relating to NASA R&D, technology commercialization, joint R&D, private-sector R&D, market, private-sector operations, NASA facilities, asset transfer, and commercial space standards to encourage U.S. private sector leadership in spacerelated commerce.

As the government has become active in promoting the commercial development of space, it has begun to, in essence, play the role of a venture capitalist and/or investment banker providing its scarce resources (the use of funds, infrastructure, and personnel) to help initiate commercial ventures that are in the public interest. As a result, government, and in particular NASA, must also act as a venture capitalist and/or investment banker by evaluating the potential impacts and benefits that will result from its "investments." NASA must ensure that its resources are allocated efficiently; i.e., resulting benefits will be maximized and benefits exceed costs. This implies that NASA must evaluate private-sector business ventures by performing financial analyses, making market forecasts, and so forth, and estimating the likely consequences of its investment, including the likelihood of achieving benefits.[14] (NASA's Office of Commercial Programs released a Request for Proposal in December 1989 with the subsequent award of a contract for the provision of financial analysis services for evaluating private-sector commercial space ventures seeking government support.) It must evaluate the appropriateness of making new investments as well as terminating previously established investments.

In order to understand the role of public-sector programs in encouraging the commercial development of space, it is necessary to have insight into private-sector investment decisions, how government actions may affect these decisions, and how the public sector should evaluate its investment opportunities. These are the main subjects of this article: to describe the ways that government programs can influence private-sector investment decisions and to develop a structure for assessing the likely effects of government programs specifically undertaken to influence private-sector investment decisions. To accomplish this, private-sector investment attitudes must be understood. Results of a survey of investment attitudes are described. The investment attitudes are summarized quantitatively in terms of the likelihood of investment which is a function of expected return on investment (ROI), risk as measured as the variability or standard deviation of ROI, expected magnitude of investment, and expected payback period. A concept is also developed for assessing the impact of government programs through the investment likelihood function. Finally, the effect of various government actions on private-sector decisions is discussed in terms of the likelihood functions.

For purposes of this discussion both commercialization and privatization are encompassed in the concept of the commercial development of space. *Commercialization* refers to the establishment of business ventures, having private capital at risk, which exploit the attributes of space in order to produce and/or market products and/or services. *Privatization* refers to the establishment of business ventures based on the transfer of care, custody, and control of government space-related assets to the private sector to perform the same or very similar function(s). (Privatization does not include those situations where government assets are managed by a private contractor.) In the following paragraphs, no distinction is made between commercialization and privatization unless explicitly stated otherwise.

Role of Government

The National Space Policy issued by the White House on November 16, 1989, and the policies of the previous Administration, seek to encourage and foster private-sector investment in space as a major goal of the federal government. Several approaches for achieving this goal have been developed and are being applied. One approach is to encourage federal agencies to buy commercially available space products and services; another is to avoid conducting government activities that compete (or potentially compete) with private-sector products and services. Continuing government-wide pressure to reduce spending has provided added incentive to identify government programs that are likely commercialization candidates and can be undertaken efficiently off-budget by the private sector.[10,15,16] Overall, the general approach has been for the government to foster policies and practices which eliminate obstacles to investment and to provide to the private sector the facilities, products, and services which are only available through the government, and are necessary to the viability of a commercial venture.

These and other policies attempt to affect the fundamental factors of investor perceptions of risk and return that determine the extent of private-sector innovation and investment relating to the commercial development of space. Space business ventures are characterized by large capital investment, long payback period, and high risk. The large capital investments will be made by the private sector only if the *private sector perceives* that adequate rates of return are likely to be achieved at acceptable levels of risk. Thus, a potential role of government relates to reducing these impediments to investment, thereby increasing the likelihood of private-sector investment.

To illustrate the effects that public sector programs can have on private-sector investment decisions through the reduction of private-sector perceived risk and by shifting the burden of funding from the private to the public sector, consider the example indicated in Fig. 1. This example illustrates the impact of a government demonstration program on the private sector. Figure 1 illustrates a not uncommon situation wherein the private sector is contemplating the investment in a business venture that requires a significant development effort culminating in a feasibility demonstration. Figure 1A illustrates the effect of full investment, including the demonstration program, by the private sector. The degree of variability of exposure (maximum of the indebtedness curve—negative of the cumulative cash flow), payback period, and ROI is indicated. A large part of the variability is due to the uncertainty of the cost and outcome of the demonstration program. Figure 1B illustrates the impact on the private sector of NASA undertaking the development and demonstration program (for example, the development and demonstration of the Earth Observation Satellite, Landsat, and the Advanced Communication Technology Satellite, ACTS) with no transfer payment from the private sector to NASA. Note that private-sector exposure is reduced as is the variability of exposure, payback period, and ROI. At the same time the expected ROI is increased. This is due to the uncertainty in the cost and outcome of the demonstration program being eliminated (private-sector decisions can await the outcome of the NASA program) as well as the private-sector demonstration program funding requirement being eliminated. Figure 1C indicates the impact of a NASA demonstration program when a payment is made to NASA from the private sector to cover the cost of the demonstration, for example, repayment of the cost in return for using a NASA-funded demonstration satellite (i.e., Landsat, ACTS) in an operational system. The impact of the payment is to increase expected exposure and payback period and to reduce expected ROI. Note that there is no change (relative to Fig. 1B) in the variability of the performance measures. It is obvious that the business venture under scenario B is more desirable than under scenario C, which is more desirable than under scenario A. It is also obvious that the course of government action can affect the likelihood of the private sector undertaking the hypothetical business venture. Since

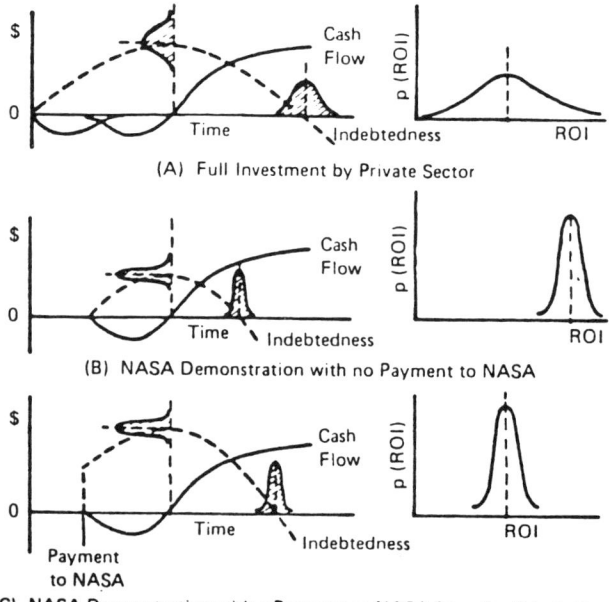

Fig. 1 Impact of a public sector demonstration program on the private sector. (Reprinted by permission of the publisher, from the AMA research study entitled, *Investment Decisions: The Influence of Risk and Other Factors*, ©1982 Research and Information Services, American Management Associations. All rights reserved.)

NASA or public-sector benefits are a function of private-sector investment decisions, the benefits will differ for each of the scenarios.

In general, the government can encourage the commercial development of space by selectively affecting perceived levels of risk (business and technical) as well as the potential for return on investment in space-related ventures. The government can accomplish this by (Fig. 2):

1) Creating new opportunities for investment (for example, through the development of new infrastructure elements and favorable policies)

2) Reducing private-sector perceived risk

3) Increasing the potential for return on investment by reducing the initial capital burden associated with high technology ventures by the private sector and/or increasing product/service markets

4) Creating an awareness of the opportunities, technology, and policies that are favorable to private-sector investment

Current government programs in each of these areas are discussed in following paragraphs. While the government is undertaking programs to encourage the commercial development of space, the government must ensure, as the steward of public resources, that the results of its actions are likely to be positive, tangible, and of benefit to the nation.

The federal government is not, of course, a single unified actor. With regard to commercial space, the government consists of many entities (refer to Fig. 3), among them the Department of Transportation, through its Office of Commercial Space Transportation; NASA, through its Office of Commercial Programs and other elements within the Agency; the Department of Commerce, through its Office of Space Commerce and the National Oceanic and Atmospheric Administration; the Department of Defense, particularly the Air Force; and the National Space Council. Each agency has a different charter and a correspondingly different set of roles with respect to space policy and commercial space policy. Different agen-

---

♦ CHARACTERISTICS OF SPACE BUSINESS VENTURES:

♦ LARGE CAPITAL INVESTMENT
♦ LONG PAYBACK PERIOD
♦ HIGH RISK

RESULT: NEED FOR GOVERNMENT ASSISTANCE TO ATTRACT NECESSARY CAPITAL

♦ ROLE OF GOVERNMENT:

♦ CREATING NEW OPPORTUNITIES FOR INVESTMENT.
♦ REDUCING THE INITIAL CAPITAL BURDEN ASSOCIATED WITH NEW PRIVATE-SECTOR INVESTMENTS.
♦ REDUCING PRIVATE-SECTOR PERCEIVED RISK.
♦ CREATING AN AWARENESS OF THE OPPORTUNITIES, TECHNOLOGY AND POLICIES FAVORABLE TO PRIVATE-SECTOR INVESTMENT.

Fig. 2   Role of government in encouraging the commercial development of space.

EXECUTIVE: WHITE HOUSE

- ♦ NATIONAL SPACE COUNCIL
- ♦ SENIOR INTERAGENCY GROUP/SPACE
- ♦ ECONOMIC POLICY COUNCIL
- ♦ COUNCIL OF ECONOMIC ADVISORS
- ♦ OFFICE OF SCIENCE & TECHNOLOGY
- ♦ OFFICE OF MANAGEMENT AND BUDGET

EXECUTIVE AGENCIES

- ♦ NASA
- ♦ DEPARTMENT OF COMMERCE
- ♦ DEPARTMENT OF TRANSPORTATION
- ♦ DEPARTMENT OF DEFENSE
- ♦ DEPARTMENT OF JUSTICE
- ♦ DEPARTMENT OF STATE
- ♦ DEPARTMENT OF TREASURY
- ♦ U.S. TRADE REPRESENTATIVE
- ♦ FEDERAL COMMUNICATIONS COMMISSION

LEGISLATIVE

- ♦ CONGRESSIONAL SUBCOMMITTEES
  - • SENATE AUTHORIZATION
  - • SENATE APPROPRIATIONS
  - • HOUSE AUTHORIZATION
  - • HOUSE APPROPRIATIONS
- ♦ CONGRESSIONAL ORGANIZATIONS
  - • OFFICE OF TECHNOLOGY ASSESSMENT
  - • CONGRESSIONAL BUDGET OFFICE
  - • CONGRESSIONAL SPACE CAUCUS

Fig. 3 Commercial development of space: a number of the players.

cies, and different parts of the same agency, may act variously and sometimes simultaneously as regulator, payload owner, vehicle operator, developer of new technology, and commercial space advocate.

Despite diverse roles, however, these agencies, in pursuing commercial development of space objectives, are all responding to two major forces: the need to limit government expenditures thoughtfully, and to respond effectively to international competitive challenges. Each agency responds to these forces in light of its own perspective and expertise, as well as its own programmatic and institutional objectives.

Private-sector investment decisions are normally based upon the consideration of a number of financial performance measures including expected ROI, payback period, capital requirements, and considerations of risk as indicated by variability of ROI or variability of other measures. In general, private-sector investment decisions are concerned with evaluating potential business opportunities and considering tradeoffs between such factors as expected ROI, risk, and magnitude of investment, which must fall within acceptable ranges; for example, the expected ROI must exceed a cutoff rate of return. However, this cutoff rate of return or hurdle that must be exceeded is related to the level of perceived risk with the hurdle increasing to offset increased risk. In addition, both the expected rate of return and risk hurdles are normally also related to the magnitude of the required investment. Thus, government programs can affect or alter private-sector investment decisions through their impact on perceived risk (i.e., through risk reduction) and required investment. Efficient government programs aimed at affecting private-sector investment decisions are aimed at those specific aspects of an investment opportunity that are impeding the investment—programs to increase ROI if return is perceived as too low, programs to reduce risk if risk is perceived as too high.

For example, if risk is perceived as too great, then programs that are specifically aimed at risk reduction (which may be more specifically related to market uncertainty, technical performance uncertainty, schedule uncertainty, and so on) are normally required rather than those aimed at increasing expected ROI sufficiently to offset the perceived risk. In such a case, the impact on the private sector is through a reduction in private-sector perceived risk and/or exposure with the increased likelihood of the private sector developing and marketing beneficial products and/or services.

Clearly, benefits can be achieved as the result of government actions altering private-sector investment decisions. It should be noted that the benefits will result, in most cases, only if investment decisions are altered and the desired investments are made and ventures are ultimately funded and the ventures prove successful. Thus, government, in allotting its scarce resources, must consider the likelihood of altering business investment decisions and the viability of the resulting business.

The government has initiated many programs and policies aimed at encouraging private-sector investments that relate to the commercial development of space. These include the following:

1. *Creating New Opportunities for Investment*—New investment opportunities have been and will be created through the development of new infrastructure elements and the adoption of favorable policies. The development of the Space Shuttle has led to new opportunities for new infrastructure-related businesses, such as Spacehab, that will capitalize on the unique microgravity environment provided by the Shuttle's orbital and return capability (Ref. 3, pp. 18–24). The Space Shuttle's jettisoned external tanks are likely to be used as the basis for new space business ventures (Ref. 3, pp. 41–47). The Space Station will likely provide unique opportunities for private-sector investments. The government adoption of policies such as the elimination of commercial payloads from the Space Shuttle and government purchase of commercial launch services have provided opportunities for a commercial space transportation industry, including the development of small-capacity, and suborbital, expendable launch vehicles.

2. *Reducing Private-Sector Perceived Risk*—Private-sector perceived risk has been affected by government programs and related policies. The government policy of not flying commercial payloads on the Space Shuttle (except in special circumstances) is aimed at reducing market uncertainty while increasing the magnitude of commercial markets for launch services,

as is the policy of government not competing with, and utilizing wherever possible, commercially provided products and services. The 1988 Amendments to the Commercial Space Launch Act relating to facility and third-party liability insurance were aimed at reducing commercial space transportation provider's risk associated with the consequences of accidental damage and allowing the U.S. commercial launch providers to compete on more equal terms with ArianeSpace (which is indemnified by the French government against significant third-party damage). The government entering into long-term contracts reduces market uncertainty. Recently NASA entered into an agreement with Spacehab wherein NASA became an "anchor tenant," thereby both increasing the magnitude of the market and reducing associated market uncertainty. NASA research and technology programs are aimed at reducing technical, schedule, and cost uncertainties. A good example of this was the NASA technology program that led to the development of the Earth observation sensors employed on the Landsat satellites being used in the Earth Observation Satellite remote sensing business venture, EOSAT.

3. *Reducing the Initial Capital Burden*—A number of government programs have been undertaken that are aimed at reducing the initial capital burden associated with private-sector investment opportunities. These programs range from tax credits, such as the investment tax credit and the historic redevelopment tax credit, to programs specifically created for the space program. These latter programs include NASA's Joint Endeavor Agreements (JEAs) and Space System Development Agreements (SSDA).[17] The JEAs, as entered into by NASA with McDonnell Douglas, Fairchild, and other organizations, provide for free transportation services utilizing the Space Shuttle.

GEOSTAR entered into a SSDA with NASA wherein transportation charges were to be deferred and repaid to NASA as a percentage of GEOSTAR revenue. This arrangement, which placed the government at risk (i.e., if anticipated revenue did not materialize for GEOSTAR), played a major role in making it possible for GEOSTAR to achieve its necessary initial financing. Well in excess of $100 million was eliminated from the capital requirements of the startup venture. In essence the government provided venture capital to the firm at a cost significantly below market rates that would have been required to achieve this financing from the private sector. (Recently, GEOSTAR, because of financial setbacks, went out of business. Thus government investment is unlikely to be paid back with the ultimate loss borne by taxpayers. Lack of accountability of government employees relating to the provision of "venture capital" is a problem yet to be addressed.)

Spacehab has also entered into a SSDA with NASA wherein Space Shuttle transportation charges are delayed until the Spacehab flight and customer payment. The result is that Spacehab capital requirements are reduced by not requiring the payment of Shuttle transportation charges until Spacehab receives payment from its customers. Again, NASA is, in essence, providing funds at below market rates in order to assist a startup situation.

NASA's Office of Aeronautics and Space Technology (OAST) is in the process of creating a technology program specifically oriented toward satisfying the identified needs of the commercial space transportation industry. This will shift funding requirements from the private to the public sector and will help to maintain the competitiveness of the U.S. commercial space transportation industry since foreign governments currently provide this type of support for their industries.

NASA and the Department of Defense (DOD) are developing the New Launch System, the next generation family of expendable launch vehicles. This technology development and demonstration program again shifts the burden of funding from the private to the public sector and (among other things) is aimed at maintaining the competitive position of the U.S. commercial space transportation industry. However, there is some concern that government development of a new launch vehicle could significantly alter the structure of the U.S. commercial space transportation industry. Space Shuttle lessons need to be reviewed.

4. *Creating an Awareness of Opportunities*—NASA has initiated the CCDS program aimed at encouraging both space and nonspace-related industries to join in a government/industry cooperative space-related research program. The program's objective is to create an awareness of space related opportunities by creating a means for conducting low-cost experimentation in space and coupling this with mechanisms that encourage the commercial exploitation of the research results. The CCDS is a jointly funded program with the government and a large number of industry participants providing the financing. The intent is for the government to withdraw its funding support as additional private-sector partners join the program and increase the private-sector funding. The CCDS is a very conceptually appealing program and appears to be generating a number of flight experiments. However, it currently suffers from not having established quantitative goals and objectives against which the program's success or failure can be measured.

NASA also has initiated a program aimed at communicating the results of its technology programs to the private sector. This is accomplished in a number of ways, including the publication of the highly regarded *Tech Briefs* and *Spinoffs* documents which describe NASA-developed technology that is likely to have commercial application.

Finally, NASA, DOD, and other government organizations fund the IR&D program. The IR&D program is jointly funded by industry and government (in FY 1989 government contributed approximately $4 billion and industry $1 billion) with the program planned, managed, and executed by industry. The program is to serve industry by encouraging R&D innovative concepts, developing technical competence and promoting a strong technical base, assuring competition and enhancing competitiveness, and promoting better government/industry communications. Government agencies present their overall program and mission plans and technology programs to industry and review industry technology programs. This interchange provides a mechanism for making industry aware of the technology requirements of the government programs and allows for a better focusing of industry R&D resources aimed at satisfying future government needs and requirements.

As has been discussed, commercialization is defined as a process that will lead to private-sector investment that will result in the marketing and sale of goods and/or services by the private sector. This process may extend from and encompass the many phases of product development from R&D through market development and sales. During this commercialization process, many investment decisions are faced. The public sector may affect these decisions, and hence the commercialization process, through programs (research, development, and demonstration programs, joint endeavor agreements, tax incentives, subsidies, patent reform, etc.) that reduce private-sector perceived performance, cost and market uncertainties (i.e., risk reduction), and/or shift the burden of funding from the private to the public sector. This is the "carrot" approach. The public sector can also wield a "big stick" in the form of regulation and the setting of standards. These programs tend to force the private sector to take actions that they would otherwise not take.[18] Certain regulatory and standards programs can also serve to reduce uncertainty (for example, mandating hookup to the hot water pipes of a district heating system tends to reduce market uncertainty and can also speed up cash flows).[19] As will be seen it is necessary to clearly identify the role of the public-sector program in the private-sector decision process.

Through the use of the carrot and/or the big stick approach, the public-sector can influence private-sector investment decisions. The basic tools of the trade or the methods available to the public sector for influencing private-sector investment decisions include research and development, demonstration, information dissemination, and incentive programs. To these must be added regulatory programs and the setting of standards. Other types of programs may also be undertaken such as establishing patents/proprietary rights, clarifying and simplifying institutional arrangements, setting tariffs and import quotas, issuing job credits, establishing private enterprise zones, and so forth.

The role(s) and major areas of influence of a number of the tools which may be used to influence private-sector investment decisions are summarized in Tables 1 and 2. Indicated are the primary and secondary roles and influences. It should be noted that a program that has one major role is also likely to have other roles. The primary roles and influences are indicated as are other somewhat lesser important (referred to as secondary) roles or areas of influence. A detailed understanding of the specific problems that must be overcome in order to achieve private-sector investment goals and objectives will allow the government to select and apply its tools in a most efficient manner.

In order to provide government planners with insight into the effect that their policies and programs may have on investment decisions, a relationship between key financial performance measures and the likelihood of investment was developed. In essence, a multiattribute utility function was developed relating the likelihood of investment to several of the more important financial performance measures. These included expected ROI, expected payback period, expected size (exposure) of investment, and a

Table 1 Relationship of government programs to role of government

Government program	Reduce risk	Ease capital form.	Create new opportunities	Create awareness
R&D	♠			
Demonstration	♣	♠		
Development of new infrastructure	♣	♣	♠	
Investment tax credit		♠		
R&D tax credit		♠		
Depreciation rules		♠		
R&D limited partnerships	♣	♠		
Subsidization				
Joint endeavor agreements				
Cost reduction	♣	♠		
Market commitment	♠	♣		
Revenue subsidization	♣	♠		
Recoupment policies		♠		
Pricing policies	♣	♠		
Low-interest loans/bonds		♠		
Insurance				
Regulation	♠			
Indemnification	♠			
Loan guarantees	♣	♠		
Standards	♠		♣	
Information dissemination				♠
Patents/proprietary rights	♠	♣		
Institutional arrangements (international competition)	♠	♣		
Policies				
Procurement	♠	♣	♣	
Pricing	♠	♣		
Scheduling	♠	♣		
.				
.				
.				

♠ = Primary impact; ♣ = secondary impact.

Table 2 Relationship of government programs to areas of influence

Government program	Major area of influence				
	Uncertainty				
	Performance	Market	Schedule	Cost	Exposure
R&D	♠	♣	♣	♣	
Demonstration	♠		♣	♣	♠
Development of new infrastructure	♣		♣		
Taxation					
Investment tax credit					
R&D tax credit					
Depreciation rules					
R&D limited partnerships	♠				
Subsidization					
Joint endeavor agreements					
Cost reduction				♣	
Market committment		♠			♣
Revenue subsidization		♣			
Recoupment policies					
Pricing policies		♣			
Low-interest loans/bonds					
Insurance					
Regulation				♣	
Indemnification				♣	
Loan guarantees				♣	♠
Standards	♠				
Information dissemination					♠
Patents/proprietary rights		♠			
Institutional arrangements (international competition)		♠			
Policies					
Procurement		♠		♣	♣
Pricing		♠			
Scheduling		♠			
.					
.					
.					

♠ = Primary impact; ♣ = secondary impact.

measure of risk. The specific risk measure chosen was the variability of ROI expressed as the standard deviation of ROI. It should be noted that all of these parameters are readily available from a financial risk analysis using simulation techniques.[19-21] Thus,

$$\text{Likelihood of Investment} = f\{m, \sigma, \text{PB}, \text{IND}\}$$

where m is the expected ROI, σ is the standard deviation or variability of ROI, PB is the expected payback period, and IND is the indebtedness but is not utilized directly; a proxy is used in its place. The proxy is the ratio of the magnitude of the investment to the investment budget under the control of the decision maker, expressed as a percentage.

In order to establish the desired functional relationship, the key to establishing the effect of risk and other financial performance measures on investment decisions, a survey was conducted under the auspices of the American Management Associations (AMA).[22] Persons queried generally had the titles of Vice President (Finance), Controller, or Director of Corporate Planning. The questionnaire at first glance appeared rather complex and imposing, yet considerably different than most questionnaires. The AMA mailed approximately 3000 questionnaires to their members selected at random but having the above titles. To these were added a number of specifically selected individuals in decision-making positions. From all of these, in excess of 300 useful responses were obtained. A very large fraction of those who responded identified themselves and indicated that they would be available for further discussions. A number of these were queried to clarify responses. Nearly all of the returned forms were filled out correctly and were internally consistent. Thus only a very small number had to be discarded.

The questionnaire consisted of two basic parts, with the former being of the standard type, collecting information on size of organization, type of business, investment decision criteria, and cost of capital. The latter part was concerned with risk avoidance attitudes.

Figure 4 illustrates the instructions that were provided for completing the risk-related aspects of the survey. The basic approach for gathering the risk data was to present a series of pictures as in Fig. 5 (a page from the questionnaire). Each picture contains four risk profiles of ROI. Within each picture the four risk profiles all have the same expected ROI but each has a different standard deviation of ROI. The pictures are arranged in columns with five pictures per column. Each picture or set of four risk profiles has a different expected ROI. These are shown relative to the firm's cost of capital R. Thus, expected ROIs of R, $1.1 \times R$, $1.4 \times R$, and $1.8 \times R$ are considered. Each column has associated with it a statement of payback period (2, 4 or 8 years) and a statement of the size of the investment (less than 1% of the capital budget, 1–10% of the capital budget, and greater than 10% of the capital budget). This approach standardizes the questionnaire to the perceptions of the respondent, using his or her own data as the general frame of reference. With each picture is a query pertaining to whether or not it is anticipated that the firm would make the necessary investment given the risk profile (A, B, C, and D), the payback

INSTRUCTIONS

It is intended to establish the liklihood of private sector investment in terms of expected (average) ROI and risk, where risk is a measure of the variability or unpredictability of ROI. The variability of ROI is illustrated in the form of graphs which indicate the chance that ROI will exceed specific values for different business situations. These situations will be described in terms of expected payback period and the maximum expected investment relative to your capital budget. In answering the following questions, please assume that the business venture is consistent with the objectives and general goals of your firm and that the basic skills, facilities, etc., are all available and that all basic non-financial criteria are satisfied. It is the intent of this questionnaire to consider only the impact of financial criteria upon the investment decision assuming all other criteria are satisfied.

Investment decisions are to be stated in terms of return on investment (ROI) which is described in the following pages in the form of risk profiles. This is illustrated where the chance of ROI exceeding different values is indicated. Of specific importance is the vertical line indicating the cost of capital (R) of your firm. In each picture in the following pages, a family of risk profiles (A → D) is presented. Each member of the family has the same expected value, but each has a different variability. For the case illustrated, the expected ROI is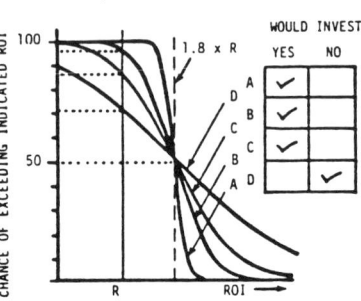
1.8 times your cost of capital (i.e., if your cost of capital is 10% then the expected value is 18%. Given this information and a statement of payback period (for example, 4 years) and a magnitude of investment (for example, less than 1% of your annual capital budget, do you think that your firm would invest given situation A?, situation B?, etc. There is a 100% chance that situation A will yield an ROI greater than your cost of capital (R) and a 0% chance that it will yield an ROI less than your cost of capital. There is a 73% chance that situation D will yield an ROI greater than R and a 27% chance that it will yield an ROI less than R. These percentages are obtained as the intersection of the risk profiles with the vertical line through R.

In the facing page an example is presented to further clarify the approach. In following pages, please check the appropriate "yes/no" box for the "would invest" decision for each of the risk profiles (A → D) and for each of the families. There are five (5) families for each of the three (3) "situations" on each page. Each page is for a different payback period; 2, 4 and 8 years. Each page therefore requires responses to 15 "would invest in A, B, C, D" questions. The "would invest" question refers to your estimate as to whether or not your firm would invest under the specified conditions.

Thank you.

Fig. 4 Instructions for responding to survey questions. (Reprinted, by permission of the publisher, from the AMA research study entitled, *Investment Decisions: The Influence of Risk and Other Factors*, ©1982 Research and Information Services, American Management Associations. All rights reserved.)

AN EXAMPLE

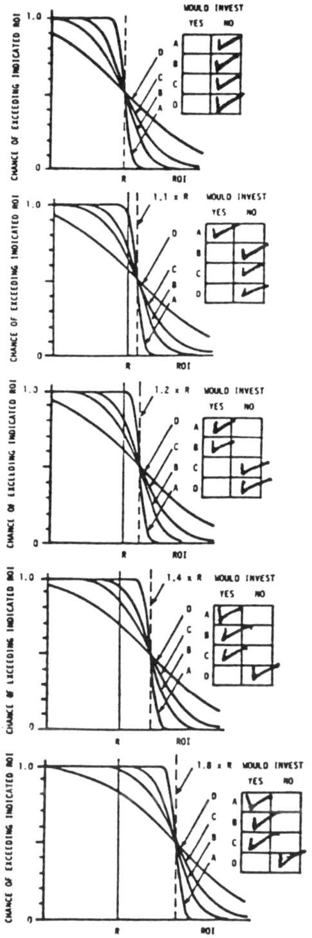

	SCENARIO	CHANCE THAT ROI IS GREATER THAN COST OF CAPITAL (R)
	A	50%
	B	50%
	C	50%
	D	50%
	A	90%
	B	67%
	C	60%
	D	55%
	A	100%
	B	80%
	C	70%
	D	61%
	A	100%
	B	95%
	C	83%
	D	70%
	A	100%
	B	100%
	C	97%
	D	82%

Fig. 5 An example of a completed survey page. (Reprinted, by permission of the publisher, from the AMA research study entitled, *Investment Decisions: The Influence of Risk and Other Factors*, ©1982 Research and Information Services, American Management Associations. All rights reserved.)

period, and the size of the investment. It is these data that lead to the determination of the investment likelihood functions.

The collected risk-related data (for the particular combination of 4-year payback period and an investment that is between 1 and 10% of the capital budget) is summarized in Fig. 6 which indicates investment likelihood (i.e., the chance that an investment will be made given a specified set of conditions) in terms of the expected ROI and level of risk as measured by the standard deviation of ROI. The scales are normalized to the firm's cost of capital. Thus the vertical scale represents the expected ROI as a fraction of the cost of capital (1.0 indicates that the ROI is equal to the cost of capital). The horizontal scale represents the risk or standard deviation of ROI as a fraction of the cost of capital. The curves represent contours of equal likelihood or probability of investment. The line marked 0.5 indicates that those investments that are characterized by points (m, σ) that fall on this line have, a priori, a 50% chance of receiving funding. There is a 50% chance of an investment when the decision makers perceive an expected ROI that is 1.73 times the cost of capital and a level of risk (standard deviation of ROI) that is 0.5 times the cost of capital.

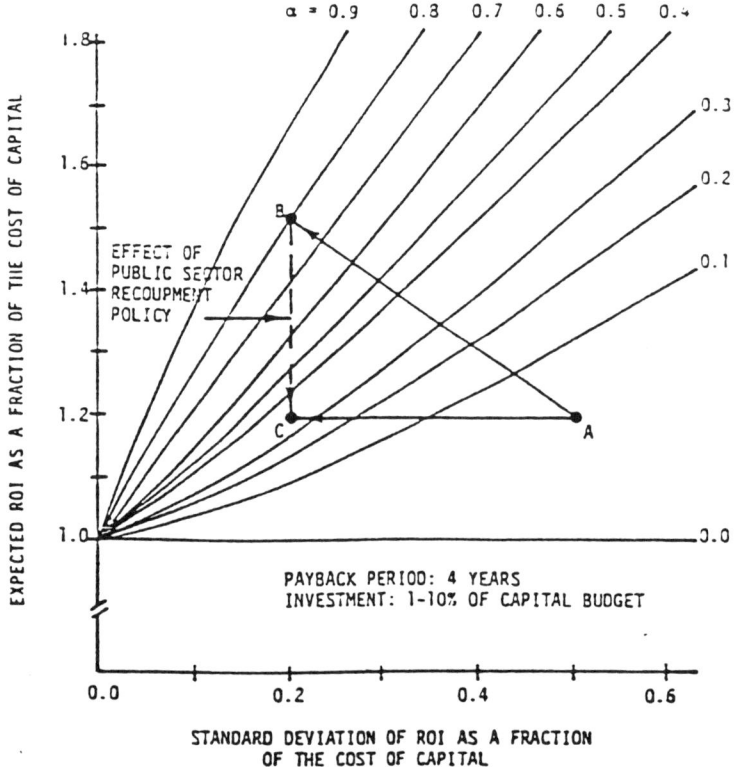

Fig. 6 Investment likelihood (α) in terms of expected ROI and risk: impact of a public-sector demonstration program.

The zero likelihood line is based upon the assumption that, except in special circumstances, it is unlikely that an investment would be made when the expected ROI was less than the firm's cost of capital. Since all points that fall on the same contour have an equal chance of being funded, the contours indicate the general risk avoidance preferences of an industry sector.

To illustrate the effects that public-sector programs can have on private-sector investment decisions, the data from Fig. 1 are superimposed on Fig. 6 as the points A, B, and C (i.e., "A" indicates the private-sector-perceived consequences of pursuing a program without government support; "B" indicates the same business situation but with the government funding the demonstration program; and "C" indicates the same business situation but with the government funding the demonstration program but requiring recoupment of its investment from future proceeds of the business). These points demonstrate the role of the public sector in affecting private-sector investment decisions through perceived risk reduction and shifting the burden of funding from the private to the public-sector. It should be noted in passing that the effect of a recoupment policy (i.e., payback to NASA for services rendered) is to drive point B toward point C and thus reduce the likelihood of private-sector investment. It is, of course, desirable for a government agency to be paid for services rendered, but this must be tempered by its effect on total benefits—the true objective should be maximization of benefits.

The significance of the investment likelihood curves is that they provide the mechanism for indicating and measuring the impact of government programs and policies on private-sector investment decisions. The value of these programs and policies results from altering private-sector decisions and stems from the benefits that would be foregone if the government program were not undertaken. The expected benefits from a public-sector program undertaken to influence private-sector investment decisions are given by

$$\text{Benefits} = \alpha_b \times \text{NPV}_b - \alpha_a \times \text{NPV}_a - \text{PVC}$$

where NPV_b and NPV_a are the expected public-sector benefits with and without the government program (i.e., development, demonstration, incentive, or other form of government program), respectively. The values of NPV_a and NPV_b can be determined using benefit/cost techniques to establish public-sector benefits. PVC is the expected present value of the cost of the public-sector program. α_b and α_a are the probabilities of private-sector investment with and without the public-sector program, respectively, as obtained from the investment likelihood curves. It is immediately apparent that the difference between α_a and α_b has a major effect on public-sector benefits. It is also evident that the value of α should not arbitrarily be taken as 1.0 with the public-sector program and 0.0 without the public-sector program. The appropriate value of α can only be established by government planning and evaluating private-sector business ventures explicitly taking into account quantitative estimates of uncertainty and the effect of the public-sector program on these uncertainties.

The affects of different government programs are summarized conceptually in Fig. 7. The general objective is to pursue policies and programs which move financial performance measures into an "acceptable region"—that is, there is a high likelihood that the private-sector will make desired investments. Fig. 8 provides examples of the impact of specific government policies and programs on private-sector investment decisions. Remote sensing (point A) of Earth resources has not been successfully commercialized since there is inadequate market to support the cost of the combined ground and space systems. However, there is an adequate market to commercialize the marketing and sales of remote sensing information products (B). This could be accomplished in a manner such that the government owns and operates the space and ground systems and a commercial business is established to market and sell the information products. In return, the market and sales organization would guarantee minimum annual payments to the government set at a level such that a significant initial investment would be made (i.e., capital at risk) with a reasonable return on investment likely—the achieved rate of return would of course depend upon the aggressiveness of the marketing and sales organization. In addition to the minimum annual guarantee, the marketing and sales organization would make significant royalty payments to the government.[7] Such an arrangement, ground-ruled out of consideration by the government in 1984 because it did not achieve total off-budget financing, would result in private capital at risk, allow the private-sector to achieve a rate of return commensurate with its marketing aggressiveness, and allow the government to share in the up-side potential as new and/or additional markets develop with increased sales. In addition, the government could, at an appropriate time in the future, still commercialize (or privatize) the space and/or ground segments.

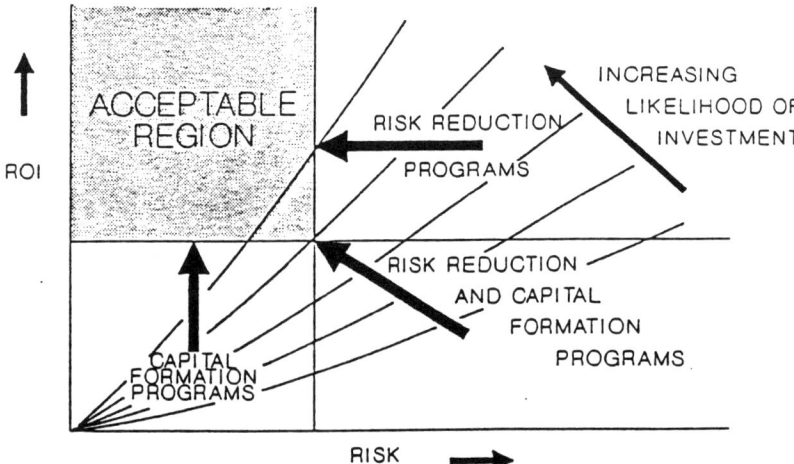

Fig. 7 Risk/return relationship and impact of government programs.

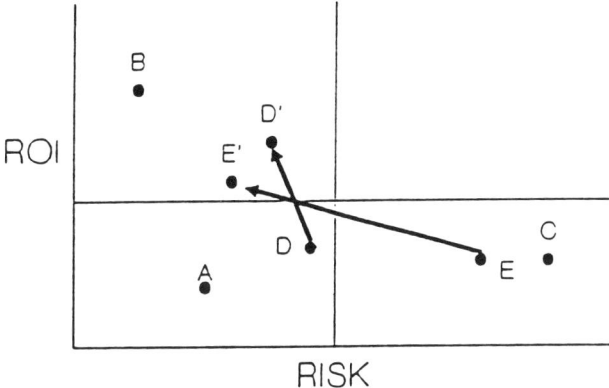

Fig. 8 Example of impacts of government programs. A, remote sensing (inadequate market); B, remote sensing market/sales organization; C, industrial space facility (ISF); D, Spacehab (without government as anchor tenant and SSDA); D', Spacehab (with government as anchor tenant and SSDA); E, commercial ELV industry (prior to 1988 amendments); E', commercial ELV industry (after 1988 amendments which encompassed government noncompetition, insurance indemnification, etc.).

Financial analyses of business plans for the Industrial Space Facility (a proposed business venture based on a free-flying space facility for long-duration space experimentation and space processing) indicated a likely low return on investment coupled with a high level of risk (C) and would have required significant government commitments to move this venture into the "acceptable region." Government anchor tenancy commitments in combination with a SSDA delaying repayment of Space Shuttle transportation fees were required to move Spacehab (D) to the "acceptable region" (D'). Government commitments not to compete with the U.S. commercial expendable launch services industry, to utilize commercial space launch services whenever feasible, to impose reasonable launch insurance requirements coupled with facility and third-party indemnifications against large losses, to continue technology development and to seek to establish a level playing field relating to international competition were required to move the space transportation launch industry from an unacceptably low-return/high-risk situation (E) to a more reasonable situation (E') that has now resulted in a number of commercial organizations making capital investments and long-term commitments to provide commercial space transportation services.

Government Planning and Evaluation of Private-Sector Ventures

Government has received, and is likely to continue to receive, proposals from the private-sector that relate to the commercial development of space. These include proposals to commercialize or privatize government programs/infrastructure elements and proposals that require government support to move a commercial venture into the "acceptable region" (as per Fig. 7). In addition, government entities need to analyze continuously their programs/infrastructure elements seeking candidates that can be called to the attention and offered to the private-sector for commercialization/privatization. In both cases, since the commitment of government resources is at stake, it is necessary for government agencies to plan and evaluate private-sector business ventures with the objective of maximizing benefits obtained from the utilization of its resources. It is important to note that benefits will normally result only when successful private-sector ventures have developed—thus, the need for government to invest in winners!

Commercialization/Privatization Proposals

As already discussed, the private-sector frequently solicits help from government agencies in order to establish viable business ventures relating to the commercial development of space. The help sought has ranged from specific programs such as government becoming an anchor tenant (Spacehab) and delayed repayment for services rendered (GEOSTAR) to general policies such as government not competing with the private-sector and government utilizing commercial space transportation services. These types of requests for government assistance (i.e., joint agreements) are likely to continue in the future. In order to respond to these requests and efficiently allocate scarce resources it is necessary to understand the private-sector decision process and the financial performance measures that are important to the investment decision.

New business ventures resulting from the joint agreement process may be internally and/or externally funded. Internal funding implies that the bulk of the funds are provided by the business organization (e.g., McDonnell Douglas) desiring the government assistance. External funding implies that a significant amount of the funds necessary for the business venture are to be provided by third parties (e.g., funds are to be obtained from capital markets as in the case of GEOSTAR). In the former case it is necessary to understand the decision process and decision criteria of the corporate entity. In the latter case it is necessary to understand the decision process and decision criteria of both the corporate entity and the capital markets.

As previously discussed, many performance measures are used by the private-sector in assessing investment decisions with the more important measures being return on investment, payback period, magnitude of exposure, net present value of cash flow, and risk. The capital markets add to these return on capital, first year of profit, first year of net cash inflow, management ability, position in the marketplace, and other quantitative and qualitative factors.

Thus government agencies should consider the above factors in the evaluation of the likelihood of commercialization. In addition, the evaluation of requests for government assistance should consider such factors as:

1) The likelihood of private-sector investment with and without the government assistance

2) The expected net present value of benefits with and without the government assistance

3) The present value (and the timing) of the cost of the assistance to government agencies
4) The types of benefits to be obtained
5) The relative importance of the different types of benefits
6) The level of risk and exposure of the private-sector relative to the public-sector
7) Business sectors (i.e., product/service types) involved
8) Recoupment of the government contribution
9) Proprietary data furnished and rights in data to be granted

There are also other factors to be considered such as:
1) Contribution to be made to the maintenance of technology superiority
2) Impact of the government assistance on an industry
3) Conflict with other agreements already entered into
4) Potential conflict with other likely agreements

In summary, there are many questions that should be answered concerning proposed government assistance among which are the following: Does the contemplated private-sector business venture (that will result from the assistance) make financial sense without the government commitment? With the government commitment? What is the chance that the private-sector will initiate the business venture without the government investment? With the government investment? Does the change in the investment likelihood warrant the commitment requested of government? What is the appropriate level of government commitment/investment? Are the resulting benefits to government worth the government investment? In order to answer these types of questions it is often necessary to perform an independent assessment of the proposed business ventures. This entails, in some cases, independent market assessments and sales forecasts and the formulation of business plans including estimates of revenues, expenses, costs, and capital expenditures and resulting financial performance measures such as annual profit (loss), cash flow, magnitude and timing of investment, ROI, payback period, and net present value of cash flow. (It is important to utilize the appropriate cost of capital for the business venture. This is a function of the firm's capital structure, market rates of return, and other factors.) In particular it is important to understand the areas of uncertainty associated with the venture, how the government assistance will lead to a reduction in these uncertainties, and how the government assistance may affect the capital structure of the venture.

The evaluation of the various forms of government assistance involves many complex analyses. This is illustrated in Fig. 9 for requests for NASA assistance received in the form of proposals from the private-sector. Upon receipt of a proposal, NASA should perform a preliminary review in order to establish those proposals that merit further analysis. The further analysis should proceed in steps, at the conclusion of each an assessment should be made by NASA as to whether further analysis is necessary or the proposal should be returned to the proposing entity with an explanation as to why it has been returned.

The first step in the financial analysis is the assessment of the likelihood of investment in the proposed entity in the absence of any NASA or other

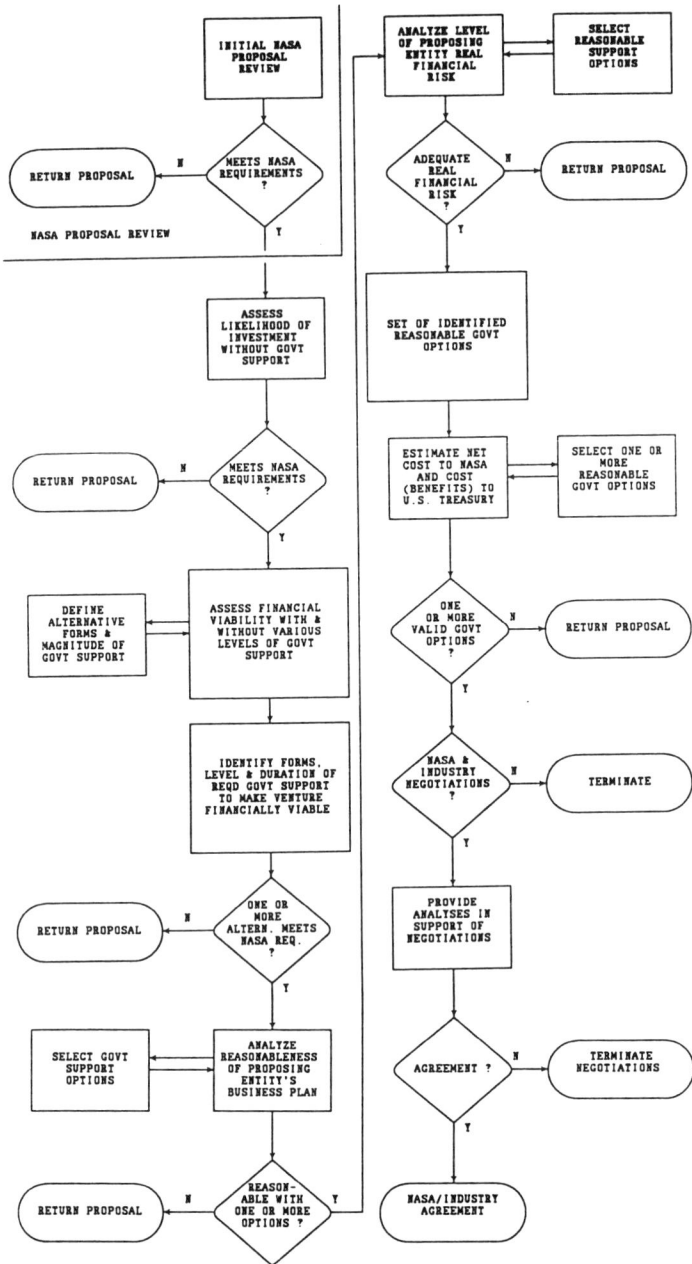

Fig. 9 Typical government analyses required to evaluate commercialization and privatization proposals: a NASA example.

government agency support. If the likelihood is too low or too high (a judgment made by NASA), the proposal should be returned to the proposing entity with an explanation of why it has been returned. If the likelihood is acceptable to NASA, then additional analyses should be performed. The next step is the assessment of the financial viability of the proposed venture with and without various levels of government support. The basic form of the support should be as requested by the proposing entity but variants of that should be investigated (such as magnitude and timing) and the financial feasibility of the venture should be assessed with each of the considered variants of government support. The particular variants of government support that will lead to financially viable business ventures should be identified in terms of the level of government support required as a function of time and the overall duration of government support.

If one or more of the identified variants of government support meet NASA requirements and are likely to lead to financially viable business ventures, then detailed financial analyses should be performed to establish the reasonableness of the proposing entity's business plan. This should entail a detailed review of such factors as the contemplated capital structure, availability of financing and the firmness of financing commitments, the expected cost of capital, availability and cost of private insurance for various risks, projected non-U.S. government market, analysis of competitive market forces, relevant management and technical capabilities, experience of the proposing entity's personnel, and other related information.

As a result of the above detailed analysis, the reasonableness of the business plan with one or more government support variants should indicate whether additional analyses of the proposed venture are in order. Given that the proposal has merit to this point, it is necessary to establish the extent to which the proposing private entity is undertaking real financial risk. This is important since public policy has been to encourage and support those business ventures that will assume real financial risk (an indication of the seriousness of the proposer's commitment and because, in many agreement forms, the government is likely to become a partner in the sharing of risk).

If NASA deems that adequate financial risk will be assumed by the proposing entity, then estimates should be made of the net cost to NASA and to the U.S. Treasury (it should be noted that returns to the Treasury may exceed the direct outlays made to support the proposed venture). In addition, the associated budget impact on NASA of the various levels and timing of U.S. government support should be made and minimum levels of government support identified.

Private Financing of Government Programs

In keeping with the thrust toward the commercial development of space and the desire to reduce government budgets by encouraging private-sector investment in areas of significant interest to government, private-sector financing of government programs has been encouraged.[16] This first requires an identification of government programs that are likely commercialization/privatization candidates and then a determination of the benefits and costs to government of various private-sector financial arrangements. As a result, it is also necessary to identify financing mechanisms which would enable the government to obtain off-budget financing for needed services in the absence of adequate federal budgets or to extend government activities within specified budgets by obtaining "additional" financing from the private sector. Off-budget financing must be accomplished in a manner such that government is not worse off than with a normal procurement. It should be noted, however, that if the benefits from a project exceed its costs, then foregoing the project because of budget constraints may be less desirable than pursuing an off-budget financing approach that in actuality is more costly than a normal procurement.

In general, the *benefits* of entering into initiatives to commercialize/privatize government programs/assets include:

1) Bringing private investment capital into the national space effort
2) Providing for shared government/industry use of a facility, thereby spreading costs among all users
3) Transferring some of government's operational responsibilities to the private sector

The principal *drawbacks* include:

1) The possibility of attempting to enter into initiatives that are poor commercial ventures because of pressures to achieve off-budget financing
2) Depending on the financing mechanisms, committing the government to sizeable out-year expenses to fund even good initiatives (a result to which the Congress might be expected to object).

The evaluation of initiatives for private financing of U.S. government programs must consider traditional procurements, off-budget financing (private financing), and commercialization (private investment). In broad terms these are as follows:

1) *Traditional Procurement*—A financing arrangement whereby the government funds a project through the normal budget process by committing budgeted funds and entering into a contract with a private company. The government maintains ownership and complete control over the design, development, and use of the project or facility. There is little or no risk to the private-sector contractor under cost-plus contracts and some risk under fixed-price contracts.

2) *Off-Budget Financing (Private Financing)*—A financing arrangement whereby the private sector provides upfront money for a project or program that is primarily for government use. In this arrangement the government controls the design, development, and management of the project or facility and pays the private entity back in the long run for its initial financial investment. Therefore, there is little risk to the private financier who is guaranteed to be paid back by the government. The intent is to relieve current budget pressures.

3) *Commercialization (Private Investment)*—A financing arrangement whereby the private sector has partial or total long-term control over the design, development, and use of a space project or program. The private sector commits capital (at risk) to an endeavor and shares in the risk and return. The implication is that commercialized space projects or programs

will be more market driven and consumer oriented than space projects under the government's control.

To date, major emphasis has been placed upon leasing options as a mechanism for obtaining off-budget financing. However, other financing schemes, including leasing variants, may be applicable and, in some cases, more desirable, depending on the program initiative under consideration. These include:[10]

1) *Private-Sector Financing/Government Ownership*—This alternative is appropriate when government desires to own the resulting facility or infrastructure element developed with private-sector financing. It includes a broad range of options with the general characteristics that the private sector finances the development of the infrastructure element, the ownership of which reverts to the government upon a payment to the private-sector. The private-sector firm may be maintained as a marketing/sales organization with appropriate incentives to encourage the development of nongovernment sales.

2) *Private-Sector Financing, Ownership, and Operation/Government Leasing*—This alternative is appropriate when it is not necessary for government to take title to a facility or infrastructure element developed with private-sector financing and there is a reasonable likelihood that nongovernment markets will develop. It provides for government lease of the facility. It includes a broad range of options with the general characteristics that the private-sector finances the development which is then leased by both government and nongovernment customers. In addition to the lease payments, the government may provide other payments (i.e., related to the private-sector investment) that will affect the negotiated government lease rate.

3) *Shared Ownership*—This alternative provides a normal partnership or corporate structure established to develop, own, and operate space infrastructure elements with the government being a partial owner of the firm. This alternative is appropriate when it is not necessary for government to take title to a facility developed with private-sector financing and there is a reasonable likelihood that nongovernment markets will develop. A partnership structure has the advantage of allowing the government to acquire the facility if future conditions and actions necessitate government ownership. A corporate structure has the advantage of broader range of private-sector participation and marketability of holdings but the disadvantage of the government not easily acquiring ownership of the infrastructure element if future conditions and actions so dictate.

Commercialization candidates may be identified through 1) private-sector efforts with resulting unsolicited proposals for government/industry joint ventures submitted to government (e.g., NASA); 2) other government agency initiatives (e.g., OMB recommendations); and 3) government agency (e.g., NASA) identification of activities that are appropriate for commercialization considerations. In other words, proposals for commercialization opportunities may be initiated from both outside and inside government and may be for new activities and infrastructure elements or may be for current or planned government programs. Thus, government agencies must have the capability and resources to evaluate proposals that they receive from outside the agency as well as to be able to identify and evaluate their own programs that may be appropriate candidates for commercialization.

A major problem has been to identify those projects, or segments of projects that may be good candidates for private-sector financing. The identification of good commercialization candidates can benefit from a screening process which addresses the following questions for each commercialization candidate:

1) Is it likely that the project will lead to space infrastructure development?
2) Is the project critical to agency objectives?
3) Is the project important for national security reasons?
4) Is significant R&D required to develop the project?
    a) Is the technology sufficiently developed to be applied on a commercial basis?
    b) Is this government agency the only organization with the necessary technological skills to adequately develop the project?
    c) Is the technology mature enough to lead to desired end products within agency scheduling, contract, and cost constraints?
5) Does the project require government agency to have extensive design and quality control specifications?
    a) Can agency priorities and needs be contractually specified a priori?
    b) Is there a contractor with a proven track record or experience with this type of project?
6) Does the project's success depend on the performance of other projects or are other critical projects dependent on it?
    a) Does the project depend on the completion or operation of a large number of other projects?
    b) Is the project dependent on a single project for its total usefulness?
    c) Are there a large number of government scientists/employees involved in the project?
    d) Will the scheduling of this project interfere with the efficient production or usage of other projects?
    e) Do many projects depend on its timely completion before they can be begun or finished?
7) Does the project face significant political obstacles?
8) Does the project face significant policy obstacles?
9) Does the project face significant legal obstacles?
10) Does the project have significant potential for nongovernment use?
    a) Does the project have secondary applications?
    b) Are there many alternative private-sector applications?
    c) Will the U.S. government require exclusive use of the infrastructure?
    d) Does the technology offer innovative "breakthroughs?"
    e) Can transportation assurances be made by the government?

11) Is the project likely to be eligible for government indemnification if insurance is not available?
12) Are there any other reasons for not commercializing the project?

Answers to the above questions should establish the need for further consideration.

Government cost reductions from off-budget financing are only possible if: 1) a decision has to be made that the project or program is to be undertaken and funded; 2) there is a nongovernment market for the program's output (either goods or services) that is identified, real, and large enough to provide sufficient revenue to yield a competitive ROI to a private investor; and 3) government is willing to share in the investment and/or risk. These criteria present significant hurdles to cost reduction, particularly in risky R&D ventures. In addition, they may require special legislation to overcome long-standing procurement regulations.

The cost reductions discussed above primarily affect the sponsoring agency's budget. Other types of cost reductions can occur from shared private financing arrangements that only affect the overall federal position: that is, public benefits may be foregone. A delayed program start due to lack of funds might cost the government more in the long-run because of lost benefits (that may or may not be quantified). This line of reasoning will not benefit the immediate problem of an agency's budget constraints, however it can be important to central budgeting organizations such as the OMB.

There are many examples of federal programs where user fees and other charges do not cover the costs. Virtually all programs consider the R&D portions as sunk costs and only charge a fee for operations. NASA wind-tunnel usage is an example where the cost of developing and building the facilities was made for government mission purposes and is not factored into charges to industry for facility use.

The government can also accrue surpluses on its programs. The Social Security program is an example where current charges (i.e., taxes) exceed current outlays and the surplus goes into a special trust fund. The Highway Trust Fund is another example where the money collected is then spent for a particular purpose, but where revenues exceed expenditures.

Government financial commitments are subject to many complicated regulations. There are some general principles that highlight the differences in planning and executing government expenditures from corporate expenditures. These stem from both Congressional actions and from Executive pronouncements. Changes and exceptions to the rules can be made; however, any significant deviation will involve a major effort on the part of the requesting agency and will be subject to intense political scrutiny. The time delays alone in getting special legislation or regulatory action could jeopardize a financial "deal" between government and industry.

Congress has enacted a variety of appropriations laws, ranging from the broad principles of the Antideficiency Act (first passed in 1870) to the Competition of Contracting Act. Judicial interpretations over the years have added further interpretations to these laws.

The overriding principle of the Antideficiency Act is the concept of pay-as-you-go. No government contracts are allowed that commit funds that have not been appropriated. Because of the annual budget cycle, commitments extending beyond 1 year are prohibited. There are exceptions, but they commonly focus on the question of whether the payments are "contingency." Only by having Congress pass special legislation permitting multiyear funding for specific projects (or classes of projects) can an agency enter into contracts that promise payment over more than 1 year. The pay-as-you-go principle restricts the flexibility of government officials in entering into leases, in negotiating "market guarantees," in agreements for "termination liability," and in many other areas that can influence the completion of a private financing arrangement.

Additionally, the OMB has issued regulations such as A-104 which sets out specific conditions and analyses required for entering into leases of capital assets. A-104 is designed to make agencies use the least cost method of financing capital assets. A-104 does not preclude (but weighs heavily against) the use of a lease, and it permits agencies to use nonmonetary considerations in decision making.

In general, leases are more expensive than upfront acquisition because they involve paying the private sector for the use of its capital. However, there are exceptions where leases can be financially advantageous. First, if there is a proven commercial market in addition to government use, then there is a basis for shared ownership or use of the assets with cost reductions to the government.

Second, if significant program delays will result from lack of financing, there could be long-run savings through a lease arrangement. These savings will not appear in the lease/purchase analysis required by A-104, but will show up in future years by lowering program budgets from what they would be if there were delays.

Third, there could be financial advantages to the government of an early start in some programs that would appear as social benefits. Such benefits to the public will not appear in any government financial budget, and in fact, may be nonquantifiable (that is, in the form of improvements in quality of life, etc.).

Finally, a lease could be financially beneficial if the project is of short duration or there are convenience factors to the government. For example, in the case of the Jet Propulsion Laboratory (JPL), the government was leasing commercial space at high rents and needed additional office space. By consolidating the space in one building and allowing the JPL to provide the new facility with a lease-back to the government, the government was able to save money and avoid upfront financing. (However, the arrangement was cheaper only in comparison to the cost of commercial leases; it was still more expensive than if the government provided upfront capital and built the facility on land it owned.)

In summary, there are many questions that should be answered concerning private financing of U.S. government programs among which are the following: How does the proposed financing option compare with the option of total U.S. government funding? Is there adequate budget to achieve government funding? If it is unlikely that government funding will be forthcoming, what costs (benefits foregone) are associated with delays? What costs and cost savings may accrue to the concerned government agency and Treasury from private financing? What is the budget impact on the concerned agency? Are there alternatives to the specific proposed financing arrangement? What is an appropriate cost of capital to be used in the analyses? What is the minimum level of government commitment that is likely to be required in order to maintain private-sector interest? Will legislative action be required? What is the true level of risk being assumed by the private sector? By whom? Is the proposing entity's business plan realistic? What other government commitments are likely to be required? Is it possible for government agencies to systematically identify projects that are good candidates for private financing? What methods/procedures should be used? What criteria should be used?

An overall process for identifying commercialization/privatization opportunities and analyzing/evaluating resulting proposals is indicated in Fig. 10.

Summary

Both government (public sector) and industry (private sector) undertake programs aimed at influencing private-sector investment decisions. These investment decisions range from capital investment decisions and corporate strategy decisions of firms to capital investment decisions of individuals. The private sector undertakes research, development, demonstration, and advertising/promotion projects that lead to increased private-sector investments and to the formation of commercial ventures. To a large extent, these programs are undertaken with the expressed intent of maintaining or increasing the value of the firm. A vast array of government programs and policies are initiated with the specific objective of developing technology and/or creating the environment which will lead to increased private-sector investment and the formation of commercial ventures which are in the public interest. These encompass research, development, and demonstration projects as well as incentive, regulatory, standards, information dissemination, and other programs and policies aimed at increasing the competitiveness of U.S. industry.

Private sector RD&D programs are undertaken with the specific intent of reducing the risk associated with investment opportunities. The risk reduction increases the likelihood that investments will be made. Many public sector programs are undertaken with the specific intent of reducing the risk as perceived by the private sector as well as partially shifting the burden of funding (reducing capital requirements, exposure) from the private sector to the public sector. The intent of these government programs is to influence or alter private sector investment decisions through the combined effects of risk and exposure reduction.

The benefits from RD&D and incentive, regulatory, and other programs and policies (the former undertaken by both government and industry and the others undertaken by government) are the result of altering private-sector investment decisions. If investment decisions are not altered or likely to be altered, then there are no benefits from the government programs and policies. This then leads directly to the fact that the benefits or value

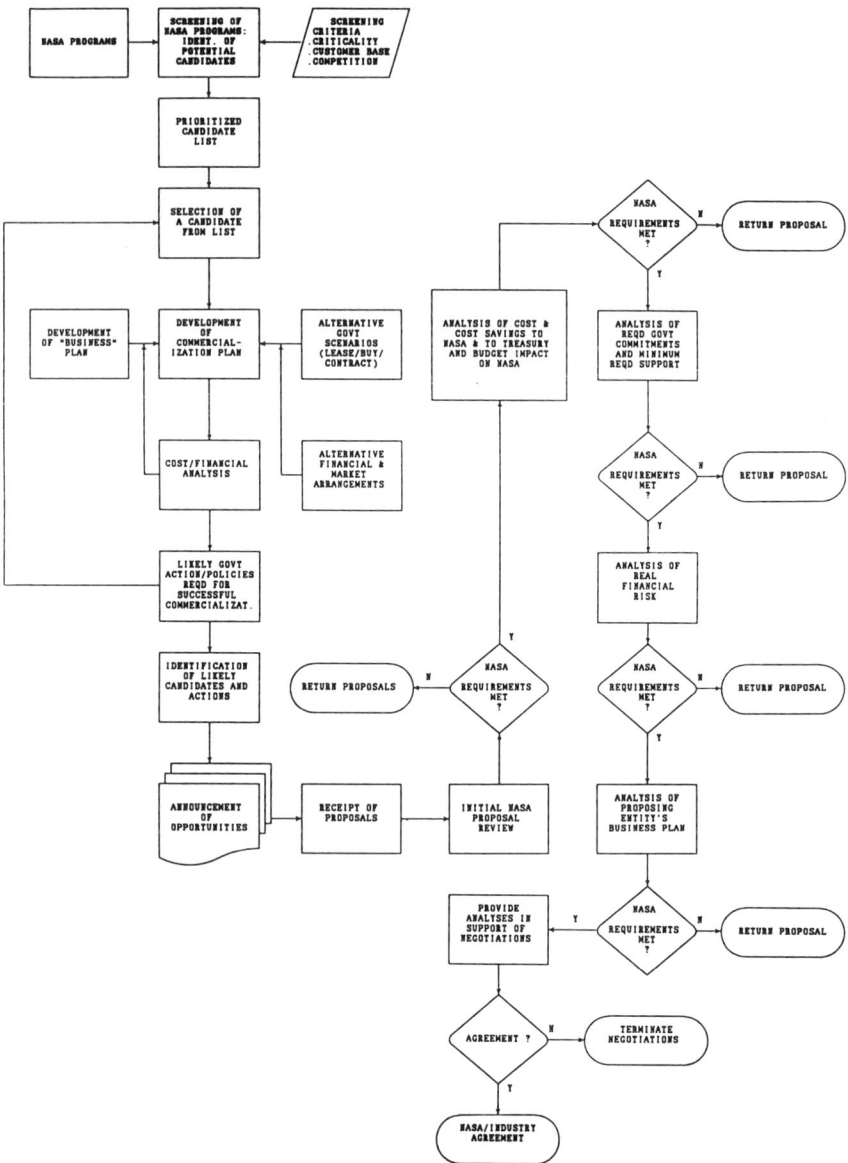

Fig. 10 Typical government analyses relating to commercialization/privatization candidates and proposals (a NASA example).

of an RD&D program to a firm or government and an incentive, regulatory, or other program to the government, must explicitly take into account the likely effect of the program on private-sector investment decisions. In order to predict the effect of these programs on private-sector investment decisions, it is necessary to establish functional relationships between important investment performance measures or investment criteria, evaluate the investment opportunities with and without the program, and then establish the effect of the program upon the investment decisions through the change in the investment performance measures.

Individuals in the public sector do not have direct control of investment decisions necessary to capitalize upon the results of the RD&D or other programs and policies. They can influence these decisions but they do not make them. Thus when planning, evaluating, and justifying programs aimed at altering investment decisions, the benefits which can be obtained from these programs can only be predicted or estimated. This prediction has a significant effect upon the value of the program since the value derives from the effect it has on altering investment decisions. The predictive mechanism is the investment likelihood functions.

Investment decisions are based upon multiple attributes—that is, an investment decision takes into account all available information such as profitability, timing, return on investment, and many other factors. Some are more important than others. It was found that in many instances the most important factors affecting an investment decision are expected ROI, risk as measured by the variability (standard deviation) of ROI, expected payback period, and expected exposure. A survey was conducted and an analysis of returns resulted in the establishment of functional relationships between expected ROI, risk, payback period, and exposure and the likelihood of private-sector investment. These investment likelihood relationships serve as the bridge between the public and the private sectors. These relationships allow the public sector to predict in a somewhat rational manner the effect that their programs will have on private-sector investment decisions and to take this into account when evaluating and planning alternative courses of action.

The planning and evaluation of public-sector programs aimed at influencing private-sector investment decisions require an understanding of 1) private-sector investment decisions, 2) the type of firms and investment decisions that may be affected, and 3) the type of effect or impact that the program is likely to have on financial performance measures. It is the interrelationship of these factors through which barriers to commercialization can be identified and programs developed that efficiently remove the barriers. Different tools are appropriate for working on different barriers. When risk caused by market uncertainty is the principal barrier, R&D, demonstration, market commitment joint endeavor agreements, revenue subsidization, and so forth may be appropriate programs, whereas tax credit, loan guarantee, and cost reduction joint agreements are likely to be less effective programs. When inadequate ROI is the barrier, then attention should be focused more on taxation, cost reduction joint agreements, revenue subsidization, and recoupment policies. The matching of programs and policies to barriers requires an understanding of commercial business ventures and leads to the need for the public sector to plan and evaluate private-sector business ventures in order to have a clear understanding of the impact and need for supportive government programs.

References

[1]Lovell, R. R., Knouse, G. H., and Weber, W. J., "An Experiment to Enable Commercial Mobile Satellite Service," National Telecommunications Conference, Galveston, TX, Nov. 1982.

[2]"Communications Strategic Plan (Draft)," NASA, 1990.

[3]"Commercial Space Ventures: A Financial Perspective," U.S. Dept. of Commerce, April 1990.

[4]Greenberg, J. S., and Palmer, D. T., "Ship Coal Bunkering Facilities," Rept. MA-RD-920-82012, March 1982.

[5]City of Baltimore District Heating Assessment, Final report, Resource Development Assoc., Marietta, GA, Sept. 1982.

[6]Tax Equity and Fiscal Responsibility Act of 1982.

[7]Greenberg, J. S., Statement Before the Committee on Science and Technology, U.S. House of Representatives, June 28, 1983.

[8]Public Law 98-575, *Commercial Space Launch Act*, 98 Stat. 3055, Oct. 30, 1984.

[9]*Commercial Space Launch Act Amendments of 1988*, Congressional Record, Oct. 21, 1988.

[10]"Commercial Development of Space: Financing Mechanisms," Final report prepared for NASA Headquarters, Princeton Synergetics, March 1989.

[11]"Space Projects: Improvements Needed in Selecting Future Projects for Private Financing," GAO/NSIAD-90-147, U.S. General Accounting Office, Sept. 1990.

[12]Bush, G., "U.S. Commercial Space Policy Guidelines," Feb. 12, 1991.

[13]Mannix, J. G., "Space Commerce Opportunities Plan," briefing to the NASA Space Commerce Steering Group, May 1991.

[14]Greenberg, J. S., "Government Tools for the Support of Commercial Ventures," IAA-84-216, 35th Congress of the International Astronautical Federation, Oct. 1984.

[15]"A Method for Identifying Space Infrastructure Commercialization Candidates," Final report prepared for NASA Headquarters, Princeton Synergetics, Inc., July 1989.

[16]Greenberg, J. S., "Private Financing Sought for Space," *Aerospace America*, Sept. 1989.

[17]Stone, B. A., "Economics Benefits of Commercial Space Activities," *Acta Astronautica*, Vol. 19, No. 9, 1989.

[18]Greenberg, J. S., "An Assessment of the Effect of Efficiency Standards on the Appliance Industry," Rept. 80-177-1-0, Econ, Dec. 1980.

[19]Greenberg, J. S., "A Financial Risk Analysis of a District Heating Business Venture," *Proceedings of the 15th Intersociety Energy Conversion Engineering Conference*, Aug. 1980.

[20]Hertz, D., "Investment Policies that Pay Off," *Harvard Business Review*, Jan.-Feb. 1968.

[21]Greenberg, J. S., and Edelman, F., "The Assessment of Uncertainty and Risk," *Financial Executive*, Vol. 37, No. 8, 1969.

[22]Greenberg, J. S., *Investment Decisions: The Influence of Risk and Other Factors*, American Management Associations, New York, 1982.

Economics of Strategic Planning

Joseph Fuller* and Kevin Lacobie†
Futron Corporation, Bethesda, Maryland 20814

Introduction

CONVENTIONAL wisdom in government and the aerospace industry is that it is virtually impossible to do long-range strategic planning in political and advance technology environments. Politics and technological innovation, it is thought, are so unpredictable and volatile that long-range planning is not worth the investment of energy and resources. Any long-range plan will be overtaken by events, making the effort useless. Actually, the opposite is true; it is because of unpredictability and economic uncertainty that strategic planning is necessary. It is the economics of uncertainty and change that make strategic planning, a system that anticipates change and guides the organization's response, an important dimension in organization.

Nowhere are volatility and uncertainty greater than in our current national space programs. Both civil and military space programs are ideal candidates for strategic planning. Both are heavily influenced by geopolitical, technological, and economic considerations. Additionally, the time span from initiation to completion in space projects is often measured in tens of years. It is ironic that in today's environment, and in the foreseeable future, the primary considerations are economic: "Is the return on investment worth the effort?" Part of the answer lies in selecting and deploying the correct mix of programs, resources, and other assets. Strategic planning, while not the total answer, can contribute substantial insight and direction to both the government and the aerospace industry.

Copyright © 1992 by the American Institute of Aeronautics and Astronautics, Inc. All rights reserved.

*President, also a 25-year veteran of NASA and the aerospace industry. Mr. Fuller is experienced in all aspects of space flight development and operations, and has performed cost and economic analysis and strategic planning for a variety of programs.

†Economist. Mr. Lacobie is concluding a doctorate in Economics at George Mason University, Fairfax, Virginia. He has published analyses of small satellite systems, and is an instructor in managerial economics.

In this chapter we briefly examine the relationship between economic theory and strategic planning, especially in an organizational setting representative of government and industry. After establishing an economic basis, a strategic planning approach, which we believe to be extremely appropriate and adaptive to the modern firm, is discussed. Finally, examples of strategies arising from economic factors are placed in the context of strategic planning. What we attempt to demonstrate is that strategic planning is critical to succeeding in an environment of risk and uncertainty, and that even though the field lacks maturity, strategic planning is sound economics. The intent is to raise the readers' sensitivity to its value and provide a basic introduction to a general strategic planning approach.

A Strategic Planning Primer

Strategic planning is the act of the organization accommodating itself to its environment and the attendant uncertainty. *Strategy* originates from Greek military terminology, for the art of the general. In this context, strategy is to be contrasted to *tactics*, the practices of the field officer. The field commander faces a given task, a hill to charge, a flank to outmaneuver. He can execute this task with pre-established tactical practices, that is, with analytical knowledge of appropriate production given known constraints in the environment.

The strategic commander, however, is the one who chooses which hill to charge, what environment to face. The general has the responsibility to be aware of changes in the data that will affect the tactical appropriateness of his field charges. Furthermore, the general must broadly deal with changing circumstances: changes that originate from the enemy's own reaction to his strategies, (hence, the strategic behavior described in *game theory*) and other unpredictable changes (weather, internal dysfunctions, etc.) that affect the battlefield. (See, for example, Dixit and Nalebuff.[1])

Planning often connotes detailed, precise, and strict blueprints of an organization's present and future actions. The plan is conceived as the function of genius, analysis, and scientific precision. Events are meant to occur in an orderly sequence. The hubris associated with such a conception should be noted. To be realized in such form, the plan must depend on the complete predictability of future events. However, we do not attribute this kind of predictability to the social environment. In fact, the rigid 5-year plans constructed by Soviet agencies were found to be empty exercises, because real human events do not occur in such orderly sequence.[2]

We reject this definition of planning and submit that the planning in most firms has never been of this variety. In successful organizations, planning practices do account for risk and uncertainty and organizational behavior. The purpose of planning is, not to produce an exact blueprint, but, rather, to conceive a strategic plan that is an ordering principle for the firm, a point of focus and alignment. This conception of the firm as organization is elaborated below.

Change, Complexity, and Uncertainty

As the Greeks understood, any effective organization must have functions in place to accommodate changes in their organizational tasks or environment. Today, this function resides mostly within the leadership. Leaders are the entrepreneurs. They adapt to change, and they initiate it. However, the traditional economic theory of the firm tends to concentrate on other matters, underrating the entrepreneurial principle.

Within a relatively stable environment, the economic problem of the firm is mainly one of technical optimization. A firm can maximize its production and profits if the data for demand, costs of inputs, and available technological processes are all known. Since most production occurs in multistaged processes, time is a required input into any firm. Thus, the firm must accommodate time by planning its operations. However, as long as the environment remains stable, planning itself remains merely a technical problem of optimizing production.

Change in the environment is not necessarily problematic, if the changes can be anticipated. Stable change, such as demand growth due to population, or the well-planned announcements of military cutbacks, can be easily accommodated in the firm's production plans, as long as the firm is aware of these trends.

Unexpected change is the problematic aspect of firm production. In an uncertain environment, a firm faces several economic problems. A firm's specific assets are at risk should their output demand or input supply suddenly change. Furthermore, the normal accounting conventions for incorporating time become shaky in an uncertain environment, leading to possible inoptimal production arrangements. Likewise, forecasting becomes unreliable in the face of uncertainty, so the firm loses much of its long-range planning ability.

The traditional economies of yesteryear may have been much less complex, but they faced a harsh and unpredictable (to them) natural environment. While science has conquered many of nature's vagaries, space companies do face an environment unique and unpredictable on its own terms—space! Furthermore, some natural systems, although simply constituted, exhibit complex behavior on the whole, increasing our uncertainty about nature's outcomes. ("Chaos" theory is actually a theory of complex orders. See Ref. 3.)

Today's Uncertain Environment

However, it is the social environment that contributes to the complexity of the world today. Effective customer demand in a global, heavily populated economy, displays a wide variety of interests. Even space companies face this today, where their ultimate consumer could be anyone from a grieving family looking for a stellar burial place, to a covert operations organization. The global economy brings with it the rapid and unpredictable changes of regional and national politics, 24-h stock markets, currency rates, and international finance. The social, economic, and political environment that impacts the local organization today is immense and not yet completely understood.

Not that the local environment is not also complex and unpredictable. Space companies are already too familiar with the uncertainty associated with representative democracies. The U.S. government is notorious for dramatic and frequent changes in space policy, so interested space ventures have had to adapt, invest heavily in developing political connections, or drop out. Export controls, trade regulations, and rapidly changing technologies are all of major concern to any space organization. Even mundane shifts in capital gains taxes or industry/university relations can have a major impact on an aerospace corporation.

Furthermore, today's space companies are no longer just subcontractors to the federal government. *Competition* is a keyword in today's space corporation. The new environment includes competition from all corners, including backyard ventures and aggressive small firms. The rapid changes in formerly Second World countries and the continued ascendence of Third World countries, like Brazil and Indonesia, add to the uncertainty of the competitive environment.

All this uncertainty in a complex environment accumulates over time, so that the future horizon of predictability and stable planning is dramatically shortened in our highly unstable international marketplace. *Planning* is no longer a matter of staking out a detailed 5-year production schedule and sticking to it. Plans cannot depend on the stability of input prices or demand, nor even on the technology that incorporates their production functions. In fact, all of the traditional assumptions for textbook economic decision making are missing in today's long-term environment.

Yet, in spite of the omnipresent uncertainty they face, firms and individuals must still act. Uncertainty becomes a core problem, but of a grander scale than simple production economizing, in today's organization. Uncertainty, not mere optimal production arrangements based on traditional economic assumptions, will drive the need for organization within the firm.

What Is Organization?

The economic theory of the firm has always asked, "Why organize?" The textbook treatment of this question assumes a static environment for the firm. Then production becomes a matter of economic technique. Given a choice of available technologies and the costs and availability of required inputs, the firm manager can calculate the optimal allocation of all inputs (land, labor, and capital). Knowing consumer demand, a manager can calculate the price, profits, and production quantity of its output. If production is in stages and requires time, or the firm must face predictable changes in input costs or demand, the manager can incorporate the firm's time preferences and interest rates to plan future production capacity. As long as the data are available, the economic allocation problem is tractable, and the firm has a rational basis for planning.

However, this treatment of the firm does not really define why the firm is necessary. Surely, all the inputs must come together; necessary raw materials and land must be fed into capital machinery with human knowledge and labor input, but this can be arranged through appropriately specified contracts. If contracting costs are minimal, all of the firm's arrangements can also be arranged as market contracts. The textbook "theory of the firm" is less a theory of organization than a description of a locus point of transactions.[4]

Organization can be partially explained by transaction costs. In production decisions, the manager faces a trade-off between obtaining production from the market or from within the firm. Involving the market can entail high transaction costs and loss of control. However, production within the firm can entail high monitoring costs and burdensome risk. When transaction costs become too high, more production is retained within the firm.[5] This economic problem is known in aerospace organizations as the "make or buy" decision, for which guidance can be incorporated into a strategic plan.

Transactions cost theory, however, remains a static explanation of the firm. Given the costs of transacting, monitoring, and so forth, the firm can decide on an optimal arrangement of contracts and resource allocations. Given an appropriately sophisticated manager, these considerations remain part of the *tactical* problems of firm decision making.

We wish to stress that the primary need for organization comes, not from economic allocation problems, but from the economic problems caused by uncertainty. This is where *organization*, true cohesiveness within the firm, becomes important. An aligned, focused response to changes in the environment is the strategic key to firm survival and prosperity. (See Demsetz's "The Theory of the Firm Revisited," in Ref. 6, especially pp. 158–160. This concept of cohesive organization seems to have first been analyzed by Selznick.[7]) Strategic planning provides this capability. In a static, completely known world, the organization need only worry about the tactical arrangement of resources and contracts. In the uncertain environment common to today's economies, the organizational task is more complex, and solutions must be found beyond static economic models.

A Strategic Plan as a Model of the Firm

The strategic problem of organization decision making is to deal with uncertainty. Our conception of strategic planning resolves this problem by viewing the planning function as a *modeling* exercise for the organization. Through strategic planning, the organization defines itself and becomes cohesive and consistent in its pursuits. The function, in turn, empowers the organization to comprehensively tackle uncertainty.

Why a Model Concept?

Several experts have recently written of a different view of the strategic plan. Peter Senge, in *The Fifth Discipline: the Art and Practice of the Learning Organization*,[8] Tom Peters, in *Thriving on Chaos*,[9] and Stanley Davis, in *Future Perfect*,[10] all hold that it is the people within the organization that are key to its success and that it is they who must be empowered to act in the best interest of the organization. This view focuses on the idea of producing entrepreneurial behavior throughout the organization as a means of injecting vitality as a stimulus to growth and excellence. These experts and others have also written of the success of large organizations made up of numerous smaller units that behave as entrepreneurial businesses. Regardless of the form, what is desired is an organization with the economic criteria of being efficient and productive.

Two principles could increase the chances of this criteria being met: first, all members operating under a common purpose, sharing a common vision, and working toward common goals. Second, reduction of the uncertainty surrounding the decisions of the organization's members can reduce risks and improve rewards. The basic premise here is that a macro model of the organization operating as a strategic plan can resolve the economic problem of confronting uncertainty.

The approach we have taken in this work is to view the strategic plan as a model of the organization residing in an environment of uncertainty to which everyone must constantly mold and adapt. What is desired is that each member carries in his head a dynamic model of the organization that would allow the person to react to a change in the environment, be it a threat or an opportunity. This is what entrepreneurs appear to do so well. For the concept to work as well in a large organization, its leaders must strive to create an internal environment conducive to creating the commitment found in entrepreneurs.

Model Attributes for Strategic Planning

To work as a model of the organization, the strategic planning function must meet several conditions. 1) The model should be comprehensive, incorporating all significant aspects of the organization. 2) There should be a relationship among the elements of the model; the elements should influence one another. 3) The model should lend itself to continuous testing. 4) The model should be sound, responding appropriately to changes in the business environment that could potentially impact the organization. Specifically, such a model should include the following attributes:

1) Communicate a clear distinction between the present and future organization
2) Demonstrate an understanding of current environment and trends
3) Explicitly accommodate strategic issues
4) Provide for clear measurement of achievement and progress
5) Show genuine care and consideration for the human element
6) Allow ample freedom for innovation and creativity
7) Encourage similar planning and strategic thinking at all organizational levels
8) Be simple and concise

Although simple in form, the guidance from strategic planning is important to aerospace organizations. The organization faces a complex environment, and without any internal stability and cohesion, its members cannot assess their own changes as positive or negative. The strategic plan becomes an institutional guidepost, an architecture, for all organizational endeavors.

Focusing on Strategic Issues

Good strategic planning requires concerted effort and dedication, substantial information about the organization and the environment, plus a keen understanding of strategic issues; all are generally available in the organization, but usually in imperfect forms. The trick is to capture this information and assess its meaning for the organization's future. An assessment of both the internal and external environments should yield much of the basic information required about the organization in the context of the marketplace. At this point particular emphasis should be placed on identifying the critical strategic issues for resolution, including those involving organizational culture. Once strategic and cultural issues are understood, they can be combined with the other planning elements and incorporated into a long-range planning framework to produce the desired strategic planning model. Eventually, the model can be developed and refined into a form easily communicable to the work force and, as appropriate, to customers, suppliers, and other stakeholders.

During strategic assessment, the organization will identify issues of their environment whose change or outcomes could substantially affect the character and capability of the organization. These issues involving the interface between the organization and its environment, of necessity, involve risk and uncertainty. Thus, these issues become strategic in nature, and require the focus of the organization's strategic modeling effort.

An example of an important area demanding strategic thought and analyses is the international space arena, with its growing number of entrants, where competition and cooperation must exist side by side. Another strategic issue of concern are the joint civil and military space programs and the many opportunities that stem from their need for similar resources and infrastructure. In such an environment, numerous critical policy decisions will be made, with their impact affecting the character, content, and scope of the nation's space program. Such decisions usually directly affect the livelihood and welfare of the nation's citizenry and its industries. Strategic and long-range planning will reveal these strategic issues, salient information, and, ultimately, the proper course for government and industrial organizations.

A Strategic Planning Modeling Prescription

This section prescribes a model for strategic planning for the aerospace environment. The model works in government or industry, but its general nature makes it applicable to any environment. Each of the model elements—purpose, vision, mission, goals, strategies, and culture—are important to creating an organization's model, and are briefly defined below:

1) *Purpose*—This answers the questions, "What business are we in?" and "Why are we in it?" Without answers to these basic questions, testing the model would be extremely difficult. As with the rest of this prescription, it is not anticipated that the answers will remain static.

2) *Vision*—The vision is composed of many components that collectively provide a visualization of not what the organization currently is, but what it hopes to become. The necessity for change should be self-evident and create a tension that can only be relieved by satisfactorily progressing in the direction of the vision. Is this a true prescription for every organization, even the successful ones? Emphatically yes! No organization can afford to rest on its laurels in a constantly changing environment. Not only must organizations constantly test the hypotheses and assumptions of their models,

but they must continually scan the environment in anticipation of positive and negative change.

3) *Mission*—This is the organization's holy crusade! It is the glue that binds the organization together. Although the organization will want to allow much variety and freedom, it must define a mission as the one cause that is universally understood and applied. The mission is the end to which all the organization's output—its products and services—are applied.

4) *Goals*—These are the means through which the organization's progress toward its vision is measured and assessed. Goals will break down vision into its component parts, creating for each a minivision, milestones, and means to qualitatively and quantitatively measure and assess progress. Again, with goals in place as an internal measure to real success or failure, the organization's strategic plan model can also be tested for appropriateness and validity.

5) *Strategies*—Strategies stake out the battlefields that have been determined to be critical to the success of the organization. Strategies will identify obstacles to victory and methodically plot their demise. The assumption being that careful attention to changes in the environment, plus the application of certain methods, technologies, and behavior, will facilitate the removal or moderation of obstacles, therefore facilitating the attainment of the goals and vision. Strategies are directly traceable to strategic issues, in that, one measure of whether an issue is strategic is whether or not it is an obstacle or a contribution toward the accomplishment of an organization's vision and goals. If these conditions are met, the issue is deserving of being addressed in a strategy.

6) *Culture*—Culture is a composite of organization values and norms, and can be influenced to enhance strategic objectives. That is, values and norms, selectively emphasized, can facilitate the development and attainment of an organization's aims. While certain values are formed early in one's life and are difficult to change, the ability to influence normative behaviors can result in a cohesive and effective organization. An organization's tasks and its environment will influence its cultural makeup. Culturally, an advertising agency will almost certainly differ from a heavy manufacturing firm. One critical aspect of the approach proposed here is the inclusion of a planning component in the organization's culture. ("Corporate culture" has become the popular subject of research and education in management theory. Deal and Kennedy[11] provide a thorough managerial perspectice on culture, whereas Casson[12] provides a fresh new look from an economist's point of view.) A necessary condition for this approach to succeed is for the entire organization to be imbued with the strategic planning model and to be dealing with uncertainty within the context of the principles described herein. (cf. Casson[12] and Milgrom and Roberts.[13])

The general model described here must be adapted to particular situations and environments. A user might have to add to the model for his or her own application. For example, the model for a firm in its early development stage might be substantially different than for a firm in the peak of its prowess in the marketplace. However, the model is fairly universal: it will work for government, industry, and nonprofit organizations. It will also work for units and teams within the organization. Actually, if it did not function effectively at all organizational levels, the basic approach would be defective. It is also important to note that the entity needs to be in continuous touch with its internal as well as external environment, that members communicate and share this information effectively, and that genuine care and consideration of the human element is included in the design.

According to Senge,[8] for a system like our planning model to function, what is needed is not merely employee acceptance, but employee commitment. Leaders in the organization must provide an internal environment that encourages personal development and the attainment of individual goals. The strategic planning model calls for the free sharing of information. In such an organization, members and teams are equipped to act in the best interest of the firm. In this scenario, an organization with informative, accurate, and timely channels of communications is better prepared to cope with uncertainty, including unforeseen opportunities.

This automatically raises concerns about communicating the model itself. Although what is being modeled is extremely complex, the model should be simple and concise. Too much information will saturate the cognitive senses and probably will not provide sufficient latitude for innovation and creativity. Furthermore, in the planning culture created by this strategic planning approach, much is left to business and other tactical planning. Besides, we must also avoid the bane of all organizations far and wide—micromanagement.

Economics and Strategic Guidelines

The principles and values described for a strategic plan are inspired from economics, information theory, and organizational development literature. However, the body of economic theory is not cohesively arranged for prescription to the strategic organization. This section seeks to highlight the direct economic theory that is available as a guideline to strategy.

The concern is to find substantiation for specific strategic plans within the economic literature that deals with uncertainty. What has economic analysis concluded to be valuable actions in an uncertain environment? The organization may already have ideas for its own strategic model, but these economic models, described in economic literature and carried out in actual business practice, can be useful benchmarks.

It is important to view the following examples as focused, not haphazard, guidelines to firm behavior in the face of uncertainty. Many firms could be pursuing plans that appear to be valid uncertainty-coping strategies, but which really are the incidental outcomes of unmanaged growth and decentralized aspirations. Focus is an important part of maintaining a survivable strategy.

Profitability is the most commonly accepted short-term decision strategy. Profit calculations based on return on investment (ROI) and other accounting tools are acceptable guides, as long as the data relevant to their forecasts remain stable.[14] ROI is easily buffeted by unpredicted environmental changes, and thus can lead a firm to rapidly alter its practices. While profits remain the bottom line for the survival of any firm, many do not like to base their decisions on the assumptions required of profit forecasting.

ROI and other profit estimators tend to hedge against the long term, so are not recommended by firms that desire more stability in their decision making. A more commonly accepted long-term strategy is market share. Firms that strive for market-share are staking their survival on the strength of their industry's demand. For long-term commitments, companies can gauge their success by comparing their results to the competition.

Profit estimators can be broadened to include nonpecuniary profit as well. Cost–benefit and social cost–benefit analysis can guide the firm through the confusing success indicators of the marketplace. By explicitly building the organization's values into its cost–benefit tools, members are provided with a continual focus on the firm's purpose and culture. Ample background information, and company-wide and consistent application of such decision-making tools are imperative for all these profitability strategies.[15]

Many space companies seem to use market-share as their major strategic model. Especially common is share growth in defense or government contract markets. The market-share corporation stakes its survival in the value of industry, and seeks to gain a larger share of that industry (see Cooper and Nakanishi[16] for an analytical view). Consistency is a hallmark of the share-driven company, since it is pursuing a well-identified product category. The product is well known, the attributes of the industry well understood, and the company simply aims to gear its efforts to doing the best within this specified environment. (See Buzzell and Gale,[17] whose findings have suggested that cohesion, not market share, is the cause of experience curves, scale economies, and profitability. These findings, supported by years of data from the PIMS data set (operated by the Strategic Planning Institute) do put market share-driven approaches in a new light.)

Advanced textbook economic theory does not directly suggest either profitability or market share as adequate uncertainty-coping strategies. Instead, the theoretical suggestion is that the best manner to handle risky or uncertain conditions is portfolio spreading and hedging. Risk management strategies assess the probabilities of different potential futures, and then devise a product mix, or *portfolio*, according to the firm's desired risk posture. Simply put, risk management suggests not to put all the firm's eggs in one basket, nor to judge single events as indicative of the firm's survivability. (Laffont[18] is a introductory, although technically demanding, text on the subject. Eatwell et al.[19] collect numerous introductory articles on capital allocation decisions and portfolio management.)

It is possible that some space companies view their competitive strategy as just a portfolio management problem. Holding a mix of defense, civilian, and commercial contracts signifies that a corporation is strategically diversifying.

Continuous innovation is yet another strategic model, one with which most space companies are familiar. Continuous innovation can take many forms, from constant customer awareness, for which successful service firms are known, to constant technological growth, which represents most space companies. By introducing change via new technologies, the innovating strategic firm hopes to stay on top of change.

Innovation, however, should not be looked upon as purely a matter of new technology, i.e., hardware. Innovation is a generic strategic idea and appears in such forms as niche building, "uniqueness" (see Ref. 9, p. 167), and product differentiation (see Ref. 20). Changes in internal organization, in methods of distribution, and in new product features should all be considered forms of innovation. (This has been commonly labeled as the technology-driven approach, in contrast to the market-driven approach, to firm strategy. As suggested, innovation should be looked at broadly and include changes as small as process improvements and customer enhancements. Recent research on sources of technological innovation indicates that the customers and customer requirements are often the source of new innovation. See Ref. 21.) Looked at in this way, organizational dynamics can be assessed as a matter of strategy.

The economics of uncertainty is a new field, with a very formal focus. The mathematics and applications of the formal findings have not been completely exposed. Experienced organizations today may find this abstract academic discipline helpful in establishing a consistent strategic model, or merely verification of their successful strategic goals.

Future of Strategic Planning

Can any organization afford to not have a strategic plan? The facts are that most organizations operate, to varying degrees, with most of the strategic planning elements discussed in this chapter. People and organizations do plan; the critical question is how effective is that planning? What we have shown is that strategic planning is soundly based on the economic principles of risk and uncertainty and the theory of the firm. Furthermore, the idea of an organization as a collection of individual entrepreneurs is gaining credence, and is central to our argument for strategic planning. The approach highlighted here, the idea of a strategic planning model of the organization, combines the economics of planning with recent developments in organization theory. What remains is for each organization to determine what its strategies are and to accommodate these in the context of an easily understood model of the organization.

The civil and military space programs and the aerospace industry, which exist in a rapidly changing environment full of risks and uncertainty, are especially suited to this modeling approach. Technological innovations, complex space missions, commercial opportunities, international competition, and environmental considerations must all be taken into account by strategic planning. The stakes are extremely high: the future of civilization, or, at least, the economic well-being of employees and the long-term survival of the organization.

We have provided one approach to strategic planning, an approach that relies heavily on an understanding of the environment and the organization, as well as on the interaction of one with the other. It is at the intersection of environment and organization that strategic interests and opportunities exist. The organization that is prepared to cope in an entrepreneurial fashion will be best suited to reap economic profits and rewards.

References

[1] Dixit, A., and Nalebuff, B., *Thinking Strategically: the Competitive Edge in Business, Politics, and Everyday Life*, W.W. Norton, New York, 1991.

[2] Rutland, P., *The Myth of the Plan: Lessons of Soviet Planning Experience*, Open Court, LaSalle, IL, 1985.

[3] Gleick, J., *Chaos: Making a New Science*, Viking, New York, 1987.

[4] Teece, D., "Economic Analysis and Strategic Management," *Strategy and Organization: A West Cost Perspective*, edited by Glenn Carroll and David Vogel, Pitman Press, Mayfield, MA, 1984.

[5] Williamson, O., *The Economic Institutions of Capitalism: Firms, Markets, Relational Contracting*, The Free Press, New York, 1985.

[6] Demsetz, H., *Ownership, Control, and the Firm: the Organization of Economic Activity*, Vol. 1, Basil Blackwell, Cambridge, MA, 1988.

[7] Selznick, P., *Leadership in Administration: a Sociological Interpretation*, Harper & Row, New York, 1957.

[8] Senge, P., *The Fifth Discipline: the Art and Practice of the Learning Organization*, Doubleday, New York, 1990.

[9] Peters, T., *Thriving on Chaos: Handbook for a Management Revolution*, Harper Perennial, New York, 1988.

[10] Davis, S., *Future Perfect*, Addison Wesley, Reading, MA, 1987.

[11] Deal, T., and Kennedy, A., *Corporate Cultures*, Addison-Wesley, Reading, MA, 1982.

[12] Casson, M., *The Economics of Business Culture*, Oxford University Press, London, 1991.

[13] Milgrom, P., and Roberts, J., *Economics of Management and Organization*, Prentice-Hall, Englewood Cliffs, NJ, 1992.

[14] Rachlin, R., *Return on Investment Strategies for Decision-Making*, Watts, New York, 1987.

[15] Tomkins, C., *Corporate Resource Allocation: Financial, Strategic and Organizational Perspectives*, Basil Blackwell, Cambridge, MA, 1991.

[16] Cooper, L., and Nakanishi, M., *Market Share Analysis: Evaluating Competitive Marketing Effectiveness*, Kluwer Academic Press, Boston, 1988.

[17] Buzzell, R., and Gale, B., *The PIMS Principles: Linking Strategy and Performance*, Free Press, New York, 1987.

[18] Laffont, J.-J., *The Economics of Uncertainty and Information*, MIT Press, Cambridge, MA, 1989.

[19] Eatwell, J., Milgate, M., and Newman, P., eds., *Allocation, Information, and Markets; the New Palgrave (selections)*, W.W. Norton, New York, 1989.

[20] Porter, M., *Competitive Strategy*, Free Press, New York, 1980.

[21] von Hippel, E., *The Sources of Innovation*, Oxford University Press, New York, 1988.

Remote Sensing: The Inconsistency of U.S. Space Policy

Simon P. Worden[*]
Strategic Defense Initiative Organization, Washington, DC 20301
and
Jordan S. Katz[†]
Comprehensive Technologies Inc., International, Arlington, Virginia 22202

Introduction

CIVIL remote sensing satellite systems and data applications have evolved from research-and-development-centered activities to operational systems. Satellite operation and application of various remotely sensed data have become critical to our national welfare. Applications of remotely sensed data from space are also proving to have significant economic benefit for the private sector. Remote sensing policy as applied to these issues is discussed in the following paragraphs. Specifically, the lessons learned from the LANDSAT program are examined as they will relate to the space segment (EOS, Earth Observing System) of the U.S. Global Change Research Program (USGCRP).

The EOS is intended to provide 15 years of continuous data from a variety of spaceborne instruments. This is an operational system, comprised of instruments that have evolved from a variety of precursor systems including LANDSAT. LANDSAT experience has shown that USG management of civilian operational remote sensing systems is inconsistent. This conclusion raises questions regarding the ability of NASA to efficiently manage the long-term operations of EOS and other critical civil remote sensing systems.

This paper is declared a work of the U.S. Government and is not subject to copyright protection in the United States.
[*]Deputy for Technology.
[†]Principal Analyst.

Civil remote sensing systems were pioneered by NASA in the early 1960s with the first polar orbiting weather satellite program (TIROS). TIROS produced the first civilian satellite images of the Earth from space. Today, U.S. government agencies operate a diverse range of third-generation civil remote sensing systems. Large user communities have developed within U.S. government agencies, universities, and the private sector.

Overall there are three fundamental needs for remote sensing data: national security, Earth science research, and commercial applications. U.S. remote sensing systems are owned, almost entirely funded, and developed by the U.S. government.

Many potential applications of land remote sensing were envisaged in the mid-1960s when high-resolution images were being returned from NASA's Gemini and Apollo programs. The U.S. Geological Survey Office was the first to recognize that orbital remote sensing data could reveal information about large-scale geological features that might take days or weeks to cover with aerial mapping.[1] Another benefit was that these images showed extensive large-scale features that had gone unnoticed from the ground or had been lost in aerial mosaics.[1] Soon thereafter, the Department of Agriculture also became interested in the use of remote sensing data to survey crops and forests.[1]

More recently, in the past decade there has been increased recognition of the potential implications of man-induced changes to the global environment (global change and, in particular, global warming). In response to this threat, the United States has started the largest scientific program it has ever undertaken, the USGCRP, to develop a comprehensive database, from which to better understand environmental changes. This is a $50 billion, 20-year effort. Its largest single requirement is to acquire a comprehensive, long-term imagery data set of the Earth from a network of operational remote sensing satellites.

USGCRP imagery requirements differ from other remote sensing needs in that they generally do not require high spatial resolution on the ground. Typically, USGCRP data will have resolutions of 100 m with at most a 30-m resolution requirement. Conversely, USGCRP data must be as accurately calibrated as possible and maintain that calibration over long periods of time. It will consist of global data sets taken as frequently as possible. This will require very large and sophisticated data storage and handling systems.

Ad Hoc Remote Sensing Policy

Today's operational remote sensing systems are the result of an ad hoc approach to space policy. Civil remote sensing capabilities initially emerged from NASA research programs. NASA was responsible for demonstrating polar orbiting and geostationary spacecraft capable of imaging the Earth 24 hours a day. NASA's charter directs it to be the primary U.S. government agency for research and development of space systems and applications. The Space Act[2] made no provision for NASA to provide operational space services, nor did it provide for a transfer of maturing space systems technology to operational organizations. Therefore, NASA's remote sensing programs focused mainly on research systems and not operational systems or applications.

In contrast, communication satellites were originally demonstrated by NASA and the Department of Defense (DoD) and then turned over to COMSAT, a quasiprivate corporation chartered by the U.S. government. COMSAT originally held a monopoly over all U.S. operational satellite communications systems. Later, domestic communication satellites were fully deregulated. Therefore, any private-sector firm could develop and operate communications satellite systems, not just COMSAT.

Weather satellites, considered a type of remote sensing system were also demonstrated in the 1960s and 1970s by NASA. In the early 1980s operational weather satellites were transferred to the Department of Commerce (DoC) National Oceanic and Atmospheric Administration (NOAA). Today NOAA is responsible for procuring and managing operational weather satellite systems.

The LANDSAT System: A Lack of Consistency in Policy

NASA began the Earth Resource Technology Satellite (ERTS), later LANDSAT, satellite program to validate the sensor technology needed to field operational remote sensing satellites. The first three LANDSAT satellites were considered research and development (R&D), not operational, satellites. LANDSATs 1 to 3 carried a Multi-spectral Scanner (MSS) which took 185 × 185-km images of the Earth with 80 m resolution in three spectral bands. NASA assumed that once it developed the satellites, the user community would fund the development of applications, so little federal funds were spent on data applications. Small amounts of money were reserved for data analysis, mostly for ensuring data quality.[1]

Ever since LANDSAT's inception, remote sensing data were thought to have significant commercial application. LANDSAT data have allowed private-sector firms to create a "value-added" information market. These firms enhance, merge, and analyze raw LANDSAT and other remote sensing data to produce finished information products for paying customers. These enhanced data are used by private firms to monitor renewable resources such as agriculture, forestry, and fisheries; to perform site analysis for engineering and construction, mineral extraction, utilities, and waste management; as intelligence sources by the media; and for mapping.[3] One of the growth applications that has recently emerged is the Geographic Information Systems (GIS), a computer database that produces geographical information based on multiple types of data, and space remote sensing data.[4]

The success of LANDSAT's research phase fueled demand for a next-generation operational system, one that would be capable of distributing data in days or weeks, instead of weeks and months. Data users demanded continuity of MSS data, but also desired higher resolution. The Carter Administration noted the applications development going on in the remote sensing research community and the private sector. They were also aware of the contradiction of NASA, a development agency, under pressure to produce LANDSAT 4, an operational remote sensing system. The solution for the Carter Administration was to privatize the new operational LANDSAT system. As commercial markets for the data emerged, they expected market development to take as long as 10 years.[5,6]

This approach to commercializing LANDSAT was never realized. The Reagan Administration rejected the previous administration's approach and accelerated the LANDSAT commercialization. Congress passed the Commercial Land Remote Sensing Act of 1984 that gave the DoC the responsibility to fund LANDSAT operations. The DoC was also given the power to regulate and license nongovernment commercial remote sensing systems. The Act further mandated DoC to produce a commercial transfer plan. According to the plan, LANDSAT would be operated by a private corporation that would have exclusive rights to all LANDSAT data. In exchange the private company would pay for all data marketing and dissemination costs. Further, Congress stipulated that data must be made available to all users on a nondiscriminatory basis, an extension of U.S. "Open Skies" policy. EOSAT Corporation, a joint venture of General Electric (at that time, RCA, the LANDSAT satellite contractor) and Hughes Aircraft (LANDSAT instrument developer), was selected to operate LANDSAT and assumed operations in 1985.

As of early 1992 two LANDSAT satellites remain in operation, LANDSAT 4 and LANDSAT 5. LANDSAT 4 was launched in 1982, and LANDSAT 5 was launched in 1985. As part of the transition to the private sector, Congress committed funding subsidies for the building of LANDSAT 6, and concept definitions for LANDSAT 7. In 1987 both remaining LANDSATs had outlived their expected lifetimes. NOAA anticipated they would cease operating, and made no commitment to fund LANDSAT operations subsidies beyond the satellites' life expectancy. LANDSATs 4 and 5 continued to operate even though NOAA did not request funding for FY 1989. Congress appropriated $9.4 million to the NOAA for FY 1989 to cover half the year. After April 1, 1989 NOAA did not have any funding available. This problem caught the attention of the newly formed National Space Council chaired by Vice President Dan Quayle. The Space Council brought together the agencies that depend on LANDSAT data and persuaded them to provide funding to NOAA to continue the operations subsidy to EOSAT. The Space Council also recommended that funding should continue as long as LANDSATs 4 and 5 remained operational. For FY 1990 and 1991 the funding arrangement continued whereby Congress appropriated the first 6 months of operations funding and the user agencies reprogrammed the remaining funds. In addition, the National Space Council and President Bush issued a National Space Policy Directive that the U.S. government would ensure the continuity of LANDSAT-type data (multispectral, 30-m data).

In FY 1991, EOSAT's contract with the federal government ended. LANDSAT 6 development and its launch costs have been paid for. The U.S. government paid for the operation of LANDSATs 4 and 5 from 1986 through 1991, and funding was requested for continued operation of LANDSAT in FY 1992, until LANDSAT 6 is launched and declared operational. The complete operations cost of LANDSAT 6 is to be the responsibility of EOSAT. LANDSAT 6 has a 5-year life expectancy.

The remaining issue is what to do about a follow-on (LANDSAT 7). The LANDSAT 7 question was addressed by the National Space Council in 1991. A deal was struck between the DoD and NASA, the two largest U.S. government users of LANDSAT data. DoD and NASA agreed to set up a joint program office. DoD will be responsible for procuring the satellite and NASA will be responsible for data distribution and satellite operations. In addition, NASA will be responsible for working out, with the assistance of the DoC, the pricing of LANDSAT data for commercial users. Further details are spelled out in the President's FY 1993 budget request. The joint program offices will allow DoD- and NASA-sponsored users to obtain data at the marginal cost of reproduction. This satisfies the national security and Earth sciences communities' data requirements through the 1990s. Commercial users' requirements and needs are to be examined further by NASA.

Congress is considering legislation[7] to deal with the continued support of the LANDSAT program. The House of Representatives Subcommittee on Space, chaired by Rep. Ralph Hall (D-Tx), introduced legislation that amends the 1984 Commercial Remote Sensing Act to accomplish the following:

1) Create a NASA/DoD joint program office that would have overall responsibility for the LANDSAT system.

2) Start procurement of LANDSAT 7, to provide continuity to LANDSAT 6, and possibly allow for some advanced sensors to be flown as well.

3) Re-negotiate data rights with EOSAT to allow U.S. government and nonprofit data users to receive data at the cost of reproduction as opposed to the "commercial" cost.

4) Establish a 5-year sensor technology development plan to develop and demonstrate advanced sensor, component, and system design technologies.

5) Instruct the joint program office to come up with options for a LANDSAT 7 follow-on.

6) Revise data policy to give nonprofit users access to unenhanced data at the marginal cost of reproduction as long as the data are being used for noncommercial applications. This is the same for private nongovernment remote sensing satellite systems (that currently don't exist). The DoC is responsible for preventing nonprofit users from using data for commercial purposes.

Private-sector firms have been held captive to this process. Being entirely dependent on LANDSAT data for certain applications puts the private sector at the mercy of this process. The bottom line has been that data sales from LANDSAT to non-U.S. government customers reached a record amount of $15 million in 1990.[8] This revenue stream is not capable of supporting LANDSAT operations let alone the cost of building additional satellites. Therefore, the private sector has been a secondary player to the needs of U.S. governments sponsored researchers and DoD users.

NASA's Mission to Planet Earth

One of NASA's primary scientific initiative for the 1990s is Mission to Planet Earth (MPE). MPE is designed to obtain long-term (15-year) comprehensive remote sensing data sets from which to assess the complex interplay of man-induced and natural mechanisms on the global environment. MPE is NASA's portion of the USGCRP, a 20-year, $50 billion dollar effort involving many government agencies. NASA's MPE is currently estimated to cost at least $35 billion, thus it is the dominant part of the USGCRP.[9]

MPE, and in fact the entire USGCRP, grew out of NASA conferences held in the early and mid-1980s. In the mid-1980s NASA formulated the Earth Observing program as part of the Space Station effort. Large polar platforms, designed to be compatible with the Station truss structure and Shuttle support systems, were to be the basis of MPE. Instruments would have been serviced and changed during Shuttle visits to these platforms.

After the Challenger disaster and subsequent Shuttle redesign, it was not possible to launch the Shuttle from the west coast launch facility into polar orbit. NASA then renamed the Station Polar platforms the Earth Observing System and moved them onto Titan IV expendable launch vehicles. In 1990, with the Presidential announcement of Mission to Planet Earth as one of NASA's two major initiatives (the other is now called Mission from Planet Earth aimed at manned planetary exploration), NASA moved EOS completely out of the Space Station program.

As envisaged by NASA in 1991, EOS would consist of two large polar satellites—EOS-A and EOS-B, each launched into a polar orbit. The first A platform was to be launched in 1998, and the first B platform 2½ years later. Since each platform would be designed for a 5-year life, three of each would be needed to ensure a 15-year continuous data set.

With about 18 instruments each and postulated data rates of close to 10^{12} bits ("terabits") per day, EOS would be the largest, most complicated, and most expensive scientific program ever undertaken. Each EOS satellite would weigh close to 15,000 kg, the largest scientific satellites ever built. About half of the instruments, and 99% of the data was to be in the form of moderate resolution (30 to 100 m resolution) multispectral calibrated imagery. In addition to the U.S. EOS platforms, there were to be a smaller European Earth observation platform and two small satellites from Japan.

The data were to be handled by the EOS Data and Information System (EOSDIS). EOSDIS was to have seven separate data centers and would itself absorb 40% of the EOS budget.

In late 1989 it became clear that Mission to Planet Earth, NASA's ongoing scientific and manned programs, and a strong start on manned solar system exploration could not be accommodated within a realistic NASA budget through the 1990s. Although MPE usually held top priority in statements by key members of Congress and public commentators due to its environmental relevance, it began to undergo close scrutiny as well.[10]

Due to international and domestic pressure, the United States found itself faced with urgent calls for responses to global warming. Some of these, such as capping carbon dioxide emissions would have immense effects on the U.S. economy, and could cost trillions of dollars. For this reason, many politicians turned to the scientific community for answers regarding how bad global warming would be and how soon it would be a problem. Since EOS would not begin taking comprehensive data sets until well after 2000, it was clear that EOS could do little to answer the immediate and urgent questions.

For these reasons, the White House [National Space Council, Office of Management and Budget (OMB), and Office of Science and Technology Policy] asked, in early 1989, the National Research Council (NRC) to investigate EOS as part of the overall USGCRP. They were to recommend near-term missions that could address global warming and assess whether EOS might be done faster and more efficiently with smaller satellites in place of the giant EOS-A and EOS-B platforms.

The NRC report, released in September 1990, made some amazing conclusions.[11] Even though its only data input were briefings supplied by NASA, the NRC contradicted NASA on several key points. They concluded that near-term missions could begin to answer global warming and other key questions soon and placed higher priority on these small missions over EOS platforms. NASA hastily renamed some small scientific missions "Earthprobes" and claimed they were answering the criticisms. However, the total cost of these Earthprobes in the 1992 budget is 10% of EOS's cost. However, the key issue was one of simultaneity. NASA had claimed that the primary reason for placing all instruments on large platforms was the scientific requirement for obtaining different types of data simultaneously. The NRC disagreed. They recommended that the EOS-B platform be broken into three smaller platforms. They further stated that, while NASA had refused to present any information on alternatives, they saw no reason that a set of small satellites flying in formation could not provide the simultaneity that was required.

Since the NRC report raised a variety of engineering issues, the OMB mandated in November 1990 that NASA submit their EOS designs to an external engineering review. This review, chaired by Scripps Institute Director, Dr. Edward Frieman, reported to the White House in September 1991.

The Frieman Committee reached even more surprising conclusions than the NRC. First, they recommended that NASA break the two EOS platforms into at least five medium-sized platforms launched on Atlas II or smaller boosters—the first in 1997. While NASA had insisted that no such launchers existed for west coast polar launches, the Frieman group found that Air Force plans, apparently unknown to NASA, would provide such options by the mid-1990s. The Frieman Committee further found that new "smallsat" technology and instrumentation developed by the Departments of Defense and Energy could provide critical global warming data by 1995. This could even include Synthetic Aperture Radar imagery which NASA had said they would be unable to get until after the year 2000 in the EOS program. Finally, the Frieman group concluded that NASA's EOSDIS approach was not well conceived and that it should be completely reworked.

The key results of the Frieman panel[13] clarified the fundamental policy issue facing the United States in global change research. On one hand, if the question was whether, and if so how much and how soon, global warming would be a problem, then this question could be answered relatively soon with small, cheap satellites. On the other hand, if the question was not whether global warming existed, but how to manage the environment on a microlevel, valley by valley, city by city, farm by farm, and so forth, then a comprehensive data set and expensive, long-term system such as EOS would be necessary.

The decision seems to be that one must address the former policy issue before launching the latter. In addition to scientific and philosophical issues, EOS, particularly EOSDIS, raises serious issues for the commercial remote sensing community. Clearly, the quality and quantity of EOS data will far outstrip any data currently available for commercial use. However, NASA's current approach is to provide EOS data, at a nominal fee, only to users willing to publish their results in open scientific literature. This is obviously incompatible with commercial use. Recent White House data policy on the USGCRP recognizes that something must be done to make such data available to the commercial user, but does not suggest any way to accomplish this.[13] Various suggestions, such as a two-tiered pricing arrangement, will clearly be a serious policy challenge to the U.S. Congress and Administration in the years ahead.

New Technology Approaches

In the above cases the requirements for remote sensing data are centered on several key instrument types. For example the prime LANDSAT issue is the long-term continuity of TM and MSS data. But the debate is over the continuity of operational spacecraft. Because there is an operational demand for data, NASA is extremely reluctant to deviate from conservative spacecraft or sensor technologies. They maintain that their proposed architectures for LANDSAT and EOS are the only possible options, despite the advances of small satellite (smallsat) technologies over the last 10 years.

Despite the serious policy problems in fielding civilian remote sensing space systems, there are some promising developments in technology, including the development of smallsat systems. Coupled with low cost, small satellite launch vehicles such as Orbital Science's Pegasus makes fast-turnaround fully commercial systems feasible for both scientific and commercial remote sensing applications.

The evolution of capabilities of smallsat and small launch vehicles could commercialize remote sensing services. The only example of such an activity is NASA's recent procurement of ocean color data (SeaWiFs). (The SeaWifs instrument, built by Hughes Aircraft, was sold to Orbital Sciences Corporation (OSC). OSC will launch the instrument on its SeaStar commercial environmental monitoring satellite in 1993. This project will be the first privately funded and operated remote sensing satellite system. NASA has signed an "anchor tenant" contract with OSC, guaranteeing to purchase a specific amount of SeaStar data for 5 years.) NASA has a validated requirement to fly the Wide Field Ocean Sensor as part of its Earth probes program, but is procuring the data from a private operator instead of developing and operating the spacecraft and sensor itself. NASA has contracted with Orbital Sciences to procure this ocean color data over 5 years. Orbital Sciences is responsible for constructing the spacecraft and sensor, and for all launch and on-orbit operations. Orbital Sciences retains exclusive rights to the data while it's less than 2 weeks old. After 14 days

NASA is entitled to the data for open research purposes. Both the commercial and research value of the data is maintained. This type of arrangement is possible because of advancement in small spacecraft, sensor, and launch technology. These advancements have brought total mission costs down somewhere between $50 and $100 million, as opposed to $500 and $800 million for LANDSAT 7. Furthermore it demonstrates that the necessary capabilities reside in industry as well as in U.S. government agencies. This is far more cost effective and less time consuming than NASA's traditional acquisition process.

U.S. government agencies independent of NASA have been in the spacecraft, sensor, and operations development business for over 30 years. Remote sensing systems driven by the unique requirements of the Strategic Defense Initiative and national security community have created new technologies that show promising civil and commercial applications. The Department of Energy's (DoE) National Labs have been developing miniaturized spacecraft propulsion, avionic, structures, and sensors. These systems, when deployed in a distributed orbital architecture could provide the same capabilities as large singular satellite remote sensing systems.[14] This relationship is analogous to a large company that has many personal computers connected to a local area network compared to having one large mainframe computer. The distributed system is more flexible, can be customized to specific requirements quicker, can be developed piecemeal (pay as you go), and is more robust.

Other sensor technologies that the LANDSAT and EOS programs are attempting to develop already exist off the shelf in other agency programs. DoE labs have developed a small, low-power, synthetic aperture radar (SAR). The Defense Advanced Research Projects Agency (DARPA) has also been a pioneer in developing more capable smallsats and launch vehicles. DoD has also made advances in TM-type sensor with higher resolution and more spectral bands.[12]

These technology developments have been largely lost on NASA and the civil remote sensing community in general. It seems that the arguments regarding LANDSAT are not concerned with the continuity of data but the continuity of spacecraft. Both civil and commercial requirements could be met with small distributed systems incorporating advanced technology. Because these new systems are smaller and less expensive, development of operational systems are within the grasp of industry, although some of this technology still needs to be demonstrated by the U.S. government. The application of this new technology coupled with the NASA procurement of data service (as opposed to satellite acquisition) would save valuable time and money in answering the critical questions of global change. Thus far, NASA continues to resist application of these technologies in the LANDSAT and EOS programs.

A low-cost smallsat could solve the current dilemma with LANDSAT. While non-U.S. government revenues could not pay for acquisition and operation of a $500 million satellite, they could pay for a $30 million one. Indeed, the DoE and DoC have proposed just such a solution to the need for continuity of LANDSAT data.[14]

Conclusions

The preceding cases of remote sensing development have demonstrated the inability of U.S. space policy to provide consistency. Because of the nature of the process the U.S. government has been unable to create long-term stability in the LANDSAT program. EOS, which will not be launched until the late 1990s, is starting to go through similar program and design gyrations.

The LANDSAT program has come full circle in the last 20 years. It began as a NASA development program, was declared operational, and went through several short-term privatization attempts. Now, future LANDSATs will land back in the hands of the U.S. government as an operational system. Privatization is a failure not in theory but in the inconsistency of the applied implementation on the part of the Reagan Administration and Congress. As the players in the Administration, Congress, and the private sector debate the future of LANDSAT, it has been forgotten that the short-term abandonment of previous plans has created the current problem. Now a new arrangement is being proposed. Joint operations by DoD and NASA, marginal pricing for U.S. government users, and the like are all significant changes over the previous plans. But, there is still no consistency here.

The largest policy contradiction is NASA's EOS program. For the last 20 years the debate over LANDSAT was fueled by the inappropriateness of an R&D agency (NASA) providing operational data services. Years of Band-Aid policy-making privatization are being scrapped.

LANDSAT will be run by NASA. Now NASA is embarking on the operational EOS program. This is also a contradiction to the one goal that existed in the late 1970s: NASA is becoming the mother of all remote sensing system operators to support global change research.

NASA views the EOS program as an R&D program as far as the development of instruments are concerned, but operational from a data standpoint. NASA is also providing a substantial amount of funding for most of the EOS data applications as part of USGCRP. This support for analysis is a departure from the earlier LANDSAT applications debates. The concern over the contradiction of NASA as an operational agency has disappeared, and the fears of the past forgotten. NASA will be the main benefactor to contractors building the space segment, ground segment, and data segment; and to the data research community as well.

In effect, NASA has become the developer, builder, operator, and primary consumer of civil remote sensing data. This was the situation that policy makers wanted to avoid in the decision to take LANDSAT to an operational status. It seems that the political implication of global change and environmentalism has eclipsed the roles and missions of U.S. government agencies. NASA has taken advantage of the environmental issue to create a large, long-term operational science mission. When examining the history of EOS, it's clear that the majority of driving requirements (spacecraft size, simultaneity, data systems, etc.) were all based on use of its infrastructure (Shuttle, Space Station, etc.). NASA retrofitted its EOS program to fit a new large and encompassing global change environmental research agenda. But, near-term political realities dictate the need to address specific areas of global change as a priority.

There are issues that will constrain NASA's approach to EOS: timeliness of information, and funding. Recent measures to control U.S. government spending have effectively capped NASA funding growth. Congress has made it clear that NASA will not get all the funding necessary to fund its original EOS and has directed NASA to reorient its program to near-term problems (global warming). The Administration has also told NASA to focus first on politically pressing global change problems within reduced funding envelopes. NASA continues to resist, citing the need for an encompassing understanding of global change to get answers to specific environmental problems.

NASA still fails to remotely sense the forest from the trees. The money does not exist for EOS as currently envisioned by NASA. Congress and the Administration have both expressed the unwillingness to support such expenditures. New technologies and launch capabilities exist to develop and field alternative remote sensing systems to potentially meet all near-term global change requirements. As EOS and LANDSAT are structured there is very little incentive for U.S. government agencies to embrace innovative alternatives. The private sector is discouraged from utilizing new technologies for private systems because of subsidized data available from EOS and LANDSAT. Further, markets for data have not expanded as rapidly due to the program uncertainty and political risk of LANDSAT. These are exactly the types of situations feared by early space policy makers when attempting to limit NASA from being an operational agency.

To correct this situation, NASA and other remote sensing agencies should embrace a strategy based on the following attributes:

1) The 10-year goal should be that all civil remote sensing systems should be commercially developed and operated, buying data not spacecraft.

2) The U.S. government should assume the role of "anchor tenant" as outlined in U.S. Commercial Space Policy (Feb. 1991).

3) The U.S. government should demonstrate (jointly with potential commercial operators), as a research program, by 1995, a distributed "lightsat" remote sensing system. As a long-term function, new technology development and validation should remain a U.S. government function.

Until such a consistent, long-term policy is implemented across all U.S. civil remote sensing agencies, commercial potential will remain unrealized. Unless we move rapidly, other nations will fill the vacuum that we are leaving.

References

[1]"Imaging the Earth (I): The Troubled First Decade of Landsat," *Science*, Vol. 215, March 26, 1982, p. 1600.

[2]The National Aeronautics and Space Act, PL 85-586, 85th Congress, H.R. 12575, July 29, 1958.

[3]"The Future of Land Remote Sensing Satellite System," CRS Rept., Library of Congress, Sept. 16, 1991, p. 39.

[4]Jordan, E. L. III, President, ERDAS Inc., Testimony Before the Committee on Science, Space and Technology, U.S. House of Representatives, Nov. 26, 199.

[5]Radzanoski, D., "The Future of Land Remote Sensing Satellite System," Congressional Research Service, Library of Congress, Sept. 16, 1991, p. 2.

[6]Greenberg, J. S., "Commercialization of the Land Remote Sensing System: An Examination of Mechanisms and Issues," *Hearings on the Commercialization of Meteorological and Land Remote-Sensing*, Committee on Science and Technology, U.S. House of Representatives, 98th Congress, No. 53, April 14, 1983.

[7]The National Land Remote-Sensing Policy Act, H.R. 3614, U.S. House of Representatives, 1991.

[8]U.S. Department of Commerce, "Space Business Indicators," Economics and Statistics Administration/Office of Space Commerce, June 1991, p. 31.

[9]U.S. Executive Office of the President, "Our Changing Planet: A U.S. Strategy for Global Change Research," Rept. by the Committee on Earth Sciences to Accompany the U.S. President's Fiscal Year 1990 Budget, Office of Science and Technology Policy, Washington, DC, 1989.

[10]Statement by Sen. Albert Gore (D-Tn), Chairman Senate Subcommittee on Science, Space and Technology, May 1990.

[11]National Space Council, *An Assessment of the U.S. Global Change Research Program*, National Academy Press, Washington, DC, 1990.

[12]Rept. on the Earth Observing System, Engineering Review Committee, Sept. 1991.

[13]National Space Council, Data Policy Directive, 1992.

[14]"BRILLIANT EYES: Highly Proliferated, High Performance Lightsats Supporting Department of Energy Leadership in the USGCRP," Livermore National Laboratories, Nov. 1990.

Engineering Design and Decision Making

George A. Hazelrigg*
National Science Foundation, Washington, DC 20550

The views expressed here are strictly those of the author and do not necessarily reflect the views or policies of either the federal government or the National Science Foundation.

Overview of Engineering Process

ENGINEERS frequently think of themselves as problem solvers. Indeed, engineers are taught problem-solving skills as the major element of their education and, throughout their lives, they seek to hone and improve these skills. To be sure, problem-solving capabilities are important to engineering. Yet, it is the contention of this chapter that problem solving is not the principal activity of engineering; rather it is decision making, and to accept this premise opens a door to a vast new array of engineering skills that are amenable to being taught.

An approach to engineering is discussed that will seem foreign to many experienced engineers. First, a set of definitions is presented. This is necessary in order to discuss the subject matter of the chapter clearly. Not all readers will agree with these definitions, nor is it the objective here to seek agreement. But the definitions presented are crisp and unambiguous, and they will enable the presentation of new concepts.

> **Science** (physical science) is the process of rationally and methodically seeking to understand nature, with the principal objective of developing a predictive or problem-solving capability.

This paper is declared a work of the U.S. Government and is not subject to copyright protection in the United States.
*Deputy Division Director, Electrical and Communications Systems.

Science deals with problem solving.

> *Problem*—In the physical sciences or engineering, a problem is a question posed in the following form: given a state of nature at one time, determine or predict the state of nature at another time, or given elements of the state of nature, determine others. For example, given the loads placed on a structural element in a space system, determine the stresses in the element. Then determine whether the element will support the loads for ten years.

Concept: To solve a problem, one must: 1) know the relevant state of nature, which engineers sometimes refer to as the initial condition or the boundary condition; 2) know the laws of nature that apply; and 3) have the necessary computational capabilities to apply the laws of nature.

The scientific method is referred to as research.

> *Research* is the activity by which a scientist (that is, one who practices science) seeks to understand nature and improve our predictive capability. Basic research is a methodical process that builds on the existing knowledge base, mainly by a process of formulating testable hypotheses and by testing these hypotheses by means of repeatable experiments. Applied research is a process of measurement and modeling aimed at the development of accurate predictive models.

Engineering is different:

> **Engineering** involves the manipulation of nature to create systems for the benefit of at least some segment of mankind.

It follows that engineering involves the allocation of resources. The allocation of resources is decision making.

> *Decision*—An irrevocable allocation of resources. The selection of design parameters for an engineering system such as a computer or an automobile constitutes an allocation of resources. Design is a decision-making process, and the design parameters represent decisions.

Concept: To make a decision, one must: 1) have options or choices from which to choose; 2) have for each option an expectation of possible outcomes, that is, what will happen; and 3) place values on the outcomes to determine which outcome is most desired. Virtually all decisions are made under conditions of uncertainty, that is, under the condition that the outcomes cannot, in advance of the decision, be determined precisely. Whenever there is uncertainty in the outcomes, the decision maker is exposed to the possibility of making a decision that will result in an outcome that is not the most desired outcome.

ENGINEERING DESIGN AND DECISION MAKING

This enables a concise definition of technology.

> **Technology** is the ability to make good engineering decisions and to carry them through to their desired conclusion.

Concept: The technology to do something is the ability to create a workable design, to carry out that design into a workable product, device, or service, and to place it successfully into service or action. Each of these things requires that appropriate decisions be made at all levels of design, manufacture, and use.

The distinction between problem solving and decision making leads to a distinction between knowledge and information.

> *Knowledge*—An agreed upon set of "facts." Knowledge is essentially the contents of a data base, with elements such as $F = ma$. For purposes of this discussion, knowledge can be defined as an understanding of the laws of nature and the ability to apply those laws to predict the behavior of physical systems.

Concept: Knowledge gives us the ability to create choices. For example, the knowledge that $F = ma$ enables one to conceive of reaction propulsion: a rocket.

> *Information*—Information relates to a specific decision. Quantitatively, it can be measured as the probability that the preferred choice in a decision will lead to the most desired outcome.

Concept: Information enables engineers to make effective choices in engineering design. It is virtually never the case that one would make a decision without some expectations on the outcome, that is, without some information. Furthermore, with perfect information, one would never make a "bad" decision. But, even with perfect information, a better decision might be made if better choices are available.

It is also possible to distinguish between different types of research.

> *Basic Research*—Research that seeks to improve knowledge, that is, man's understanding of nature, and man's ability to predict the behavior of physical systems.

Concept: Basic research seeks to provide a fundamental understanding of nature that leads to better predictive capabilities and, hence, to an ability to create and use new options for engineering design.

> *Applied Research*—Research that seeks to improve information, that is, to improve the ability to make choices in decision making.

Concept: Applied research enables the engineer to create designs that better achieve the goals set for the system. For example, better information reduces the need for overdesign that has weight, cost, and performance penalties.

There is an additional activity that might be classified as research that does not fit under these definitions of either basic or applied research. It may be called *look–see research*.

> *Look-See Research*—Research that is performed as a process of simply trying things, not to better understand nature nor to improve information for decision making. Look–see research consists of trying things, more or less at random, but perhaps with some insights, to accomplish a goal. Look–see research frequently occurs immediately following the invention of a new measurement device, as scientists are eager to take a new look at nature afforded by the device. Quite frequently, the view of nature that new instruments give leads to the discovery of new phenomena. As new phenomena are found or new results obtained from look–see research, scientists turn quickly to basic research in order to understand them.

Concept: Look–see research is driven by technology, both by technological needs and by technological advances. Engineering achievements frequently drive look–see research. It is the mode of research that is performed when either no other opportunity presents itself or when special new opportunities for look–see research appear, for example, as in the case of the invention of a new instrument. Look–see research leads to new areas for basic and applied research, and it can create new options. But it does not provide information for decision making. It must be followed by some amount of applied research to enable application.

As defined above, science is not a process of decision making; engineering is. Thus, by these definitions, research ends where decision making takes over.

The engineering process begins with an understanding of the objectives of a project. The engineer must develop an understanding of the benefit that the project will provide, and how this benefit will depend on the engineering design itself. In essence, the engineer must create a set of values by which engineering decisions will be made.

Next, the engineer seeks conceptual designs to accomplish the relevant objectives. Conceptual designs include a general description of the system, but generally do not include specifics such as dimensions. In the case of an airplane, for example, the conceptual design variations might include high wing vs low wing, jet vs turboprop, aluminum vs composite structure, and so on. In the case of a space vehicle, the conceptual design might include the nature of the power source, propulsion, and structure. Some people refer to the creation of conceptual designs as engineering synthesis.

Each conceptual design will contain a number of design parameters. Again, for an airplane, these could include the wingspan, engine thrust, gross weight, and so on. The choice of these parameters is a decision-making process that engineers often refer to as optimization. The goal for each conceptual design is to pick those values of the parameters that lead to a design that maximizes the net benefit of the system. But uncertainties cloud both the choice of the values for each parameter and the computation of the benefit that these choices produce. One goal of research is to reduce these uncertainties. But some forms of uncertainty cannot be resolved entirely by research. For example, the fatigue life of an airplane depends on the weather through which it flies and the turbulence it encounters, the extent to which it is loaded each time it flies, and the care with which the pilots handle it. These parameters can be known only within limits, and no amount of research will resolve them any better.

On the other hand, research can resolve many uncertainties. For example, the finest details of the design of a load-bearing member of a structure can affect stress concentrations that could lead to failure. Improved theories and methods of stress computation may improve prediction of such conditions and enable better design. In the absence of such theories, the engineer typically chooses to be "conservative." That is, the engineer overdesigns the member to allow for uncertainty. This author knows of no comprehensive theory of overdesign to compensate for uncertainty, but engineers make occasional reference to concepts such as statistical design, factors-of-safety, and "fudge factors" that implicitly allow for uncertainty.

Once the engineer has obtained an adequate state of information for decision making, both the system conceptual design and its design parameters are chosen. Frequently, this is an iterative process. It is generally expensive to obtain accurate information for an engineering design, so many conceptual designs may be eliminated at an early stage in the process, and only those remaining will be compared with a high degree of accuracy. Then, once a specific design concept is chosen, it may be subject to much more in-depth analysis.

The engineering process described here relies heavily on the results of scientific research to produce design options that emerge as part of conceptual designs and to select values of design parameters and predict the performance of the resulting system. But it includes also synthesis and decision making, which are not processes that can be addressed by the physical sciences. We can add one more definition.

> **Systems engineering** is the treatment of engineering design as a decision-making process.

Science rarely produces results that, in themselves, hold value to society. Results that themselves are of value to society relate to advances or discoveries that are called *cosmological*. The Copernican theory of the solar system, that the Earth revolves around the sun and therefore is not the center of the universe, is an example of a cosmological advance. All other scientific advances lend value to society only through their implementation, typically in engineering systems. *Engineering is the process that makes science of value to society.*

Historical Perspective

The above concepts and definitions deviate rather substantially from those taught to and perceived by most engineers. It is interesting to consider from where the present notions derived.

Engineering has been practiced for millennia, but until rather recently it was taught through a process of apprenticeship. The current, formalized approach to engineering education may be attributed largely to Napoleon. In his youth, Napoleon was an artillery officer, and he was awed by the potential power of artillery. Yet he noted that artillery was an art, little understood by most artillery officers. When he rose to power, he sought to do something about this, and he provided support to the major French mathematicians of the day to develop a theory of artillery amenable to being taught. He then sought to incorporate an artillery curriculum into higher education. At the time there existed a well-developed university system, based on the Greek system. But the university system did not deem it appropriate to incorporate the new, applied curriculum.

The word *university* derives from the Greek *universe*, meaning all or everything. And the Greeks thought of the university as the place where one would learn about everything. But, oddly, apprenticed skills such as engineering were not thought to be included in the topics appropriate for a university. Thus, Napoleon created the French polytechnic. The word *polytechnic* derives from the Greek words *polys*, meaning many, and *technik*, meaning art or craft, hence, many arts. Engineering was taught in the French polytechnic, but, interestingly, not agricultural engineering.

Toward the end of the Napoleonic wars, the military engineers who had been educated in the polytechnic found themselves unemployed. They felt that their education made them useful to society, yet there was a stigma attached to them that made it hard for them to find jobs. So they found a new name to call themselves; a name that drew as great a distinction between them and the military as they could mange. They called themselves *civil engineers*.

The French polytechnic system was copied in the United States as the land-grant college system, formalized by the Morrill Act of 1862, which included curricula in both military engineering and agriculture, hence, the A&M colleges. Rensselaer Polytechnic Institute became the first U.S. institution to award a degree in engineering in 1824. From that time through World War II, there was little change in the teaching of engineering. Engineering was taught largely by engineers educated to the Master's level.

But World War II changed forever the perception of engineering. The war was widely recognized as a technological battle that was won by engineers and scientists working on "mega-projects" such as radar and the atomic bomb. These projects advanced technology so rapidly that the Allies gained an awesome advantage in military power, and the possibilities afforded through concentrations of scientific and engineering research and development became entirely apparent. In 1950, the National Science Foundation was created, and together with other federal agencies, provided a continuing stream of funds for advancing technology.

Around 1950, Compton, a famous Nobel physicist, became the president of the Massachusetts Institute of Technology (MIT). He observed that technology had begun to advance too fast to pass it down from generation to generation by a process that still had a substantial element of apprenticeship. In 1952, Gordon Brown, chairman of the electrical engineering department at MIT implemented Compton's idea, and began the turn to basics—mathematics, physics, and chemistry. Other major schools followed immediately, and the rush to a science base for engineering was on. This rush intensified through the 1960s and 1970s, and during this time engineering faculties turned to the hiring almost exclusively of Ph.D.s. The new Ph.D. engineers were specialists and researchers, and they were in kind more scientists than engineers. Some engineering schools even eliminated the word "engineering" from their names and became schools of "applied science." Today the influence of science on engineering is profound. Most engineering institutions teach applied science as their engineering curriculum. This education ends with the notion that, upon graduation, an engineer must obtain experience (that is, become an apprentice) to round out his or her education.

This chapter seeks to dispel the myth that engineering is applied science or problem solving, and provides a basis for formally teaching much of engineering that is yet thought of as an art.

Decision-Making Process

The three key elements of decision-making processes are identified above: identification of options or choices, development of expectations on the outcomes of each choice, and formulation of a system of values for ordinally ranking the outcomes and thereby obtaining the preferred choice. Two things complicate almost all engineering decision-making processes: first, uncertainty on the outcomes and, second, knowing what to include in the universe of phenomena that need to be considered in determining the outcomes of the options. If indeed systems engineering is the treatment of engineering design as a decision-making process, then it will be seen that systems engineering is an activity that knows no disciplinary bounds. This is in part what makes engineering decision making so difficult to teach.

About 30 years ago a turboprop airliner was taking off from Logan Airport in Boston, Massachusetts. During the takeoff role, the engines ingested some 30,000 starlings, which resulted in a substantial loss of power and a crash involving considerable loss of life. During the investigation, it was easy to determine that the loss of power was the probable cause of the accident, and that the loss of power was the result of starling ingestion. But why did 30,000 starlings fly into the engines of the airliner? It was eventually determined that a small, power takeoff shaft on the engines, used to drive hydraulics, electrical, and other systems, developed a vibration when the engine was accelerated to takeoff power. This vibration, although inconsequential structurally, had a spectral signature very similar to that of the mating call of the male cricket, a favorite food of the starling. Redesign of the shaft to eliminate this vibration eliminated bird ingestion problems. Here is an example that shows how far afield from the "conventional" engineering disciplines an engineer might need to go to fully analyze a design. The lesson is that engineering design should always be thought of as a multidisciplinary or even omnidisciplinary activity.

To further illustrate the nature of engineering design and decision making, an example problem is presented. Consider a jar that is filled with beans and the question is asked, "How many beans are in the jar?" As in the case of most engineering decisions, there are many methods by which one could arrive at an answer. First, and most obvious perhaps, is to pour the beans out of the jar and count them. It might appear that this method could give an exact solution, but this is not necessarily the case. In counting several hundred or thousand beans, it would be easy to make a mistake. Also, beans break in half. Should each half be counted separately as a full bean, as half a bean, or not at all? And bags of beans frequently contain small, beanlike stones. Should they be included? Then, upon pouring the beans back into the jar, they might not fit. Even by counting, it may not be possible to obtain a unique and "correct" solution.

At the other end of the spectrum, it is possible simply to guess. Although this method is less likely to be as accurate as counting the beans, it is simple, easy, and takes little time. In between these extremes lie a wide range of methods that could be broadly referred to as modeling. Before beginning to examine models, however, it is useful to define the option space.

The option space is the space of possible designs or decisions. In the case of the bean-jar example, the option space is easy to define: it is the set of real, positive integers. Knowing this greatly clarifies the issue. One is simply to pick a single number from this set. Generally, defining the option space for an engineering design is a more complex task. But it can be broken into two subtasks: design synthesis and parameterization. Design synthesis considers the broad configurational possibilities. Again, for an airplane, these could include the general configuration of the airframe— high wing, low wing, T-tail or straight tail, number of engines, wing-mounted or fuselage-mounted engines, types of materials, and so on. The second part is parameterization. This means identifying the parametric alternatives. Some parameters are very limited. For example, if an existing engine is to be used, one is limited to available engines. Other parameters, such as the wingspan, may take on a range of parameter values.

In engineering, defining the option space requires two steps: design synthesis or the definition of design configurations, and parameterization or the identification of the parameters and their permissible ranges that define the specific implementation of a design configuration.

Regardless of the ease or difficulty in establishing the option space, it is the proper first step in a decision-making process. Later, it will be necessary to evaluate all options and to make a choice. This is a much easier process if the number of options to be evaluated is fixed at the outset. Indeed, the level of analysis necessary may well be determined by the nature of the option space.

Next is the problem of determining for each option the outcome space. Although determination of the outcome space is what most of engineering education focuses on, it is rarely presented as such, and thus the results are lacking. The outcome space is the set of possible outcomes for each potential design *and their associated probabilities of occurrence*. Here the example of beans in a jar is particularly simple. The outcome space is simply the set of real, positive integers, *with a probability on each*. Again, one could simply guess at the number of beans in the jar, and extend the guess to a probability distribution on the number of beans. But a more professional approach is modeling. A simple model for the number of beans in a jar is the following:

$$N = \frac{V_j}{V_b} f \quad V_b = 4/3 \, \pi r_b^3$$

V_j is the volume of the jar, V_b is the average volume of a bean, and f is a "packing" factor or the fraction of the jar that is filled with beans as opposed to air. Assuming the beans to be spherical, the average volume of a bean is given in terms of the average radius r_b. Of course, much more complex models could be used. For example, the volume of a bean could be computed assuming that beans are oblate spheroids, and models could be developed for the computation of f. Such models could even account for edge effects, so that f is dependent on the size and shape of the jar.

In the above model, V_j, V_b, and f can be estimated with reasonable accuracy. What the model has done is to "disaggregate" the variable N into three variables (in this case r_b is used in place of V_b), each which can be estimated with an accuracy such that their aggregation via the model into N leads to an estimate of N that contains less uncertainty than a direct estimate of N itself. The process of modeling is shown in Fig. 1. Direct estimation of N leads to considerable uncertainty. The development and

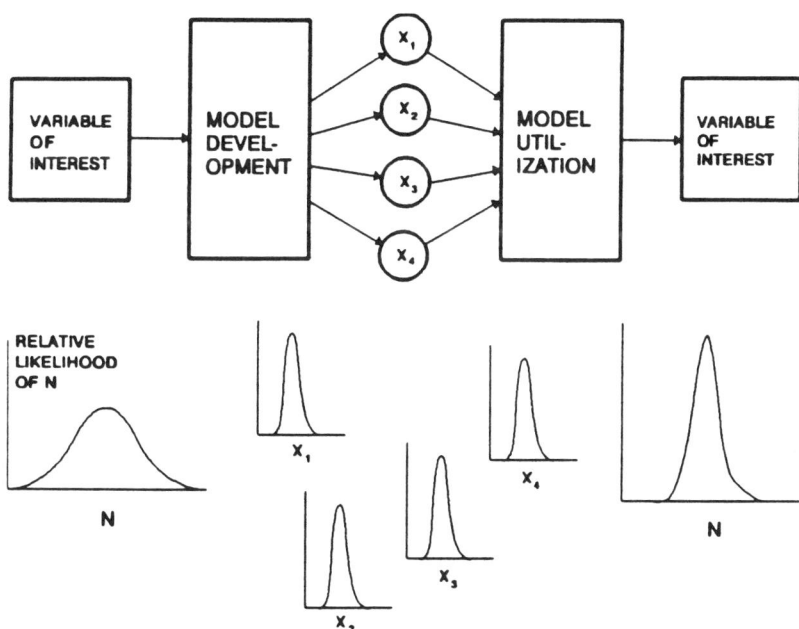

Fig. 1 **The notion of a model.**

use of a model leads to an estimate of N that has less uncertainty. We could say that a "good" model is one that reduces the uncertainty in the variable(s) of interest.

Despite the use of models, however, there almost always remains uncertainty in the estimate of the parameter(s) of interest. This is certainly the case in the bean jar example above. Indeed, as noted earlier, many engineering problems have an inherent uncertainty that no amount of analysis could remove. Consider again the stress imposed on the wingspar of an airplane. This depends on loading of the airplane, the weather through which it flies, and the manner by which the pilot flies the airplane. If a manufacturer produces 100 similar airplanes, each will be subjected to different stresses through its life. A valid question is: "Given the wide range of uncertainty on the actual, life-cycle loading of the wingspar, how strong should it be?" Too weak, and there will be failures. Too strong, and its excess weight will cause economic loss because of reductions in parameters such as payload, range, and speed.

Clearly, it is not possible to decide on how strong the wingspar should be without some measure of the economics, including the risk of failure and accident, of the airplane. The same is true of guessing beans in a jar. Why do we want to guess how many beans are in the jar? Is there a payoff? What are the rules?

The importance of the rules can be demonstrated by means of a trivial example. Suppose that, in the bean jar example, a prize of $10 is to be offered for a "good" guess, defined as a guess that is lower than the actual number of beans in the jar. A guess of one bean satisfies this criterion, and wins the prize. There is no need to model beyond the knowledge that there is at least one bean in the jar. What is more, the decision to choose an answer of one bean is a good decision independent of how many beans are actually in the jar. In many cases of engineering decision making, it is permissible to err on the side of conservatism, so long as the economic penalty for doing so is small.

We ultimately wish to set upon the derivation of an approach that combines engineering analysis of design options with values to obtain preferred design choices. First, however, an observation is in order:

The results of an engineering analysis, that is, the outputs of a model, are of no value to a decision maker unless they include some measure of their accuracy.

For example, a model that produces the answer, "There are 1037 beans in the jar," without an assessment of the accuracy of this number, provides no insights of value. It is only when a decision maker assigns a level of confidence to such an answer that it is useful in a decision-making process. Herein lies a major fault in the development and teaching of engineering models. To be useful in engineering design and decision making, they must include, implicitly or explicitly, an assessment of their accuracy. Yet, virtually no models make this inclusion. Thus, the first step we must face is to extend the modeling techniques most engineers have been taught so that they provide more than an answer. We want them to provide *information*.

There are two basic sources of modeling error: model error and parameter error. A mathematical model may be defined as an abstraction of reality. Mathematical models of the physical world are abstractions of nature. As an abstraction, no model is either a complete or perfect representation of the thing it represents. The simple bean counting model provided above is no exception. Beans are modeled as spheres, and many physical phenomena are compressed into one parameter, the packing factor. This model uses three parameters: the volume of the jar, the average radius of an (assumed spherical) bean, and the packing factor. None of these parameters can be determined precisely, although measurements and experience may lead to fairly accurate knowledge of their values. Again, information on the values of these parameters takes the form of probability distributions—no information is contained in a single number estimate.

In the case of the bean jar example, the model given could produce answers across any reasonable range of values provided that the input parameters are chosen appropriately, with the knowledge of the approximations they represent. Thus, it is possible to use this model both to estimate the number of beans in the jar and to estimate the accuracy of the estimate. There are many approaches one could use to do this, but the one that will be discussed here is generic in that it tends to work for nearly all problems of this sort. This is the Monte Carlo method. The Monte Carlo method is a simulation technique that builds a probability distribution of the output variable by repeated application of the model, each time with a different set of input parameters. The method works as shown in Fig. 2.

The state of information on each input parameter to the model is expressed as a probability distribution, typically a probability density function pdf. These functions are integrated across the parameters to obtain cu-

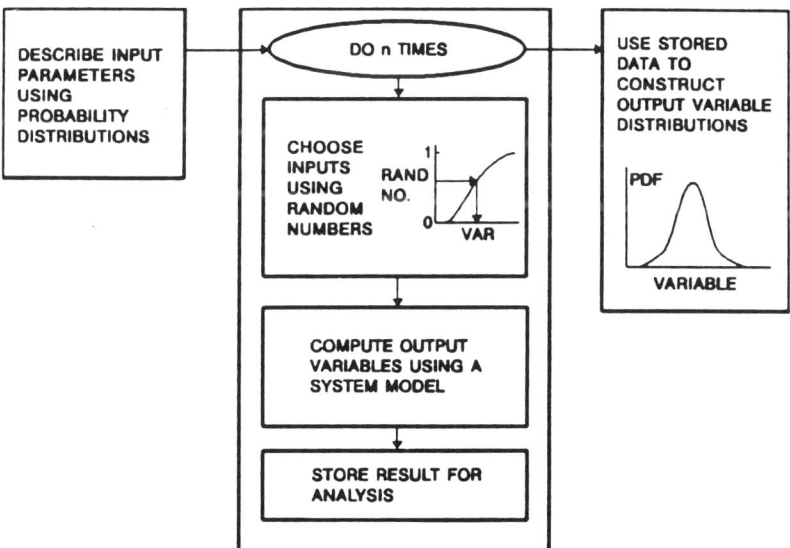

Fig. 2   The Monte Carlo method.

mulative distributions that range from zero at the minimum values of the parameters to one at the maximum values. Then, for each parameter, a random number is chosen in the range zero to one. The random numbers are entered on the ordinates of the cumulative distributions to select specific values of each parameter. The selected set of parameters is then entered into the model and the output variable is calculated. By repeating this procedure many times (perhaps 1000 times), a probability distribution on the output variable may be generated. This distribution contains the information available on the output variable.

There are many potential pitfalls of the Monte Carlo method of which one must be aware and avoid. First, the random numbers used might not be random. Some random number generators produce a set of numbers that are correlated with lags of a few to several numbers. This can be very bad if the correlation lag is equal to the number of variables in the model, such that successive values of the output variable are highly correlated. Second, some parameters have extremely unlikely values that produce significantly different consequences. It may be unlikely that these values will be sampled in proportion to their likelihoods, thus severely distorting the results. Procedures exist for dealing with this situation. They basically consist of adjusting the probabilities to assure adequate sampling, and then adjusting the output probabilities accordingly. Despite such problems, the Monte Carlo method, in the proper hands, is an extremely powerful approach to the solution of many problems.

The application of the Monte Carlo method to the model of the bean jar example might produce a result similar to that shown in Fig. 3. This distribution notes that there is a minimum number and a maximum number of beans, and that it is a virtual certainty that the actual number of beans in the jar lies within this range. The distribution also shows that there is a most likely number, this is the guess that is most likely to be correct, and there is an expected or mean number, which is generally different from the most likely number. The most likely number and the expected number are almost always different if the distribution is nonsymmetrical, and this is often the case if the variable cannot range below zero, as is the case here (there is no such thing as a negative bean).

To further illustrate the need for careful development of a value space for the selection of an option, we will continue with the bean jar example.

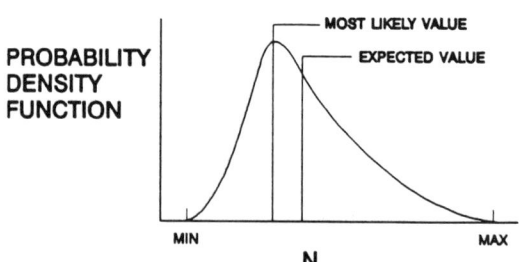

Fig. 3 The notion of uncertainty.

Assume that there is for a good guess a payoff of the form shown in Fig. 4. This function provides for a prize of $250 for an exact guess, diminishing to zero for a guess that is low by 250 beans. There is no prize for a guess that is high, even by one bean. Such a payoff function is nonsymmetrical, as are many payoff functions in engineering. There is also a cost C for making a guess. Let N denote the actual number of beans in the jar, u denote the guess that is submitted, and P denote the payoff. Given the state of information contained in the pdf of Fig. 3, which we will denote $f(N)$, the expected value of a guess u is given by

$$E\{u\} = \int_0^\infty f(N)P(N - u)\,\mathrm{d}N - C$$

One might think that a logical choice of u would be the value that maximizes $E\{u\}$. This indeed would be the case for a person whose decisions are based on expected values only. The implication of expected value decision making, however, is that the decision maker has infinite resources available and can wait to average his or her winnings. A decision maker with limited resources might instead be highly risk averse. The probability of obtaining a net gain is given by the equation

$$R\{u\} = \int_u^{u+250-C} f(N)\,\mathrm{d}N$$

Here, $R\{u\}$ is the probability that the decision maker will not lose money as a function of his or her choice. A risk-averse decision maker might want to choose the value of u that maximizes this probability. Figure 5 illustrates the choices of u that maximize $E\{u\}$ and $R\{u\}$ respectively. These values of u are not the same, and indeed one could plot $R\{u\}$ vs $E\{u\}$ parametrically with respect to u, with the result shown in Fig. 6. Generally, the curve of Fig. 6 will be smooth as it passes through the maxima in $R\{u\}$ and $E\{u\}$. It follows that the expected value decision maker could increase his or her probability of winning something at (to first order) no loss in expected value, and the risk-averse decision maker could increase his or her expected winnings at (to first order) no loss in probability of winning something. Therefore, it is unlikely that any rational individual would choose a value of u that produces one of these extremes. More likely, the decision maker

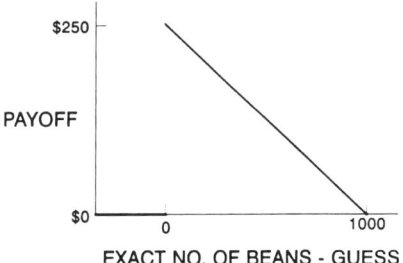

Fig. 4 **A nonsymmetrical payoff function.**

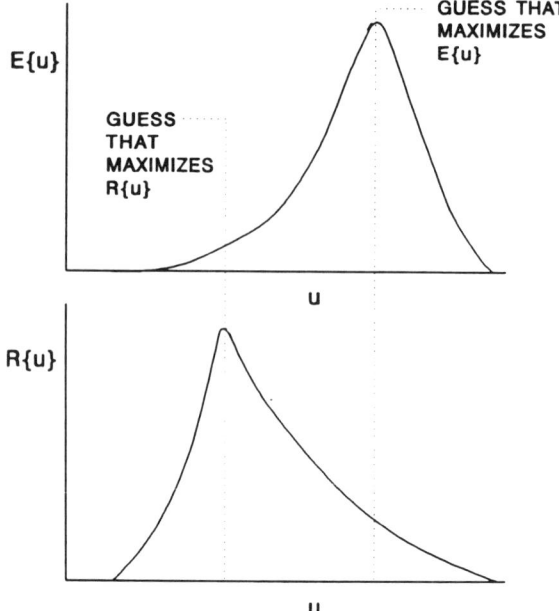

Fig. 5 Probability distributions on $E\{u\}$ and $R\{u\}$.

would choose a value of u that lies somewhere on the solid part of the curve, between but not at the endpoints of the solid portion.

The question is now, "What value of u provides the best choice?" We can go no further seeking the solution unless we turn to the decision maker's value space, and define his or her risk-return preferences. The mathematical basis for the construction of a value space is contained in an area that is referred to as *utility theory*. Utility theory has roots that date to the 18th century. But a major extension is attributable to von Neumann and Morgenstern, who added the dimension of risk through the notion of the

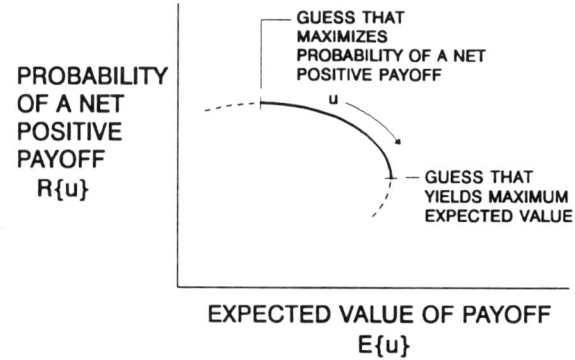

Fig. 6 The option space.

von Neumann-Morgenstern lottery. We will not dwell on a detailed derivation and explanation of utility theory here. Rather, we will merely point out the existence of the value space and illustrate its use.

A value space is a multidimensional space that describes a decision makers preferences for alternatives whose outcomes are represented as a set of attributes. Each set of attributes may be assigned a number, which can be thought of as representing the "worth" or value of the attribute collection, such that if alternative A is preferred by the decision maker over alternative B, the number assigned to A will be larger than the number assigned to B. Thus, the value space enables the ordinal ranking of alternatives.

Engineers make use of value functions quite frequently in selecting between alternatives. A popular type of value function is known as the linearly additive value function. This function represents total value of an alternative as a linearly weighted sum of the values associated with each of the attributes of the outcomes of the alternatives, including risk. The form of the linearly additive value or utility function is

$$V = \sum_{i=1}^{n} \alpha_i v_i(x_i)$$

Here, the values of each of n attributes are given by $v_i(x_i)$, and the weights are α_i.

One thing generally can be said of linearly additive value functions, despite their extensive use. They are almost always inappropriate for application to engineering decision making. A simple test usually suffices to prove this. Consider an option for which the outcome is perfect in all attributes save one, and it fails completely in that attribute. For example, consider an airplane that is fast, maneuverable, strong, durable, easily maintained, and very low cost. But the airplane has a range of 1 mile. Clearly, the airplane is utterly useless. How does it measure against other alternatives in a linearly additive value function? If it beats out more reasonable alternatives, it is clear that the value function is inappropriate.

There are several alternative forms of value functions that are more appropriate to engineering decision making. These include the multiplicative function

$$V = \prod_{i=1}^{n} v_i(x_i)$$

and the lexicographic function

$$V = \max_{i=1}^{n} v_i(x_i)$$

The multiplicative function has a logarithmic form that is much like the linearly additive function, but the nonlinearity of logarithms takes care of the problem noted above:

$$V = \sum_{i=1}^{n} \log \left[\alpha_i \, v_i(x_i) \right]$$

Note that the log (V) increases monotonically with V, such that the order of preferences is unchanged despite which form one uses. Thus, it is unnecessary to take the antilogarithm to obtain a useful value measure. Also note that it is only necessary that value functions rank alternatives. They need not rate alternatives against one another, that is to say, they need not give a measure of how much better one alternative is than another. Indeed many experts in utility theory argue that only ordinal values need exist.

Many engineering problems have value functions that can be reduced to purely economic considerations. Attributes such as performance, reliability, maintainability, and lifetime can all be reduced to a common economic measure. In such cases, it is appropriate to make this reduction rather than to treat these as separate attributes.

A value function for the bean jar example could take the form shown in Fig. 7. The curves of constant value shown in Fig. 7 are referred to as indifference curves. A decision maker would be indifferent to options whose outcomes lie on a single indifference curve, and would prefer options whose outcomes lie on indifference curves of higher value. This concept enables the optimum choice of u by combining Figs. 6 and 7 as shown in Fig. 8. The optimum choice is that value of u that maximizes value, which occurs at the point of tangency between the option space curve and the indifference curve representing the highest obtainable value.

The methodology outlined above and applied to the simple example of guessing how many beans are in a jar, unlike almost all other engineering methodologies, is a decision-making methodology, not a problem-solving methodology. It does not seek nor does it use an answer to the problem, "How many beans are in the jar?" Rather, it seeks and uses *information* regarding the question, "What is the best guess of how many beans are in the jar?" In doing this, the methodology maintains consideration of many things:

1) The allowable set of options,
2) The probability distribution on possible outcomes of each option,

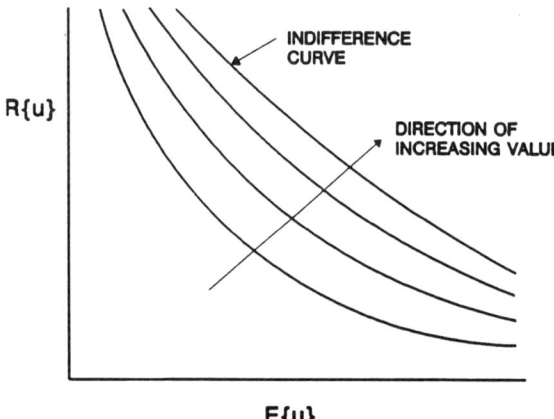

Fig. 7   A typical indifference field.

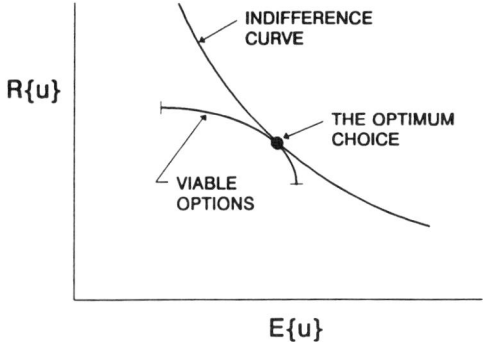

Fig. 8 Risk-value tradeoffs.

3) The cost of pursuing each option
4) The preferences, including risk, of the decision maker

In essence, deciding on a guess u requires testing the hypothesis, "u is the best guess."

Cost and Value of Information

There is one more significant difference between problem solving and decision making; a problem becomes solvable when the three elements of problem solving are available—essentially, laws of nature, boundary conditions, and computational capability. But a decision must be made when it must be made, regardless of the state of information at the time. And it is obvious that one's ability to choose effectively between options is strongly dependent on the amount of information—the state of information—that is available. Yet, information nearly always comes only at some cost, that cost increasing rapidly as the quality of the information improves, and perfect information is rarely obtainable. Thus, two questions arise: "What is the optimum amount of information to seek in support of a decision?" "How can one measure the state of information regarding a decision?"

The optimum amount of information to seek in support of a decision depends on the value obtainable from improved information and the cost of improving the information. Implicit in this statement is the notion that one never starts into a decision-making process with a total lack of information. Thus, there is always an informational status quo. To state it differently, one never makes a decision with absolutely no expectation on the future given the choice taken. It is the informational status quo that one improves upon when one "buys" information, and so it is this improvement and the value and cost of the improvement that needs to be analyzed.

First, consider the form that improved information takes. It was noted above that information takes the form of a probability distribution such as that shown in Fig. 3. An improved state of information would take the form of a probability distribution that is more peaked or less broad than that extant. Figure 9 shows how improved information might appear. Im-

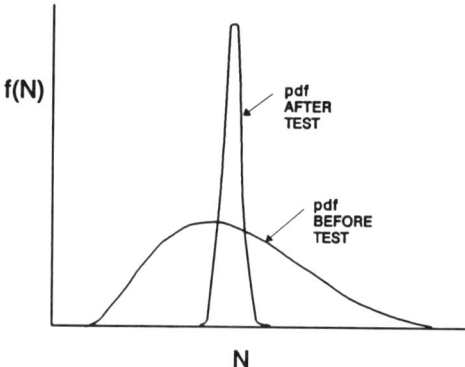

Fig. 9 The result of improved information.

proved information might be obtained by performing a test, or by the use of a mathematical model, or both. In the case of the bean jar example, one might count the beans in a jar of smaller volume and scale this answer up to the volume of the jar in question. In advance of the test or analysis, one should be able to estimate the quality of the information that will be obtained. But one cannot obtain the information itself. What this means is that one should be able to estimate the breadth of the probability distribution function after the test (in terms of, say, its standard deviation), but not its location (that is, essentially, its mean). In fact, all that can be said of the location of the after-test pdf is that its location is described probabilistically in terms of the before-test pdf. This fact greatly complicates the determination of the value of improved information, and it is beyond the scope of this chapter to provide an analytic approach to its evaluation.

Conceptually, one wishes to evaluate the values and costs of different amounts of improvement upon the current state of information, and one would then select to buy that level of information that maximizes the net value—value less cost. This process is shown in Fig. 10. Improved information provides value in a decision-making process because it enables the taking of choices whose outcomes have higher value as measured in terms of the attributes of the outcomes.

Sometimes, when it is difficult to assess the value of improved information, it might prove insightful to estimate the value of perfect information. The bean jar example is such a case. Here, the value of perfect information is quite simple to estimate: it is the maximum prize less the cost of making a guess. With perfect information, one could be guaranteed of choosing the best possible value of u, the value that obtains the maximum prize. Whenever it is this easy to estimate the value of perfect information, it is a worthwhile thing to do.

Note that the optimum amount of informational improvement one should seek to buy depends on the cost of the improvement. Thus, it is not only important to be able to estimate the cost of an engineering system, it is also important to be able to estimate the cost of obtaining improved in-

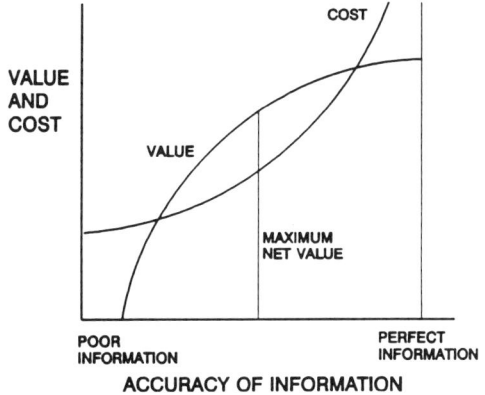

Fig. 10 Optimizing the purchase of information.

formation regarding alternative system choices. This is an area that has been largely overlooked by cost analysts. The methodology outlined here puts these estimates into a context for use similar to the context of system cost estimates.

When considering whether to buy improved information for a decision, it is helpful to know what is the current state of information. It is noted early in this chapter that information relates to a specific decision. This becomes very clear when we set out to define the state of information. Indeed, the state of information can be quantified only in reference to a specific decision: it depends on the outcome spaces of all available alternatives and the preferences of the decision maker.

Consider the generic decision shown by the decision tree of Fig. 11. The option space is defined as the choices $i = 1, 2, 3, \ldots$, and the decision for which the state of information is to be evaluated is denoted by the square. The presence of uncertainties associated with each possible choice

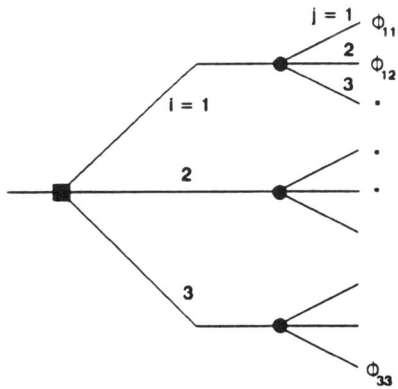

Fig. 11 A decision tree.

are denoted by the large dots, and the outcome spaces for each decision are denoted by the symbol ϕ_{ij}. Each outcome occurs with probability p_{ij}, and has associated value V_{ij}. Comparing outcomes, the values are such that if

$$V_{ij} > V_{\bar{i}j}$$

then

$$\phi_{ij} \succ \phi_{\bar{i}j}$$

which is read, ϕ_{ij} is preferred to $\phi_{\bar{i}j}$. If

$$V_{ij} = V_{\bar{i}j}$$

then

$$\phi_{ij} \sim \phi_{\bar{i}j}$$

which is read, neither outcome is preferred to the other. Now define the variable

$$\delta_{ij}^{\bar{i}j} = \begin{cases} 0 \text{ if } V_{ij} < V_{\bar{i}j} \\ 1 \text{ if } V_{ij} \geq V_{\bar{i}j} \end{cases}$$

Given these definitions, the probability that decision i will result in an outcome that is preferred to or at least as good as decision $\bar{i}$ is given by

$$P_{\bar{i}}^{i} = \sum_{j} p_{ij} \sum_{j} \delta_{ij}^{\bar{i}j} p_{\bar{i}j}$$

The state of information H is now determined as the probability that the following statement is true:

Choice i will produce an outcome that is preferred to or at least as good as any other choice.

This probability is

$$H_i = \prod_{\bar{i} \neq i} P_{\bar{i}}^{i}$$

As written, H_i is unity if one can be sure that i is the choice that will lead to the most preferred outcome. The value of H_i diminishes with degraded states of information. Thus, if H_i is close to unity, one might choose not to buy improvements upon the current state of information. But if it is well below unity, one might choose to delay a decision until a better state of information can be obtained.

Note once more the role played by cost estimation. Cost is a system attribute. Uncertainty on cost is uncertainty on the outcome space of a choice. Therefore, cost uncertainty is a valid reason for seeking improved information on a decision.

Conclusions

This article has sought to provide a new approach to engineering, an approach that treats engineering design as a process of decision making, not as a process of problem solving. We have defined this approach as *systems engineering*, and shown that, with this definition of systems engineering, it becomes amenable to being taught. Indeed, it is evident that the discipline of systems engineering includes a broad class of subdisciplines such as microeconomics, decision theory, utility theory, optimization, and cost analysis.

The approach presented here is a self-consistent theory of engineering decision making that sets the basis for the discipline of systems engineering, but leaves much to be developed. The objective is to provide a methodology for good engineering decision making, and it is shown that the essence of good decision making is good information. A precise definition of "information" is given, and a quantitative measure of its quality is developed. The quantitative measure of informational quality includes the nature of the options available to the decision maker, the information available on the outcomes of each option, and the preferences of the decision maker. In this way, the quality of information ties together all aspects of the decision.

Finally, the methodology developed here places cost analysis into a precise context for incorporation into the engineering decision-making process. It is shown that cost information is needed not only for engineering products, but also for the engineering information-gathering process as well.

Considerable extension and development of this methodology is already in the published literature. The bibliography provided below is a good starting point.

Bibliography

Arrow, K. J., *Social Choice and Individual Values*, 2nd ed., Wiley, New York, 1963.

Bernoulli, D., "Specimen Theoriae Novae de Mensura Sortis," *Commentarii Academiae Scientiarum Imperalis Petropolitanae*, Tomus V, 1738, pp. 175–192. Also, "Exposition of a New Theory of Risk Evaluation," *Econometrica*, Vol. 22, Jan. 1954, pp. 23–36. Translated by Louise Sommers.

Cyert, R. M., DeGroot, M. H., and Holt, C. A., 1978, "Sequential Investment Decisions with Bayesian Learning," *Management Science*, Vol. 24, No. 7, 1978.

Fishburn, P. C., "Utility Theory," *Management Science*, Vol. 14, No. 5, 1968, pp. 335–378.

Hazelrigg, G. A., "Economic Viability of Pursuing a Space Power Concept," *Journal of Energy*, Vol. 1, No. 2, 1977, pp. 93–99.

Hazelrigg, G. A., "Evaluation of Long-Term RD&D Programs in the Presence of Market Uncertainties," *Energy Systems and Policy*, Vol. 6, No. 2, 1982.

Hazelrigg, G. A., "The Engineering Decisionmaking Process," *Proceedings of the Cambridge Symposium on Optical and Optoelectronic Engineering*, sponsored by The International Society for Optical Engineering, Cambridge, MA, Nov. 6–11, 1988, pp. 118–127.

Howard, R. A., Matheson, J. E., and Miller, K. L., *Readings in Decision Analysis*, 2nd ed., Stanford Research Institute, Menlo Park, CA, 1976.

Keeney, R. L. and Raiffa, H., *Decisions with Multiple Objectives*, Wiley, New York, 1976.

Takayama, A., *Mathematical Economics*, Dryden Press, Hinsdale, IL, 1974.

Tribus, M., *Rational Descriptions, Decisions and Designs*, Pergamon Press, Elmsford, NY, 1969.

von Neumann, J., and Morgenstern, O., *Theory of Games and Economic Behavior*, 3rd ed., Princeton Univ. Press, Princeton, NJ, 1953.

Resource Allocation in the Congress

Terry Dawson*
U.S. House of Representatives, Washington, DC 20515

Introduction

RESOURCE allocation within the Congress has become an increasingly complex process whose outcome for individual projects is difficult to predict. Three groups of committees in the two separate Houses of Congress are involved in the process of setting the NASA budget. Superimposed on this process is the requirement to allocate resources in such a way that implementing legislation will not be vetoed by the President.

Additional constraints are imposed by such factors as: 1) the presence of overarching guidelines and restrictions (such as the Gramm-Rudman Act or the Budget Enforcement Act of 1990); 2) the organization of specific key congressional committees (e.g., the current organization of the Appropriations Subcommittee that funds the NASA budget forces NASA to directly compete for funding with such activities as veterans affairs and housing programs rather than defense spending or total federal spending for research); and 3) the particular districts and states that are represented by key leaders in Congress.

Budget Process in Congress

The budget process starts with the submission of a budget request by the President in late January or early February. This budget request covers the upcoming fiscal year that will commence on October 1. The key committees in each House of Congress that must act on this budget request include the Budget Committee, an Authorizing Committee, and the Appropriations Committee.

This paper is declared a work of the U.S. Government and is not subject to copyright protection in the United States.
*Engineering Advisor, Committee on Science, Space, and Technology.

Legislative Process

It may be useful at this point to describe the legislative process that a bill goes through before becoming law. The process is essentially the same for the authorization bill and the appropriations bill. The initial part of the process is also the same for the budget resolution that is prepared by the Budget Committee. However, because the House Budget Resolution applies only to the House of Representatives and the Senate Budget Resolution applies only to the Senate, differences between the two bills do not need to be resolved, nor is the final product submitted to the President for his signature.

Since revenue and appropriations bills traditionally originate within the House of Representatives, the following description is from the perspective of the House. The legislative process begins with a series of comprehensive hearings on the President's budget request at the subcommittee level. At these hearings, the subcommittee receives testimony from administration officials, industry and academic leaders, and other interested individuals.

Following these hearings, the subcommittee chairman will prepare a version of the bill which he or she believes strikes the best balance between fiscal reality and the competing needs of the various programs covered within the bill. This document is referred to as the *Chairman's mark*.

When the Chairman's mark is ready, the chairman schedules a subcommittee "markup" of the proposed legislation. At this markup, the members of the subcommittee discuss the proposed legislation, and offer "amendments" that they feel should be adopted. These amendments are debated and then voted on. After all of the member's amendments have been dealt with, the chairman calls for a vote on passage of the revised legislation. When a majority of the subcommittee members vote to approve the revised bill, it is "reported out" to the full committee for further consideration.

Note, in addition to the actual bill, the subcommittee also approves a legislative "report" which further explains the provisions contained in the bill. In some cases, the report also expands on the provisions contained in the bill by giving additional specific guidance. This additional guidance does not carry the same weight as the legislative language contained in the bill, but it is useful to the administration in further defining the intent of Congress. After being adopted by the subcommittee, the report follows the bill to full committee and then to the floor of the House. The report can be revised in full committee, but not on the floor.

When the full committee receives a piece of subcommittee-passed legislation, the chairman of the full committee schedules a full committee markup of the bill. At full committee, the subcommittee chairman describes the bill that is before the members and summarizes any amendments that were adopted by the subcommittee. Committee members then propose additional amendments that are debated and voted on. When this process is complete, the chairman calls for a vote on the amended bill. If a majority of the members vote in favor of the amended bill, it is submitted to the full House for action.

However, before a piece of legislation can be debated on the floor of the House, it must be granted a "rule" by the Rules Committee. A rule is a document that will govern the process and debate that will take place on the floor. The rule specifies the amount of time that will be allocated to debate the bill and stipulates the number or type of amendments that may be offered on the floor.

After a rule is granted, the bill can be placed on the legislative calendar. Though somewhat more formalized, the process on the floor is similar to that in full committee—the bill is described, debated, and amended. Generally, each amendment is debated and voted on separately. The exception is a group of noncontroversial amendments that the full committee chairman (who at this point is called the *floor manager* for the bill) may choose to have dealt with en bloc (i.e., bundled together as a single amendment).

When all amendments have been dealt with, a vote is called for on "final passage" of the bill. If a majority of the members vote in favor of the bill, it passes and is sent to the Senate for its action.

Generally, the Senate acts in parallel with the House (although somewhat behind it). In the Senate, the House-passed bill, or a Senate substitute, goes through a similar process—i.e., subcommittee action, full committee action, and floor action.

In almost all cases, there are differences in the House-passed and the Senate-passed versions of a bill. These differences *must* be worked out before the legislation can be submitted to the President. The most common way of working out these differences is for the Speaker of the House and the President of the Senate to appoint *conferees* to work on the bill. These conferees (which collectively are also referred to as a *Conference Committee*) and their staffs meet either formally or informally to negotiate compromises for the differences between the two bills.

When these differences have been worked out, the newly revised bill is revoted on by the House and the Senate. When both have passed bills with *identical* language, the bill is sent to the President for his action.

The President basically has three choices: he can sign the bill (in which case it becomes law); veto the bill (in which case it is sent back to the Congress to either attempt to override the veto, accept defeat and give up, or start the process all over); or choose to take no action. If the President decides to take no action, the fate of the bill is decided automatically after 2 weeks. If the Congress is still in session 2 weeks after he has received the bill, it automatically becomes law. On the other hand, if the Congress is not in session 2 weeks after the President receives a bill, it is automatically vetoed. This type of veto is called a *pocket veto*.

Thus, you can see, the process that a bill must follow is very long and cumbersome. And, at every step of the way, the bill is subject to amendment or revision. These amendments generally reflect the prevailing political sentiment of the members who are working on the bill at that time.

Budget Committee

The House and Senate Budget Committees were established in 1974 as a means of bringing greater discipline to the congressional budget process. Basically, the responsibility of the Budget Committee in each House is to control the "bottom line" of the federal budget—to ensure that the aggregate spending of the Congress does not exceed prior guidelines that have been established.

The Budget Committee is one of the first focuses of attention on the President's budget request. Within a couple of weeks after the President's budget has been submitted, the Budget Committee receives a report from the Congressional Budget Office (CBO) entitled "Economic and Budget Outlook for Fiscal Years XXXX to XXXX." This report contains an economic forecast and an estimate of the spending that is implied by current law for the next 5 years. These data are presented by functional category and are generally referred to as the *CBO baseline*.

Unless the Congress and the President have directed otherwise in current law for a particular project or activity (as would be the case for a "multiyear appropriation"), the CBO baseline assumes that each area of the budget increases by an amount equal *only* to inflation.

Several weeks later, congressional committees submit a report called "Views and Estimates" to the Budget Committee. This report presents an early estimate of the action that a particular committee anticipates taking on the part of the overall budget that falls within that committee's jurisdiction—i.e., does the committee anticipate passing a funding bill that is greater than or less than the "baseline" that was contained within the CBO report to the Budget Committee.

The Views and Estimates from each committee indicate how far above or below the CBO baseline the committee expects to be for each element of the budget for which that committee is responsible. *Views* are contained which explain why the committee believes that the anticipated action (to raise or lower funding for a particular activity) is warranted.

The Budget Committee has divided all elements of the federal budget into 20 specific groupings called *functions*. Most of NASA's spending falls within Function 250—General Science, Space, and Technology. This function contains funding for space research and technology, general science, and basic research not specifically covered by other functional areas (e.g., national defense and health). This includes funding for NASA, the National Science Foundation, and the high energy and nuclear physics research programs of the Department of Energy.

NASA funding not contained within Function 250 is included in Function 400—Transportation. All of NASA's aeronautical research activities are incorporated within this budget function.

At about the same time that other committees are preparing their Views and Estimates for the Budget Committee, the Budget Committee begins to hold an extensive set of hearings. At these hearings, testimony is received from administration officials, business and labor leaders, economists, academics, and members of Congress (e.g., chairmen of other committees).

After receiving the Views and Estimates from the relevant committees and completing its hearing process, the Budget Committee reports its budget plan for the coming fiscal year. This plan (called the *Concurrent Resolution on the Budget*) must be formally adopted by the Congress by April 15.

Generally, the Congress is prohibited from considering (i.e., debating or voting on the floor of the House or Senate) any revenue, spending, entitlement, debt, or credit legislation before the Budget Resolution has been adopted.

The Budget Resolution sets forth guidelines and targets for federal revenues, spending, the deficit, and the national debt that the Congress *must* follow in preparing appropriations bills, tax bills, and other spending legislation. Basically, the Budget Resolution tells the Appropriations Committee the *total* amount of funds that are available for the committee to appropriate to *all* of the activities under its jurisdiction. This funding ceiling that is imposed on the Appropriations Committee is frequently referred to as the "302(a)" or "602(a)" allocation, based upon the section number of the original enabling legislation.

After the budget process is completed and the subject fiscal year has begun, a series of analyses are conducted to determine if the revenue and spending assumptions that the Budget Committee used in compiling the Budget Resolution are still valid. If it appears that mandated deficit targets will not be met, then mandatory "sequestrations" (or reductions in the budget) are imposed.

The CBO must submit its final sequestration report within 10 days after the end of a congressional session. Within 15 days of the end of a session, the Office of Management and Budget in the White House must submit its final sequestration report and the President must issue any necessary sequestration order (i.e., the order that implements an across-the-board cut in the budget). Thirty days later, the General Accounting Office must file a compliance report on the sequestration order.

To date, sequestration orders have only had to be issued for fiscal years (FY) 1986 (with a 4.16% cut in the budget), 1990 (with a 0.0076% cut in the budget), and 1991 (with a 0.0013% cut in the budget). As it turns out, the FY 91 sequestration was subsequently rescinded by the Budget Enforcement Act of 1990 (i.e., the Act that emerged from the "Budget Summit" that was held between the Congress and the White House during the spring and summer of 1990).

Authorizing Committee

The function of an Authorizing Committee is: 1) to set guidelines (or ceilings) against which funds may be appropriated for specific programs; 2) to conduct oversight (including cost, schedule, and performance reviews) for those programs that do receive appropriations; and 3) to create and maintain the legislative and policy language that governs the various functions of government.

As an example of the latter responsibility, it was the Authorizing Committee which wrote the Acts that created NASA, the National Space Council, and the Office of Commercial Space Transportation within the Department of Transportation. The Authorizing Committee also establishes the broad elements of space policy—from the tenets of space commercialization to the guidelines that control the types of missions the Space Shuttle will be permitted to fly.

In the funding arena, the annual Authorization Bill for NASA identifies which programs and activities are authorized to receive appropriations, and sets an upper limit for the amount of funds that can be appropriated to each program or activity. Subsequently, the Appropriations Committee determines specifically how much to appropriate to these particular programs and activities. In addition, the annual Authorization Bill for NASA also establishes the guidelines that control the amount of flexibility the Agency will have to move funds from one project or activity to another during the fiscal year.

The Authorizing Committee that is responsible for overseeing NASA in the House of Representatives is the Committee on Science, Space, and Technology. Within this committee, the Space Subcommittee has jurisdiction over the space-related activities of the Agency, while the Technology and Competitiveness Subcommittee oversees all of NASA's aeronautics-related activities.

In the Senate, the relevant NASA oversight committee is the Committee on Commerce, Science, and Transportation. Within this committee, all NASA jurisdiction is vested in the Science, Technology, and Space Subcommittee.

In comparison with the Appropriations Committee, the Authorizing Committees generally have a much more narrowly defined area of responsibility and more people (both members and staff) whose responsibility it is to maintain an oversight and expertise of a particular subject matter. Accordingly, the Authorizing Committees frequently deal in more depth with those subjects than do the Appropriating Committees.

Appropriations Committee

Most of the "power of the purse" resides in the Appropriations Committee. Although the Authorizing Committee does have some say in setting priorities and determining whether or not appropriations are "authorized" for specific programs, it is the Appropriations Committee that provides the actual dollars that are spent. As such, this committee is the proverbial "500-pound gorilla" within the budget process.

Like most committees in the House and the Senate, the Appropriations Committee is divided into subcommittees: in this case, 13 subcommittees. The subcommittee that oversees the NASA budget is called the *Subcommittee on VA, HUD, and Independent Agencies*. It has jurisdiction over the Department of Veterans Affairs, Department of Housing and Urban Development, and about 23 other independent agencies, offices, institutes, councils, and other groups.

The National Aeronautics and Space Administration is among this group of "independent agencies." Other prominent members of the group include: the National Science Foundation, the Environmental Protection Agency, the Federal Emergency Management Agency, and, within the White House, the National Space Council and the Office of Science and Technology Policy.

Examples of other organizations within the group of independent agencies include: the Consumer Product Safety Commission, the Court of Veterans Appeals, the Interagency Council on the Homeless, the National

Credit Union Administration, the Selective Service System, the Federal Deposit Insurance Corporation, and the Resolution Trust Corporation.

The first step in the appropriations process is the 302(a) allocation from the Budget Committee. (Note: for the next several years, this distribution will be referred to as the "602(a)" allocation, because of superseding legislation that will temporarily take control of this process.) This allocation tells the Appropriations Committee the total amount of funds that are available to be appropriated during the coming fiscal year. In other words, it defines the overall size of the "pie" that is available to be divided up during the budget process.

In the next step, the chairman of the Full Appropriations Committee and the chairmen of each of the 13 Appropriations Subcommittees meet as a group (frequently referred to as the "College of Cardinals") to divide up this "pie" into 13 pieces, i.e., one piece for each subcommittee. This distribution is called the "302(b) allocation" (or currently, because of the temporary change in controlling legislation, the "602(b) allocation").

After a subcommittee has received its funding allocation, it is imperative that the appropriations bill which is reported out by that subcommittee (and is eventually passed on the floor) remains within the dollar ceiling that was specified in the 302(b) allocation. Accordingly, the basic process for the subcommittee members is to establish relative priorities among the various agencies or activities that fall within the subcommittee's jurisdiction. Thus, at this point, the budget allocation process becomes a "zero sum game"—every increase in one activity has to be offset by a compensating decrease in some other activity.

To keep this point in proper perspective, it is important to note that in the case of the VA, HUD, IA Subcommittee, the 302(b) allocation is normally billions of dollars *below* what the subcommittee members believe is required to fund *all* of the projects that they feel are deserving. Therefore, the level of competition between various projects within the jurisdiction of a subcommittee can be quite acute.

Selected Factors that Affect Budget Allocations

There are many factors that can have an effect on the allocation of funds to specific projects. Some of the more important of these factors are briefly discussed below.

Presence of Overarching Guidelines or Constraints

Due to a confluence of factors, deficits in the federal budget began to reach unprecedented levels during the 1980s. As the level of alarm rose among economists and the general public, the Congress realized that something would have to be done to correct the problem.

An initial step was the adoption of the Gramm-Rudman-Hollings Balanced Budget and Emergency Deficit Control Act (frequently referred to simply as "Gramm-Rudman") in 1985. This Act required the federal budget to not exceed specific deficit levels beginning in fiscal year 1986. The deficit was to be reduced over a period of 6 fiscal years leading to a balanced budget in fiscal year 1991. In 1987, the target date for achieving a balanced budget was moved to fiscal year 1993.

The requirements of the Gramm-Rudman Act were fundamentally restructured by the passage of the Budget Enforcement Act of 1990. This Act was the product of months of negotiations (commonly referred to as the "Budget Summit") between the Congress and the Administration. The Act covers 5 budget years from FY 1991 to FY 1995.

Under this new legislation, deficit targets are effectively eliminated for the first 3 years. In their place, discretionary "caps" (or budget ceilings) are established for three budget categories, each enforced independently. The three categories are: Defense, International, and Domestic. For the last 2 years covered by the legislation (i.e., FY 1994 and FY 1995), the "walls" between the categories are eliminated and caps are established for all discretionary spending as a whole.

The presence of a "wall" between defense and domestic spending for fiscal years 1991 through 1993 is significant because it effectively eliminates the possibility of applying "savings" that accrue from defense cuts (which are now possible due to reduced tensions in the world) to make additional funds available to meet domestic needs—including investments in research and development (R&D), scientific investigation, and the space program.

One of the basic requirements of the Budget Enforcement Act of 1990 is that all spending legislation for the period covered by the Act (1991–1995) must be "deficit-neutral." This is commonly referred to as the "pay-as-you-go" provision of the Act. It means that any increases in spending for one program must be offset either by decreases of an equal size in other programs or by increases in taxes.

When the requirement of the "pay-as-you-go" provision is combined with the "walls" provision and the general hesitancy of recent congresses to raise taxes, we see that during the period of 1991 through 1993, all increases in the NASA budget will generally have to be offset by equal-sized reductions in other domestic accounts. This means that if the Congress is to agree to making increases in the NASA budget, it will have to do so at the expense of funding for other high-value discretionary activities such as education, housing, public health, public works (including airports, highways, bridges, flood control), and so forth.

In this sort of environment, it is clear that it will be difficult, if not impossible, for NASA to continue to receive the sizable year-to-year increases that it has enjoyed since the late 1980s. When the wall comes down between domestic spending and defense spending in the FY 1994 budget, it will be easier to make increases in the NASA budget (i.e., by using funds that result from decreases in the defense budget). However, in that year (and in subsequent years) there will also be significant demands from other sectors to use the "peace dividend" to fund their favorite domestic programs. So, even then, it is not clear how much better off NASA will be in terms of the budget allocations that will be made available to the Agency.

It should be noted that as this chapter is being written, the Congress is considering lowering the wall between defense spending and domestic spending for the FY 1993 budget, rather than waiting until FY 1994, as required in the Budget Enforcement Act of 1990.

Jurisdictional Organization of Appropriations Subcommittees

As noted earlier, the appropriations budget for NASA is written by the VA, HUD, IA Subcommittee of the Appropriations Committee. Organizationally, this places NASA in direct competition for funds with such politically sensitive activities as veterans health care and housing for the poor and the homeless.

As an example of this direct competition, when the VA, HUD, IA Appropriations Bill for 1991 was being debated on the floor of the House of Representatives, a member of Congress offered an amendment to kill the Space Station and move the funds into the housing part of the bill. He made the argument that "we need housing for the poor more than we need it for astronauts in space." Fortunately, this particular amendment lost, but they don't always lose.

Thus, because of where NASA is placed within the jurisdictional organization of the Appropriations Subcommittees, it is very difficult to directly trade-off funding for NASA activities and those that are generically somewhat similar, such as defense spending or total federal spending for research. Typically, these sorts of distributions *between* subcommittee jurisdictions can only be made by the College of Cardinals (i.e., the chairman of the full committee and the chairmen of each of the 13 subcommittees) when it meets to make the 302(b) budget allocations, i.e., the allocations that divide up the total overall funding "pie" that has been approved by the Budget Committee into the 13 pieces that will be given (one each) to the subcommittees.

Accordingly, it turns out that in almost every step of the budget process, NASA is competing for funds with a somewhat different cast of characters. Within the Administration, NASA essentially has to compete with all other federal departments and agencies. In the review of the Budget Committee, NASA competes within Function 250 with other domestic science initiatives (such as the Superconducting Super Collider) while at the same time competing for funds with other budget functions (such as education, energy, natural resources and the environment, etc.). Within the authorization bill, NASA does not really have any competition as such; the entire bill deals only with NASA programs. And finally, within the appropriations process, NASA programs find themselves in direct competition for funding with housing and veterans programs.

In recent years, there has been some talk within Congress about revising the jurisdictional charters for the Appropriations Subcommittees, but this talk has not yet gone very far.

District Concerns by Members

Because of the nature of our representational democracy, members of Congress are highly motivated to look out for the best interests of their constituents. This means that when funding bills are being considered in the Congress, members are very attuned to the extent to which a particular piece of legislation will provide funding or jobs to the constituents in their district or state. Generally, the more funding or jobs their constituents would receive, the more supportive a particular member will be toward a bill.

Given the way that Congress is organized, some members clearly have more influence on funding legislation than do others. For example, members of Congress who have the privilege of being a member of the Appropriations Committee generally have more influence on funding legislation than do members of Congress who do not sit on the Appropriations Committee. Likewise, on the Appropriations Committee and its subcommittees, chairmen generally carry more weight than do other members of the committee.

As reported in "Part One" of the President's Budget Request for Fiscal Year 1993, "earmarking" of R&D funds in appropriations bills (i.e., the practice of specifying that particular projects be given to specific institutions or researchers, rather than being awarded competitively) has grown by 23% from FY 1991 to FY 1992. In FY 1992, a total of 566 "earmarks" totalling $993 million for facilities and research were contained in appropriation bills for R&D. Of particular note was the finding that the VA, HUD, IA Appropriations Bill experienced the greatest increase over 1991 in the dollar value of earmarks (i.e., the earmarks in that bill increased by $187 million between FY 1991 and FY 1992).

Many observers believe that this practice of earmarking funding for "pet" (and frequently unauthorized) projects in particular congressional districts has the effect of reducing the overall effectiveness of the federal R&D research dollar. Additionally, it reduces the amount of funds that are available to support more justified projects that have gone through the standard authorization/appropriation process.

On the other hand, some people argue that bypassing the normal (and somewhat cumbersome) peer review process for the distribution of research grants can at times actually be a more efficient way of allocating funds to individual researchers.

Influence of Lobbyists and Constituents

The basic function of a lobbyist is to formulate the best possible set of arguments that can be constructed in defense of a particular program or issue, and then to communicate those arguments to those in Congress who can affect the outcome. By listening to lobbyists on both sides of an issue, a member of Congress is generally able to hear the best possible arguments that can be made for and against a particular proposition. In this way lobbyists can play a very valuable service in keeping members of Congress well informed on the issues that confront them.

Likewise, it is important for constituents who feel strongly about a particular issue to communicate those views to their Senators and Congressmen. Members of Congress *do* "read" their mail. Read is put in quotes here because the volume of mail received in a congressional office is generally too great for a member to personally read each and every letter. But, every letter *is* read by a member of the Congressman's staff, and the staff do keep their Congressman fully informed on the content, count, and trends that are reflected in the daily mail.

Funding Stability

In recent years, there has been a growing debate on how to improve funding stability for federal R&D programs. It is generally recognized that there is too much instability in the funding that is made available for most federal research projects. This instability manifests itself in frequent downward adjustments in the budgets for specific projects, numerous "restructurings" during the life of a project, and the periodic cancellation of some projects due to a lack of funds. The source of some of this instability is the Congress, but some also comes from within the Administration.

Most funding bills in Congress are for a single year (although in the case of appropriations for construction or R&D projects, funds that are not expended in the specified year can generally be "carried over" into the next year, or even later years in the case of construction projects). Some observers have suggested that funding stability could be improved if the Congress would adopt the practice of passing multiyear funding bills.

During each of the past 4 years, the Science, Space, and Technology Committee in the House of Representatives has passed a multiyear authorization bill for NASA. However, that committee's counterpart in the Senate has generally not agreed that the time is right for multiyear legislation. So the "Conferenced Bill" that has been sent to the President for signature has almost exclusively been single-year in nature. (Some exceptions to this general rule have included a 3-year authorization for the Space Station beginning in FY 1989, and a total authorization for the full duration of the R&D effort for the CRAF/Cassini program in FY 1991.)

To have the greatest effect, multiyear legislation would have to be included within the appropriation for an agency or project, not just its authorization. At this time, the appropriators seem to have little interest in using the multiyear funding approach on the NASA budget. One problem is that the overall budget situation within the Congress is in just too great a state of flux. Embracing multiyear funding for some or all of the NASA budget would lock the appropriators in and remove their future flexibility for dealing with what has been a very fluid budget situation. On the other hand, some observers would argue that this is exactly what a multiyear appropriation is supposed to do.

An additional concern in some circles is the appropriateness of using a multiyear appropriation for an endeavor that is as inherently uncertain as an R&D project. Multiyear appropriations have generally been used only for production procurements, where funding requirements can be predicted with much greater accuracy than is possible for almost any kind of a developmental initiative.

In 1987, the Appropriations Committee did use the multiyear approach for funding a NASA project—the $2.1 billion replacement for the Space Shuttle Challenger. That initiative was spectacularly successful. The replacement Orbiter (Endeavour) was delivered on time, with high quality, and at a cost that was hundreds of millions of dollars *below* budget. But here again, this was a case of a "production project" whose cost requirements could be accurately estimated ahead of time.

Another source of the funding instability in recent NASA programs has been a case of too many programs chasing too few dollars. In other words, NASA has simply had too many programs "on its plate" for the dollars that are available.

In the past, NASA has pleaded with the Congress to approve a new program, saying that "it only needs ___ dollars this year" (and of course, the blank is always filled in with a small number). Frequently, this has been successful, and the Congress has funded the project. Several years later, however, when the annual funding requirements for the project have grown to a level that is 5 to 10 times what the first-year funding was, it begins to become "unaffordable," because too many other projects are in that same state and there just aren't enough funds in the overall NASA budget to support so many large or mature R&D efforts.

All too frequently in recent years, this situation has been handled by canceling a major development which is well into its R&D cycle. This clearly is not a wise way of managing scarce financial resources. On the other hand, the alternative of stretching out the schedule for *all* of the projects within the Agency is almost as bad. This has the effect of causing huge overruns in almost everything, making the cost efficiency of the dollars that are spent lower still, and causing even greater problems in the "out-years."

Another alternative is to simply *not approve* a proposed new project in the first place. This approach argues that the Congress should keep the "long term" in sharp focus, granting approval for program starts *only* when it can be clearly demonstrated that the funds that will be required throughout the life of the program are likely to be available.

In this way, the Congress would no longer take a "1-year-at-a-time" view of the federal budget. Even though the Congress would still be authorizing and appropriating funds on a 1-year basis, these funds would *only* be authorized and appropriated after the Congress had taken a multiyear view of the budget and determined that it was likely that the funds which would be required throughout the development cycle for the new project would be available.

This view seems to be crystallizing within the Congress, as demonstrated by committees and members demanding to know 5-year (rather than 1-year) funding requirements for individual projects, and the growing number of members who are asking questions like, "Where will the money come from to pay for this project in future years, if we approve it this year?"

Questions like this are particularly relevant in the case of the current NASA budget. Viewed from one perspective, the NASA budget has a large "core" of ongoing activities which include: flying the Space Shuttle, building and operating the Space Station, conducting space science missions, doing basic aeronautical and space technology research, operating a major communications and space tracking network, etc.

On top of this core of ongoing activities, NASA now wants to add three major new developmental efforts: the Earth Observing System, the New Launch System, and the Space Exploration Initiative. However, *each* of

Table 1 NASA budget trends in terms of percentage growth and cuts

Fiscal year	% Budget growth requested by NASA	% Budget growth granted by Congress	% Cut made by Congress
1960	175.8	184.1	3.0 Add
1961	84.2	84.1	0.1
1962	101.3	89.3	5.9
1963	107.5	101.3	3.0
1964	55.5	38.8	10.7
1965	6.8	2.9	3.6
1966	1.9	−1.4	1.6
1967	−3.1	−4.0	0.9
1968	2.7	−7.6	10.0
1969	−4.8	−12.9	8.6
1970	−5.6	−6.2	0.6
1971	−9.9	−11.6	1.9
1972	0.0	−0.1	0.1
1973	2.9	2.9	0.0
1974	−10.4	−10.8	0.5
1975	7.5	6.3	1.1
1976	10.1	9.9	0.2
1977	5.0	7.5	2.4 Add
1978	6.9	6.4	0.4
1979	12.9	12.2	0.6
1980	16.1	15.0	0.9
1981	6.0	5.3	0.7
1982	11.6	9.0	2.3
1983	10.4	13.3	2.6 Add
1984	4.5	6.2	1.6 Add
1985	3.8	4.3	0.5 Add
1986	4.4	2.8	1.5
1987 (Exc. 105)[a]	−5.2	10.8	16.8 Add
1988	10.2	4.7	5.8
1989	27.6	21.1	5.1
1990	21.8	12.8	7.4
1991	23.0	12.8	8.3
1992	13.6	3.5	8.9

[a]This excludes an additional $2.1 billion in multiyear funding for a replacement orbiter (OV-105).

these initiatives will require one or more billions of dollars *annually* within about 5 years of receiving a "go-ahead." In the current (and projected) austere budget environment, questions like, "Where will the money come from?" would seem to be right on target.

If the Congress approves these projects, and the out-year money does not become available, it and the Administration will be right back in the same boat of having to cancel projects that are nearing maturity, stretching out many programs to keep them all afloat, and so on.

Congressional Cuts in NASA Budget over Time

As a closing item for this chapter, it may be interesting to look at the historical score card for the cuts that the Congress has made in the NASA budget request. The data for this analysis are depicted in Table 1, which presents for the past 33 years: the percent budget growth that was requested by the Administration for NASA, the percent budget growth for NASA that was granted by the Congress, and the percent cut that the Congress made in the NASA budget request.

The common public perception is that "the Congress always makes a substantial cut in NASA's annual budget request." The data in Table 1 would seem to not totally support this common view. In particular, it is interesting to note that:

1) 18% of the time the Congress has actually appropriated *more* than was requested for the NASA budget.

2) 64% of the time, congressional cuts in the NASA budget have been *less than* 2 percent.

3) 78% of the time that the Congress has made a cut greater than 5% in the NASA budget, the budget request had been for a growth in excess of 10% (and most of the time, the budget request had been for a growth in excess of 20%).

These numbers would seem to indicate that over many decades the Congress has normally been quite supportive of the NASA budget request.

Economic Issues Facing the United States in International Space Activities

Henry R. Hertzfeld*
HRH Associates, Bethesda, Maryland 20816

Introduction

AT the dawn of the space age in the late 1950s there were only two nations, the United States and the Soviet Union, that had the technological ability as well as the hardware to gain access to space. Since both nations saw their relative power measured by military potential and since each perceived the control of outer space as a military advantage, both nations invested heavily in advancing their space capabilities. Terms such as the "missile gap" were political rallying points for the United States buildup of technology and defense capabilities.

The race to the moon in the 1960s was a United States civilian space effort. It was a politically motivated program that was intended to show United States technological superiority to our allies and to lessen the embarrassment caused by the launching of Sputnik in 1957 and the Soviet ability to launch heavier and more capable satellites before the United States. Of course, the development of heavy lift vehicles for the Apollo program as well as the knowledge gained in all areas of space activity also had potential national security uses.

With the exception of the sharing of scientific data from civilian space programs with the world, political competition with the Soviet Union was paramount. The United States government cooperated with other nations in establishing worldwide satellite communications capabilities, but the

---

Copyright © 1992 by the American Institute of Aeronautics and Astronautics, Inc. All rights reserved. This paper first appeared as a chapter in: V. Lopez and D. Vadas, *The U.S. Aerospace Industry, A Global Perspective for the 1990's*, published by the Aerospace Industries Association of America, Washington, DC, Sept. 1991. Substantial revisions have been made to the chapter for this publication. The views expressed herein are those of the author and do not necessarily reflect the views or opinions of the AIA.
*President, HRH Associates, 5208 Baltimore Ave., Bethesda, MD 20816.

sophisticated guidance systems, electronics, computer software, materials research, propulsion equipment, and so forth needed for space activities were produced in the United States virtually exclusively for United States government needs.

Today the world is very different. Any nation with the ability to pay for a launch can have access to space. At least seven nations have launch facilities and dedicated launch vehicles. (The United States, the former Soviet Union, France, Japan, China, India, and Israel have all successfully launched satellites and other payloads. Other nations such as Norway, Brazil, and Australia are either launching small sounding rockets or have plans to develop major launch facilities. However, not all nations have the capability to launch special or heavy payloads.) Many more nations have industrial and technological capabilities that permit them to manufacture both components and full space systems. The superpowers no longer have a monopoly in this industry. And, among most Western nations, industrial capabilities for possible commercial activity in space are developing into competitive businesses. In fact, a private company can now build or purchase satellites and launch vehicles as well as launch services and subsequent operations support for space activities. The government's role is becoming more regulatory in nature (safety in launches, ensuring international treaty provisions are enforced, and use charges for government facilities) for certain types of civilian activity in space.

Even though space activity is truly an international affair and even though the private business opportunities in space are increasing, space is still overwhelmingly a government sector activity. Financially, well over 95% of the world's investments in space are through government treasuries. Most of the investment is for research and development of new vehicles, new satellite capabilities, and new major space systems. Because governments are funding these systems, the initial use of the technology is for public purposes: national security, environmental monitoring, treaty verification, and fundamental research. However, because industrial firms and laboratories are performers of these activities for the government agencies, the commercial opportunities are beginning to become evident, and with continued government stimulation, the future may provide interesting and profitable business opportunities.

Space Industry

Space industry is a new industry. Because of its relatively new status, it is still a hybrid combination of large firms primarily engaged in other activities alongside very small new firms devoted entirely to space activities. Since government work in space has been paramount in funding and procurement over many years, most large firms in space are those with extensive government contracting experience. And, because the largest financial effort in space has been in getting to and returning from space, space transportation has been dominant. It is natural, then, to first define the space industry in terms of defense and aeronautics suppliers.

However, space also encompasses a multitude of other types of firms and activities. Electronics hardware companies that produce satellites and guidance systems have separate space divisions. Computer software development has become crucial to the smooth and efficient operating of space activities. Space research requires advanced and special instrumentation developments. Launch services includes traditional activities such as storage and maintenance of payloads and scheduling as well as more complex activities involving fueling of rockets and down range communications and control. There are also startup research firms that are designing and planning for special commercial research in microgravity environments. And there are firms that are developing markets in selling space to perform research on special space-based laboratories. In addition to these and other businesses specifically organized to deal in space, there are a host of supporting firms specializing in finance, insurance, technology brokerage, and other business services that have focused on space activities.

This survey views space as an industrial and service activity that spans many traditional economic sectors. Space has become an integral activity of the aerospace industry, and most of the available data on space include both the transportation sector [Standard Industrial Code (SIC) 376] and the satellite manufacturing sector (part of SIC 366). Even though the official data collection agencies may not have caught up to the present in defining a space industrial sector, this report attempts to be more rather than less inclusive in describing what now is and what will eventually become the space industry. [The Aerospace Industries Association's statistical publication, *Fact and Figures*, defines the aerospace industry to include all aspects of aircraft production (SIC 3721, 3724, and 3728), all aspects of space transportation equipment manufacturing (SIC 3761, 3764, 3769), radio and TV communication equipment (SIC 3663), and aeronautical instruments (SIC 3812 and part of 3829). Missing from that definition are important elements of computer hardware, business services (software development, financial and insurance services, management services), engineering services, and university and government laboratories. Also missing are components of the chemical industry (propulsion fuels, etc.) and instruments (e.g., optical instruments for remote sensing applications as well as medical instruments for astronaut monitoring). This is only a partial listing of the types of industries now involved in space and where specific firms have been created around space activities. However, it illustrates how pervasive and how important space has become to the economy and how difficult it is to accurately estimate the "space industry."]

The most surprising fact about space is the large amount of money invested in space and related activities each year. Worldwide, nearly $80 billion is was being spent on space and space-related activities in 1988. [Given the recent changes in the Soviet Union, it is very difficult to estimate their current space expenditures or predict their future investment in space research and development (R&D) and operations. The estimates for most other nations and their relative position in worldwide space expenditures have not materially changed during the past 4 years.]

It is extremely difficult, if not impossible, to arrive at a precise and meaningful estimate of worldwide space activity for a number of reasons. First, exchange rate fluctuations make international comparisons difficult over time, and imprecise within a particular year because of the differences in comparative purchasing power. Second, most nations invest in space for national security as well as commercial competitive purposes. Defense budgets are often classified, and even when they are published, an accurate accounting of the costs of personnel and facilities that have dual purposes is not usually possible. Third, statistical reporting methods vary across nations. For example, the European Economic Community data consider space industries as those that build completed space systems. It is not clear in their data whether propulsion and engines in their data set include related space components or not. Finally, most national space data are reported for government programs alone. Data on industrial sales of space and spacerelated equipment may not be accurately categorized, may include double counting (sales to government entities as well as government expenditures), and may be omitted (e.g., privately held companies) in some national statistics.

The worldwide distribution of space investments for the year 1988 was approximately:

United States: 42% (slightly less than $35 billion per year)
Soviet Union: 45% (slightly more than $35 billion per year)
Rest of the world: 13% (about $10 billion per year)

[As mentioned above, given the political uncertainties in the former Soviet Union, their 1992 investment in space is likely to be lower than their historical average and growth rates. However, if they are able to make the transition to a market economy and are able to sell space products to other nations, their expenditures could remain second only to the United States. Much will depend on whether they opt to join with other nations in space R&D and exploration (including the United States). If the former Soviet republics increase their level of international cooperation in space, then even the expenditures of the United States in space R&D could be reduced through efficiencies gained from these cooperative efforts.]

In the rest-of-the-world category, Europe accounts for 40%, China about 30%, Japan about 15%, and all other nations together about 15%. And, of the European investment, France is the predominant nation with just under 60% of the European total, and the Federal Republic of Germany the next largest with 21% of the total. Italy has rapidly increased its financial interest in space activities in recent years and now accounts for approximately 15% of the European investment in space R&D. The other nations, with the exception of the United Kingdom invest very little on a global scale in space. The United Kingdom, once a major investor and leader in launch vehicles and space research, has significantly decreased its relative share of European space investment. (See Ref. 1. However, the United Kingdom is still a major investor in the satellite communications industry.)

Although the common perception is that the United States is lagging in its investments in space R&D relative to the rest of the world, the data show otherwise. For example, the ratio of United States civilian space R&D (government programs) to European spending was 35:1 during the height of the Apollo program in 1965. By 1975 that ratio had fallen to 4.3:1, reflecting both the decline in the United States space program and the start of a buildup of interest in space in Europe. The ratio hit a low of 3.7:1 in 1980, but by 1988 had rebounded to 3.9:1. In other words, as measured in absolute financial terms, the United States has actually increased its space expenditures relative to Europe, even though there has been dramatic growth in European space investments in the past 10 years. And, while not insignificant, Japan's expenditures on space amount to only one-third of Europe's.

Why then, if investments are that large in space activity, and if the United States still maintains its financial leadership in space activities, is there so much worry about losing a competitive edge to other nations? The evidence, as is discussed below, is mounting that other nations are technologically ahead in many areas and that they are aiming at direct competition with the United States in space. Why are these nations able to get so much "commercial bang for their buck (yen, franc, mark, etc.)?" Is the United States doing something wrong in its space investment, or are there other trends and issues that the analyses overlook? The remainder of this article assesses current international trends in space government and business activity and discusses policy options for the future.

Major Economic and Political Issues in Civilian Space

Industrial Policy

Industrial policy can be very beneficial to the long-term growth of the economy. At one time in United States history, industrial policy was accepted as normal. The development of the railroads and the West would not have occurred without an industrial policy. Commercial aviation was encouraged from policies that included NACA R&D, post office contracts for air mail delivery, air traffic control operations, and improved weather forecasts. The economic success that the United States has enjoyed for so many years can be traced back to many important industry-specific public investments that have provided the conditions that encouraged new sectors to grow.

For a variety of reasons that are too lengthy to enumerate in this article, the United States Government and United States industries have entered a protracted period of mutual distrust. This has led to adversarial roles concerning many types of economic regulations, including tax policy, antitrust, labor relations, environmental concerns, and so forth.

These destructive and wasteful adversarial relationships carry over to research and development. In R&D a cooperative role is necessary, and, in fact, does exist in the aerospace industry because of its unique relationship with the government. However, because of the overall private/public attitude, it is often clouded by excessive rules, regulations, and restrictions.

In the 1950's, it was recognized that the nation's road system had to be improved. In order to build the interstate highway system, the enabling legislation had to be couched in terms of the needs for national defense and mobilization rather than a specific policy to stimulate a number of industrial sectors, including: interstate trucking, construction and construction materials, automobiles, petroleum, and so on. The impacts on industrial development, land use, and transportation from this system of highways has shaped our cities, suburbs, and industry over the past 35 years. These impacts far outweigh the military aspects of that legislation.

However, not generally recognized within the government are two factors: 1) having no official industry policy is an industry policy in and of itself, and 2) in the space sector, the government actually has a fairly well-defined policy of stimulating particular firms and industries through both R&D and operational funds.

Examples abound in space infrastructure and transportation of an industry policy. Sometimes it is hidden under umbrellas such as national security policy (e.g., launch vehicles). And, sometimes it is simply a response to a real or perceived threat from foreign competition (communications). Finally, there are some expenditures to develop infant industries (materials and other advanced research in space).

In the space launch vehicle industry, the government has taken very specific steps to develop a private capability that will be competitive with the systems under development in other nations. Through large military orders, it is hoped that the major vehicle manufacturers will realize economies of scale, reducing operating costs while improving methods of production, develop advanced vehicles through internal R&D efforts, and adapt the vehicles developed for the Defense Department for commercial purposes. Also, by restricting United States government payload launches to United States vehicles, markets are created for those manufacturers.

The NASA Advanced Communications Technology Satellite (ACTS) program has been characterized by funding instability. Each year, the Office of Management and Budget (OMB) turns down NASA proposals to fund ACTS and Congress restores some of the funds. This program's objective is a renewal of communications research efforts, originally curtailed by the government during the early 1970s when communications satellites became a profitable private business. At that time it was felt that United States industry would fund R&D out of profits from telecommunications revenues. However, United States firms did not invest sufficient R&D to keep the industry well ahead of foreign competition. The ACTS program is a renewed effort by the government to pump funds into R&D that may have significant civilian business implications as well as serving future governmental purposes. Its success has been hampered by the lack of guaranteed continuity in public funding.

Another example is the stimulation of research in space through the Space Shuttle and Space Station programs. NASA has made space available to industry to perform research at practically no cost to industry. (However, there are many hidden costs to this program. NASA makes space available at its convenience—a schedule is hard for industry to predict. The costs of industry to cooperate with the government are significant in that a number of regulations must be complied with, there is a large amount of paperwork involved, and there is the risk of proprietary research becoming public knowledge.) Eventually NASA will have policies of cost reimbursement and other charges for the use of the system, but initially the opportunity for industry to get a head start on foreign competitors has been offered.

Officially the government does not engage in industry policy. However, the United States government has made a commitment to developing space industry, even if officially it cannot be called an industrial policy. This represents a continuation of a long-term historical United States policy toward the governmental support of transportation and economic infrastructure.

Internationally, there are very significant cultural differences among nations, partly dictated by geography and history, and partly by economic realities. In one form or another, virtually every foreign nation has an industrial policy that is much more aggressive than that of the United States. These policies focus on aiding domestic industries to compete in the world marketplace.

Space is no exception. Virtually all other nations pump funds into the development of space with the specific intention of stimulating their own firms in world competition for related products and services. In this regard, it becomes not so much a comparison of United States subsidies to stimulate space R&D with other nations subsidies but a comparison of the underlying purpose and motivation of the investments. (One detailed investigation into alleged unfair practices of the Member States of the European Space Agency in launch vehicles was conducted by the United States Trade Representative in response to a petition from a U.S. company, Transpace Carriers, Inc. The Determination under Section 301 of the Trade Act of 1974 (Memorandum of July 17, 1985, Presidential Documents, *Federal Register*, Vol. 50, No. 140) found that "ESA practices are not sufficiently different from those of the U.S. to be actionable under Section 301." Policies studied included: government inducements, range services, loans and capital grants, hardware, protected home markets, indirect government assistance, and costs and pricing. This determination did not study the practices of other nations, nor did it establish any international guidelines concerning unfair practices. It also viewed the problem as of the mid-1980s and was specific to this particular petition. It should not be viewed, therefore, as a general finding that the U.S. subsidies to space industries are no more or less than those of other nations.)

The Convention establishing the European Space Agency (ESA), approved by the member nations over 15 years ago, specifically states in Article II (d), "The purpose of the Agency . . . by elaborating and implementing the industrial policy appropriate to its programme and by recommending a coherent industrial policy to the Member States." Article VII of the Convention further elaborates on the industrial policy issue by stating, ". . . improve the worldwide competitiveness of European industry . . . and development of an industrial structure appropriate to market requirements . . . ." (Convention for the Establishment of a European Space Agency, Paris, May 30, 1975.)

Foreign government motivations are reinforced by the particular types of investments that they permit and encourage. One key to the commercial orientation of government is reflected in the ownership equity of firms in the space business. CNES, the French Government Space Agency, owns equity in Arianespace and in Spot Image, as well as in many other French companies that are suppliers to space businesses. (See, for example, Ref.

2.) This direct government interest in the companies acts to stimulate cooperation between industry and government. In addition, equity owners include financial institutions that have the capacity to provide an infusion of money if needed. The atmosphere is one of a partnership rather than an "arms-length" relationship. In contrast, in the United States, the government rarely has equity in private companies and United States financial institutions generally lend money but do not take strong equity positions. As a result, an atmosphere of competing interests develops, which can impede smooth operations and continued long-term financial and technical support.

Increasing Foreign Dominance in Specific Technologies

A review of the pattern of expenditures of foreign government space budgets illustrates the tendency toward a focus on specialization in particular areas of space activity. Besides the highly visible development of new launch vehicles, the two most active areas of space research and development are occurring in the fields of communications and remote sensing. However, some nations are also focusing on materials processing in space (Germany) and on terrestrial ground receiving equipment (Japan).

Government activity in communications and remote sensing reflects the use of space for national purposes: security, treaty verification, land use, environmental and agricultural planning and monitoring, and communications. Some of these applications will result in economic and competitive advantages for the investing country. However, the primary purpose for government involvement remains in governmental use of the technologies.

Most major nations are currently positioning themselves for a future in space activities. Multinational alliances are being formed to study and evaluate investments in many areas, including manned space activities. Research is being conducted to develop underlying technologies in propulsion, life sciences, materials, and other areas that will be needed for the space programs of the future. With the exception of the United States and the former U.S.S.R., a large part of this work is intended to train personnel and gain knowledge, experience, and resources for major investments that are expected to be far in the future. Therefore, even though it appears that Japan, ESA, and others are gearing for a full-service space program, the near-term realization of such a plan is not likely. Moreover, these relatively small investments in study groups and enabling technological research may reveal market niches and specialities that will be essential to all nations in space and may also spin off profitable near-term business opportunities for these nations.

In addition, foreign companies and banks are investing in United States firms that specialize in space activities. Matra, a large French aerospace company, recently purchased Fairchild Industries. A consortium of six Japanese companies that includes one of the largest banks in Japan (Mitsubishi Trust), the Japan Air Line, as well as the Japanese industrial concerns that produce space hardware, announced an investment in 10% of the stock of Spacehab. Spacehab is a small United States firm established to produce and sell research facilities in the microgravity environment of outer space.[3]

Investments in materials processing and ground receiving equipment represent an investment portfolio specifically aimed at long-term economic competitive gain. When a nation invests in a technology that may generate large-volume sales at relatively low prices, the target is the consumer market. As with most consumer electronics like stereo equipment, televisions, and VCRs, the Japanese have positioned themselves to capture the satellite receiving equipment market. If direct broadcast TV materializes, their current investments in terrestrially based technologies will be in the forefront of the marketplace for consumer receiving equipment from satellite transmissions. Unlike the small-scale (but big-ticket) production of satellites, launch vehicles, and other space-based equipment, the mass production consumer equipment represents vast potential profits.

United States industry understands and performs R&D, production, and marketing of consumer products. But, in recent years domestic industries have been unable to develop competitive equipment, particularly in the field of consumer electronics. The future products that may come from space R&D efforts of United States companies offer a truly rich possibility for United States companies to capitalize on the very large government and private space R&D investment and know-how. However, foreign governments and industries are targeting economic uses of space and view space investments quite differently from the United States. The United States can expect to see vigorous competition in those areas that promise large economic profits.

The apparent United States weakness in the commercial aspects of space has occurred through a number of interacting factors. The companies (or divisions of companies) in the United States that have both the space know-how and the capacity to implement that expertise are the large defense contractors. These firms are geared to producing big-ticket, specialized products. The R&D that these firms perform is primarily done under government sponsorship or is done in anticipation of technologies needed for winning future government contracts. It should be no surprise that the products developed from this R&D effort are not consumer-oriented, reasonably priced, large-volume, high-profit-margin items.

The problem is not that the United States is not competitive, nor is it a problem of lack of R&D efforts, nor is it a problem of low-quality products. The major problem in international competition in the space sector is the system that has developed in the United States that views space as a political/military industry rather than a consumer-oriented industry. From the funding mechanisms to the distribution channels, the industry is dominated by a lack of interest in consumer products.

The lesson for United States industry is clear. If the United States is to capitalize on its massive investment in space R&D, the orientation of both the government and the private firms with the appropriate technological knowhow must change to reflect increasing foreign competence and competition in the areas that will challenge firms to be profitable in the consumer market as well as the government market. These changes are now taking place, as U.S. industry responds to the post-Cold War era.

Growing Commercial Activity in Terrestrial Operations

Another perspective in space business activity is the Earth-based operations support of research, development, and production. These opportunities are in familiar and traditional areas of business and involve relatively little risk (compared to space-based equipment and manufacturing). The majority of such services are presently provided to the government. However, growing commercial interest in space will mean a growing demand for terrestrial support services. Specific activities include the management of launch sites, communications and control services, and the preparation and storage of payloads.

United States know-how is very competitive in these services. For example, there is a plan to build a new launch site in Australia designed to use the Soviet Zenit rocket and to offer launch services at competitive prices with some United States and other launch vehicles. An American firm, United Technologies, has won the bid to manage the operations of that site, if and when it is completed.

In addition to providing business opportunities to United States firms, these services also put the firms in an on-site and hands-on position to learn the technologies and uses of space and to eventually design and produce space-based products.

Fiscal Consequences of Big-Ticket R&D Programs

One of the inevitable consequences of a successful long-term R&D program in a scientific or engineering speciality is that new accomplishments build on past accomplishments. As new technologies are integrated with the existing ones, the complexity of the research effort increases. There are two outcomes of this phenomenon:

1) The risks of success of the next project are determined by the combination of the new technology being successful multiplied by the risks of success of the integration of the two (or more) technologies working together properly.

2) The cost of new research efforts increases at an increasing rate.

In other words, the probabilities of future success decrease while the costs increase. Of course, there are many mitigating factors that can intervene to delay this phenomenon, including major scientific or technological breakthroughs that leapfrog over incremental improvements. However, sooner or later both costs and risks of continued research in a particular field increase.

Space is no exception. In space transportation, simple rockets are replaced by multistage rockets. One-payload launch vehicles are replaced by vehicles capable of launching two or more payloads. In communications, satellites with several transponders capable of lasting a few years are replaced by platforms with many transponders and other instruments capable of being refitted and lasting many years. In remote sensing, passive systems with only several spectral bands are replaced by active and passive systems with many bands. Astronomical research done through the use of terrestrial telescopes may now be possible through complicated space-based telescopes.

Recent events have clearly illustrated the cost, scope, complexity, and risks of advanced space equipment and instruments. The Challenger disaster dramatically showed how management and technological complexities can work against themselves to increase the risk of failure. The space telescope similarly illustrated how small errors can go undetected in large systems. Although failures can be corrected in future tries, the risks of other parts of these large and complex systems failing are still mathematically significant.

The implications for the next generation of large space programs is obvious. The Space Station is expensive. It is not only new, untried, and complex, but its success is dependent on the continued successful operation of the Space Shuttle system, as well as many other support systems. Any failure of a support system could jeopardize the success of the Station. Because of the high cost of the Space Station in a time of budget deficits, the present plans do not include backup or replacements for the Space Station, should a failure occur.

Plans for the Lunar/Mars program and other very long-term, very exciting, very challenging, and very expensive space research programs are similarly risky. Man has never before tried to use the Moon as a base for space operations, nor has man tried to mine and use the resources that may be found on the Moon. New equipment will have to be designed and built for the purpose. The effort will also require new and larger launch vehicles to be available for the mission.

Many risky technological projects should be initiated. Pushing the state-of-the-art can be a very rewarding (both economically and politically) governmental and industrial endeavor. However, the United States is currently in a position of budget restraint. The deficit is growing each year and the "discretionary" budget is pared down each year in an effort to control expenditures. Space and other advanced R&D have been spared significant cuts, even in this environment of restraint. But, it is unlikely that in the next decade there will be a political consensus that space research is of sufficient priority to command the resources necessary to build all of the ambitious programs that the space community may request.

Three major reasons are behind this lack of consensus. First, there is no overriding reason to speed up space exploration because the former U.S.S.R. nations are no longer perceived as a military threat. Second, the space sector is not yet that important from an economic competitive position (at least compared to other export industries). Third, recent events have underscored the risks associated with space activity. Given the high costs and potentially catastrophic failures, Congress and the public are wary of initiating new programs that could cost hundreds of billions of dollars over the next 20 years.

We often forget that the government commitment to NASA to go to the Moon was made in an era of budget surpluses and in an era of a perceived "missile gap" with our Cold War adversaries. We are simply not in the same era and cannot make the same political and economic decisions regarding new space proposals.

Finally, with the exception of government contract work and communication satellites, there is a noticeable lack of commercial interest in investing in space. Companies that have been formed to provide private research facilities in space or to operate launch vehicles have had major difficulties getting the financial community to invest. Only a very few have been successful, and their success is often hinged more on government contracts than on private purchase guarantees. As space industry matures, this may change. But it is not likely that there will be an opening of the floodgate of opportunities during the next decade. Beyond that it is anybody's guess, and it lies far beyond the financial planning horizon of today's business executives.

Internationally, the same factors are at work. Nations are questioning their investments in space. Beyond having a presence in space technology and positioning themselves for the possibility of taking advantage of opportunities as they may emerge in the future, most nations are very cautious about large commitments to space. The United States and the former Soviet Union were the only two major players in the space research game. And, it is unclear, today, whether Russia will continue to be a major player in space. However, most countries are forming alliances to gain information and access to all areas of space technology and are in a position to increase those investments in promising areas on relatively short notice.

For the United States, the major question is whether to sizably augment existing budgets for major new exploratory programs. For other nations, the question is whether the economic advantages afforded by space technology development will have competitive payoffs. Because the commercial success of space technologies is closely tied to the availability and operations of the space infrasturctures developed by the United States, these two questions are closely interrelated.

Over the next decade, as other nations' space capabilities mature and the access to space for all types of payloads is available from a variety of providers, the ties to the success of major space programs will be lessened. This will open the way for more independent decisions from other nations, made on the basis of economic criteria more than political criteria.

United States Leadership

The United States views itself as the leader in space. Challenges to this leadership are construed as challenges to the United States technological capabilities. The suggestion that the United States is not the leader is an emotional issue for many. And well it should be, given the huge cumulative expenditures of the United States government and industry over the past 30 years relative to the expenditures of other nations.

The preamble to the NASA Act of 1958 directs the agency to preserve the role of the United States as *a* leader in aeronautical and space science and technology.[4] A 1978 White House Civil Space Policy Document takes a more direct position on leadership when it declares, space policy will "assure American scientific and technological leadership in space for the security and welfare of the nation . . . ."[5] In 1988 a new Presidential Space Policy Directive was issued. It states that "a fundamental objective guiding United States space activities has been, and continues to be, space leadership."[6] This policy, recognizing the changing world situation in space, somewhat qualifies the absolute leadership idea by adding a clause that no longer requires leadership to include preeminence in all areas and disciplines of space enterprise, but only in key areas crucial to security, scientific, technical, economic, and foreign policy goals. One may wonder what else there is. However, at least the document addresses the possibility that other nations may step ahead of the United States in some areas without violating the United States policy of leadership.

Presumptuous statements about United States leadership in space accurately reflect official sentiment about United States policy toward international cooperation and competition in space. In the past, most cooperative efforts with other nations have been in the form of the United States sharing data and know-how in return for some commitment of personnel and resources to use the information in a joint research effort. In effect, most other nations have been treated by the United States as a less-than-full partner in space ventures.

Major foreign space research institutions are now flexing their own muscles. The ESA, in particular, is demanding an equal partner role in negotiations with the United States on cooperative programs.[7] Organizations such as ESA, Arianespace, NASDA, and so forth now have their own independent capabilities in space that are equal to, or may in some cases exceed, those of the United States. And, more and more, these nations are forming alliances in space endeavors that do not include the United States at all.

The concept of leadership cannot and should not remain a United States egotistical stumbling block in international space endeavors. The United States must recognize that others have the technological capability to move forward in space without United States involvement. Even if that capability is derived from prior United States help through cooperative ventures, the fact remains that the technology has transferred and that in some areas where others are now investing greater resources than the United States, their technologies may be superior.

Space is a very multidimensional activity. As the complexity of space systems increase and as budgets remain stable or decrease, specialization will become more important and more expensive. Consistent with the original 1958 NASA Act, the United States will remain *a* leader but may not remain *the* leader in all space activities.

National Security

Until the mid-1970s the major United States investment in space R&D was civilian, mainly in support of the Apollo and Shuttle programs. [Military procurement of guided missiles, advanced electronic guidance systems, and other space-related technologies is normally not included in the statistical comparisons of expenditures in space R&D between the Department of Defense (DOD) and NASA. However, if one were to include these acquisitions, the total expenditures of the government in defense-related aspects of space would have been greater than the civilian expenditures.] In the mid-1970s defense expenditures on space R&D began to rise rapidly, accounting for the majority of the government's investments in space. This trend continued throughout the 1980s.

When NASA was formed in 1958, there was a clear mandate to have a civilian space program for peaceful uses of space. This meant that the government had to develop a clear organizational break between space R&D oriented toward military purposes and that oriented toward civilian uses of space. This division of purpose and identity continues today, even though NASA and the DOD share a number of common bonds. (Many NASA field centers are located directly adjacent to military installations. Launch sites such as the Kennedy Space Center serve dual purposes: launching both civilian and military payloads. The Space Shuttle, primarily a civilian vehicle, is also used for classified launches. The manufacture of expendable launch vehicles for the DOD has been tied to stimulating a civilian industry as well. The Nationl Aerospace Plane program is jointly run between the Air Force and NASA.)

Although military applications of space technologies may have quite different purposes from civilian applications, the fact remains that the research done for one purpose has carry-over to the other purpose. And, as space research becomes more expensive, more complex, and more risky, the need to share technological information between the DOD and NASA (and vice versa) becomes more important for successful programs. Because of the historical division that was made with the formation of NASA and the resultant cultural barriers toward the sharing of some types of military and civilian research results, some opportunities for capitalizing on United States government-sponsored space research have been lost.

The NIH (not invented here) syndrome has often been attributed to industrial managers who are unwilling to look at new ideas if they originate outside their own research laboratories. However, the same syndrome can, and does, appear within the government. In a time of severe budget restraints and where the United States lead in some areas of space is eroding, it is time that both NASA and the DOD are encouraged to share their technological expertise.

In addition, declassifying selected technological information may permit United States companies to compete more effectively in world markets. Although this is a sensitive topic, one example from the history of the space program will illustrate the issue. In the area of remote sensing, NASA was permitted to develop sensors with a resolution of 30 m on the original Landsat satellites, and 10 m with the more advanced thematic mapper instrument on later Landsats. By executive agreement, NASA was not permitted to develop or use instruments that showed resolution less than 10 m. This agreement was classified information until the late 1970s. The DOD was performing research on advanced sensors with less than 10-m resolution during the same time period.

The civilian remote sensing industry was limited by the restrictions placed on the systems by defense considerations. During the 1980s, France developed its own remote sensing instruments, some with resolutions greater than that allowed to United States civilian firms. The U.S.S.R., which also developed images with a high degree of resolution, has offered them for sale to the public. The United States remote sensing industry now finds itself behind other nations in certain types of products because of the slowing down of civilian research due to defense classification rules.

The United States economic intelligence network should be more sensitive to areas where other nations introduce products that are not available to United States industry to develop because of a slowness in the system of declassifying defense technological breakthroughs.

International Agreements

The web of international cooperative government and industry research, technological development, production, and distribution of space information, products, and services has become very complex. Space is mainly an activity of the wealthier industrial nations. But, particularly through the application of communications satellite networks such as Intelsat, even the poorer nations of the world have access to and use space hardware and services. And some emerging nations such as India, China, and Brazil have put valuable scarce resources into developing advanced space capabilities in order to buy their ticket into the leading technologies of the next decade and the next century.

Virtually every nation in the world monitors the technology development that is occurring in the space arena. Many nations are positioning themselves with the knowledge and resources to enter space markets when and if they see possibilities. In order to do this, these countries rely on joint and cooperative efforts among themselves and with the nations that have major space research investments.

Even those nations and consortia that can afford the high entry fee into the space world benefit from cooperation. The U.S., the former Soviet Union, and ESA and its member nations, Japan, China, and India share research efforts and results.

Today, however, the United States is vulnerable to being left out of many agreements. One reason is that technology has matured to the point where others, working together, can expect to compete with the United States. Because of budget pressures and because of a negotiating attitude that makes other nations feel as though they are unequal partners, the United States government has a poor reputation as an partner.

The United States must accept the fact that because of its success in space ventures, the rest of the world is out to catch up. It is inevitable that the United States will be excluded from some international agreements. This phenomenon is the product of a very successful space program that has stimulated worldwide competition. It also means that the United States must pick its partners carefully (particularly where economic interests are at stake) and should negotiate agreements to provide for an equitable sharing of results.

Lack of Coordinated Government Actions

The United States space program started as a coordinated effort. NASA was created in 1958 and the Apollo program, the major initiative of the agency, was focused almost entirely within NASA, under the overall supervision of the Vice President. Since the early 1970s, there have been a number of additional agencies involved in the management of space-related programs. The Departments of Defense, Agriculture, Commerce, State, and Transportation have major stakes in space activity. The Office of Science and Technology Policy (OSTP), U.S. Trade Representative (USTR), and Office of Management and Budget (OMB) in the Executive Office of the President are involved in decision-making. Coordination among these interested parties has varied over the years, but has never been smooth. Space policy for the United States is not focused and centralized. Its locus changes depending on the particular question asked. Recently formed intergovernmental groups such as the Space Council are attempting to resolve these issues, but the problem goes deeply into the organization of the government and cannot easily be resolved by creating yet another bureaucratic level on top of what now exists.

The United States space policy with regards to international cooperation and competition has also been inconsistent. Major initiatives of NASA to jointly fund and perform R&D in space with other nations have had very mixed histories. One problem has been financial. (A recent example of this recurring problem concerns the cancellation of the Omega/VIMS project, which is a U.S.–French experiment that was scheduled to fly on a 1994 Soviet Mars mission. See Ref. 8.) Often a government agency will negotiate an agreement, only to find several years downstream that the United States Congress or the OMB has cut the budget available for that program. United States agreements in space are often in the form of a Memorandum of Understanding (MOU). This type of agreement is not as binding as a treaty and can be changed by an agency. However, abroad some of these MOUs are recognized by foreign governments as having the legal status of treaties. Other governments have made long-term commitments in reliance on the United States promise (conversation with Ian Pryke, ESA Washington Representative). The reputation of the United States as a reliable partner in international space agreements is not good.

Also hurting United States international positions in space is the lack of a consistent policy. Flip-flops have occurred many times in the past. The relationship between economic policy and space research policy may create conflicts. One example of the problem has been evident in the policy concerning the use of particular launch vehicles. United States government satellites must fly on United States made launch vehicles. United States private payload owners may choose on the open market. But, until recently, United States payload owners could not fly on Chinese or Soviet rockets. Now, a limited number of United States satellites are permitted to be launched on the Chinese Long March Vehicle. And very recently, the United States government has granted permission for private United States payloads to use the Russian Zenit launch vehicle if and when it operates from Australia's proposed spaceport.

In spite of the United States position toward free trade, restrictions still apply to launch vehicles. The government does not permit foreign firms such as Arianespace to compete on the launch of United States government payloads. In and of itself, this policy appears to be in the best interest of stimulating domestic launch vehicle demand and providing an incentive to United States industry. However, the Ariane family of vehicles are not competitive with the Chinese or Soviet vehicles because of the pricing policy differences between the market economies that exist in the United States, Japan, and Europe and the nonmarket economies in China and Russia. The pricing of nonmarket economy vehicles is less likely to be based on the costs to manufacture the vehicles and unfair competition may result.

Further, United States policy has consistently been inconsistent with regard to commercial space endeavors. Recent attempts to standardize policy have helped, but commercial space policy is still a tool of political power and pressure more than it is determined by economic market forces. When there are only 15 or 20 launches of commercial satellites per year, there cannot be a highly developed competitive market for the services. And, since communications satellites are still the only true space activities in the commercial sector, competitive space policy is largely relegated to the launch vehicle and communications satellite manufacturing sectors. Until a free, large, and robust market develops, commercial space policy will continue to be ad hoc and somewhat unpredictable.

United States Space Activities in 2001

The United States is committed to having a serious space program. Billions of dollars have been spent in research and in space-based equipment to support that program over the past 30 years. Although the emphasis may shift in cyclical ways between civilian and defense and also may shift within the technological areas encompassed by space initiatives, there will be a continued program for the foreseeable future. Expenditures of the United States government for space can be expected to remain at a level that will support current activity and provide for some programmatic growth. However, large increases devoted to major new and expensive space initiatives are unlikely to be realized due to several factors, including:

1) Continued United States budget deficit putting pressure on big ticket discretionary expenditures

2) Risk of failures in space initiatives as dramatized by Challenger, Hubble Telescope, and other recent problems

3) Lack of perception of political or military "race" with other nations

4) Other big science programs such as the mapping the human genome and the superconductor/supercollider

There are only two major stimuli that might trigger sizable increases in government space expenditures in the foreseeable future. One would involve a U.S. reaction to another nation's use of space for military purposes in gross violation of current United Nations treaties. The other could be initiated by a major discovery, such as finding the existence of extraterrestrial intelligent life, which would reinvigorate public enthusiasm for space exploration.

Similarly, the prospect of large cuts in the space program are equally unlikely because:

1) There are proven civil uses of space for communications, navigation, and weather forecasting purposes that need to be maintained.

2) Defense has committed large resources for long-term programs to space activity and there are many proven uses of space for defense purposes (remote sensing, communications, surveillance, etc.).

3) Technological spin-offs from space R&D contribute to economic development, as do the direct expenditures needed to maintain a cadre of trained scientists, engineers, and research institutions that are available to maintain the past investments and provide for reserve capacity if a time of need arises.

The possibility of large cuts in the space budget is even less likely than the possibility of large increases. What is more likely to happen is the postponement or cancellation of very ambitious long-range exploration programs that have few well-defined goals and are likely to have actual costs that far exceed estimated costs.

Conclusions

Space commerce, space R&D, and space exploration have become worldwide endeavors. The United States, although still the largest investor and the leader in many areas of space activities, influences, but no longer controls the directions other nations take in space. Space programs and projects that cross all types of borders (industrial, political, technological, etc.) and that are truly international, interindustry, and interdisciplinary will be the theme of the next decade. One major motivating factor in this trend is the very high cost of space activity. Another factor is the changing political and economic climate of the former Soviet Union and Eastern Europe. A third factor follows the lead of the private sector, where both national and multinational firms are entering into more and more joint research, production, and marketing agreements. In effect, the burgeoning world market means that companies and nations are showing a willingness to accept a smaller portion of a larger economic pie rather than taking the risks and costs involved in trying to capture the whole pie.

This report has focused on the United States space program since it has dominated the non-Soviet world for the past 30 years. Most other nations have mirrored their space programs on the organization and structure of the U.S. program. This has occurred because it makes it easier to deal with the United States if a foreign nation has similar organizational divisions, and because being first with a major civilian space program that is characterized by its openness and willingness to share information with others, the U.S. program is a natural model for others.

Therefore, it is not surprising that the current problems with the space program are also mirrored in the programs of other nations. High costs, budget deficits, large-scale programs without hard economic justifications, and so forth characterize most civilian space activities around the world. The major difference between the U.S. program and those of other Western nations are the size of the investment and the overt willingness of other nations to focus on economic opportunities in space. The U.S. government is focused on national security, general welfare, and the desire to explore the unknown. U.S. aerospace industries are primarily concerned with doing government business. Other nations focus their space efforts differently. Companies are frequently wholly or partially owned by their governments, and therefore the governments are far more attuned to industrial success in space.

Furthermore, other nations recognize that with limited funding compared to the United States, they cannot have a space program that does everything. The United States has built a capability in all aspects of space that no other nation can match. This is unlikely to change in the near future and, in fact, may become even more pronounced as Russia faces economic difficulties and reorganizes its space efforts.

References

[1] Pardoe, G., "The Selling of Space," *New Scientist*, Jan. 21, 1989, pp. 45–48.

[2] Arianespace, Annual Rept., 1988, p. 37.

[3] *Wall Street Journal*, June 5, 1990, p. A13.

[4] NASA Act of 1958, 42 USC 2451.

[5] Fact Sheet on U.S. Civil Space Policy, Oct. 11, 1978.

[6] Fact Sheet for Presidential Directive on National Space Policy, Feb. 11, 1988.

[7] Pryke, I., "The U.S.A. and International Cooperation," *AAS Space Times*, July/Aug. 1987.

[8] *Washington Post*, June 20, 1990.

Author Index

Bach, L. 171
Cassidy, D. E. 307
Christensen, C. B. 45, 207
Cohendet, P. 171
Dawson, T. 403
Fuller, J. 357
Gabler, E. 263
Greenberg, J. S. 3, 117, 323
Hazelrigg, G. A. 97, 381
Hertzfeld, H. R. 151, 417
Katz, J. S. 369
Lacobie, K. 357
Lambert, G. 171
Ledoux, M. J. 171
Lee, C. M. 293
Mandell, H. C., Jr. 57
Moore, D. 233
Simonoff, J. 35
Thibault, M. 89
Wagenfuehrer, C. 45
Worden, S. P. 369

PROGRESS IN ASTRONAUTICS AND AERONAUTICS
SERIES VOLUMES

*1. **Solid Propellant Rocket Research** (1960)
Martin Summerfield
Princeton University

*2. **Liquid Rockets and Propellants** (1960)
Loren E. Bollinger
Ohio State University
Martin Goldsmith
The Rand Corp.
Alexis W. Lemmon Jr.
Battelle Memorial Institute

*3. **Energy Conversion for Space Power** (1961)
Nathan W. Snyder
Institute for Defense Analyses

*4. **Space Power Systems** (1961)
Nathan W. Snyder
Institute for Defense Analyses

*5. **Electrostatic Propulsion** (1961)
David B. Langmuir
Space Technology Laboratories, Inc.
Ernst Stuhlinger
NASA George C. Marshall Space Flight Center
J.M. Sellen Jr.
Space Technology Laboratories, Inc.

*6. **Detonation and Two-Phase Flow** (1962)
S.S. Penner
California Institute of Technology
F.A. Williams
Harvard University

---

*Out of print.

*7. **Hypersonic Flow Research** (1962)
Frederick R. Riddell
AVCO Corp.

*8. **Guidance and Control** (1962)
Robert E. Roberson,
Consultant
James S. Farrior
Lockheed Missiles and Space Co.

*9. **Electric Propulsion Development** (1963)
Ernst Stuhlinger
NASA George C. Marshall Space Flight Center

*10. **Technology of Lunar Exploration** (1963)
Clifford I. Cummings
Harold R. Lawrence
Jet Propulsion Laboratory

*11. **Power Systems for Space Flight** (1963)
Morris A. Zipkin
Russell N. Edwards
General Electric Co.

*12. **Ionization in High-Temperature Gases** (1963)
Kurt E. Shuler, Editor
National Bureau of Standards
John B. Fenn,
Associate Editor
Princeton University

*13. **Guidance and Control—II** (1964)
Robert C. Langford
General Precision Inc.
Charles J. Mundo
Institute of Naval Studies

*14. **Celestial Mechanics and Astrodynamics** (1964)
Victor G. Szebehely
Yale University Observatory

*15. **Heterogeneous Combustion** (1964)
Hans G. Wolfhard
Institute for Defense Analyses
Irvin Glassman
Princeton University
Leon Green Jr.
Air Force Systems Command

*16. **Space Power Systems Engineering** (1966)
George C. Szego
Institute for Defense Analyses
J. Edward Taylor
TRW Inc.

*17. **Methods in Astrodynamics and Celestial Mechanics** (1966)
Raynor L. Duncombe
U.S. Naval Observatory
Victor G. Szebehely
Yale University Observatory

*18. **Thermophysics and Temperature Control of Spacecraft and Entry Vehicles** (1966)
Gerhard B. Heller
NASA George C. Marshall Space Flight Center

*19. Communication
Satellite Systems
Technology (1966)
Richard B. Marsten
*Radio Corporation
of America*

*20. Thermophysics of
Spacecraft and Planetary
Bodies: Radiation
Properties of Solids
and the Electromagnetic
Radiation Environment
in Space (1967)
Gerhard B. Heller
*NASA George C. Marshall
Space Flight Center*

*21. Thermal Design
Principles of Spacecraft
and Entry Bodies (1969)
Jerry T. Bevans
TRW Systems

*22. Stratospheric
Circulation (1969)
Willis L. Webb
*Atmospheric Sciences
Laboratory, White Sands,
and University of Texas
at El Paso*

*23. Thermophysics:
Applications to Thermal
Design of Spacecraft
(1970)
Jerry T. Bevans
TRW Systems

24. Heat Transfer
and Spacecraft
Thermal Control (1971)
John W. Lucas
Jet Propulsion Laboratory

25. Communication
Satellites for the 70's:
Technology (1971)
Nathaniel E. Feldman
The Rand Corp.
Charles M. Kelly
The Aerospace Corp.

26. Communication
Satellites for the 70's:
Systems (1971)
Nathaniel E. Feldman
The Rand Corp.
Charles M. Kelly
The Aerospace Corp.

27. Thermospheric
Circulation (1972)
Willis L. Webb
*Atmospheric Sciences
Laboratory, White Sands,
and University of Texas
at El Paso*

28. Thermal
Characteristics
of the Moon (1972)
John W. Lucas
Jet Propulsion Laboratory

*29. Fundamentals
of Spacecraft Thermal
Design (1972)
John W. Lucas
Jet Propulsion Laboratory

30. Solar Activity
Observations and
Predictions (1972)
Patrick S. McIntosh
Murray Dryer
*Environmental Research
Laboratories, National
Oceanic and Atmospheric
Administration*

31. Thermal Control
and Radiation (1973)
Chang-Lin Tien
*University of California
at Berkeley*

32. Communications
Satellite Systems (1974)
P.L. Bargellini
COMSAT Laboratories

33. Communications
Satellite Technology
(1974)
P.L. Bargellini
COMSAT Laboratories

*34. Instrumentation
for Airbreathing
Propulsion (1974)
Allen E. Fuhs
Naval Postgraduate School
Marshall Kingery
*Arnold Engineering
Development Center*

35. Thermophysics and
Spacecraft Thermal
Control (1974)
Robert G. Hering
University of Iowa

36. Thermal Pollution
Analysis (1975)
Joseph A. Schetz
*Virginia Polytechnic
Institute*
ISBN 0-915928-00-0

37. Aeroacoustics: Jet
and Combustion Noise;
Duct Acoustics (1975)
Henry T. Nagamatsu,
Editor
*General Electric Research
and Development Center*
Jack V. O'Keefe,
Associate Editor
The Boeing Co.
Ira R. Schwartz,
Associate Editor
*NASA Ames
Research Center*
ISBN 0-915928-01-9

38. Aeroacoustics: Fan,
STOL, and Boundary
Layer Noise; Sonic
Boom; Aeroacoustics
Instrumentation (1975)
Henry T. Nagamatsu,
Editor
*General Electric Research
and Development Center*
Jack V. O'Keefe,
Associate Editor
The Boeing Co.
Ira R. Schwartz,
Associate Editor
*NASA Ames
Research Center*
ISBN 0-915928-02-7

39. **Heat Transfer with Thermal Control Applications** (1975)
M. Michael Yovanovich
University of Waterloo
ISBN 0-915928-03-5

*40. **Aerodynamics of Base Combustion** (1976)
S.N.B. Murthy, Editor
J.R. Osborn,
Associate Editor
Purdue University
A.W. Barrows
J.R. Ward,
Associate Editors
Ballistics Research Laboratories
ISBN 0-915928-04-3

41. **Communications Satellite Developments: Systems** (1976)
Gilbert E. LaVean
Defense Communications Agency
William G. Schmidt
CML Satellite Corp.
ISBN 0-915928-05-1

42. **Communications Satellite Developments: Technology** (1976)
William G. Schmidt
CML Satellite Corp.
Gilbert E. LaVean
Defense Communications Agency
ISBN 0-915928-06-X

*43. **Aeroacoustics: Jet Noise, Combustion and Core Engine Noise** (1976)
Ira R. Schwartz, Editor
NASA Ames Research Center
Henry T. Nagamatsu,
Associate Editor
General Electric Research and Development Center
Warren C. Strahle,
Associate Editor
Georgia Institute of Technology
ISBN 0-915928-07-8

*44. **Aeroacoustics: Fan Noise and Control; Duct Acoustics; Rotor Noise** (1976)
Ira R. Schwartz, Editor
NASA Ames Research Center
Henry T. Nagamatsu,
Associate Editor
General Electric Research and Development Center
Warren C. Strahle,
Associate Editor
Georgia Institute of Technology
ISBN 0-915928-08-6

*45. **Aeroacoustics: STOL Noise; Airframe and Airfoil Noise** (1976)
Ira R. Schwartz, Editor
NASA Ames Research Center
Henry T. Nagamatsu,
Associate Editor
General Electric Research and Development Center
Warren C. Strahle,
Associate Editor
Georgia Institute of Technology
ISBN 0-915928-09-4

*46. **Aeroacoustics: Acoustic Wave Propagation; Aircraft Noise Prediction; Aeroacoustic Instrumentation** (1976)
Ira R. Schwartz, Editor
NASA Ames Research Center
Henry T. Nagamatsu,
Associate Editor
General Electric Research and Development Center
Warren C. Strahle,
Associate Editor
Georgia Institute of Technology
ISBN 0-915928-10-8

47. **Spacecraft Charging by Magnetospheric Plasmas** (1976)
Alan Rosen
TRW Inc.
ISBN 0-915928-11-6

48. **Scientific Investigations on the Skylab Satellite** (1976)
Marion I. Kent
Ernst Stuhlinger
NASA George C. Marshall Space Flight Center
Shi-Tsan Wu
University of Alabama
ISBN 0-915928-12-4

49. **Radiative Transfer and Thermal Control** (1976)
Allie M. Smith
ARO Inc.
ISBN 0-915928-13-2

50. **Exploration of the Outer Solar System** (1976)
Eugene W. Greenstadt
TRW Inc.
Murray Dryer
National Oceanic and Atmospheric Administration
Devrie S. Intriligator
University of Southern California
ISBN 0-915928-14-0

51. **Rarefied Gas Dynamics, Parts I and II** (two volumes) (1977)
J. Leith Potter
ARO Inc.
ISBN 0-915928-15-9

52. **Materials Sciences in Space with Application to Space Processing** (1977)
Leo Steg
General Electric Co.
ISBN 0-915928-16-7

53. **Experimental Diagnostics in Gas Phase Combustion Systems** (1977)
Ben T. Zinn, Editor
Georgia Institute of Technology
Craig T. Bowman, Associate Editor
Stanford University
Daniel L. Hartley, Associate Editor
Sandia Laboratories
Edward W. Price, Associate Editor
Georgia Institute of Technology
James G. Skifstad, Associate Editor
Purdue University
ISBN 0-015928-18-3

54. **Satellite Communications: Future Systems** (1977)
David Jarett
TRW Inc.
ISBN 0-915928-18-3

55. **Satellite Communications: Advanced Technologies** (1977)
David Jarett
TRW Inc.
ISBN 0-915928-19-1

56. **Thermophysics of Spacecraft and Outer Planet Entry Probes** (1977)
Allie M. Smith
ARO Inc.
ISBN 0-915928-20-5

57. **Space-Based Manufacturing from Nonterrestrial Materials** (1977)
Gerard K. O'Neill, Editor
Brian O'Leary, Assistant Editor
Princeton University
ISBN 0-915928-21-3

58. **Turbulent Combustion** (1978)
Lawrence A. Kennedy
State University of New York at Buffalo
ISBN 0-915928-22-1

59. **Aerodynamic Heating and Thermal Protection Systems** (1978)
Leroy S. Fletcher
University of Virginia
ISBN 0-915928-23-X

60. **Heat Transfer and Thermal Control Systems** (1978)
Leroy S. Fletcher
University of Virginia
ISBN 0-915928-24-8

61. **Radiation Energy Conversion in Space** (1978)
Kenneth W. Billman
NASA Ames Research Center
ISBN 0-915928-26-4

62. **Alternative Hydrocarbon Fuels: Combustion and Chemical Kinetics** (1978)
Craig T. Bowman
Stanford University
Jorgen Birkeland
Department of Energy
ISBN 0-915928-25-6

63. **Experimental Diagnostics in Combustion of Solids** (1978)
Thomas L. Boggs
Naval Weapons Center
Ben T. Zinn
Georgia Institute of Technology
ISBN 0-915928-28-0

64. **Outer Planet Entry Heating and Thermal Protection** (1979)
Raymond Viskanta
Purdue University
ISBN 0-915928-29-9

65. **Thermophysics and Thermal Control** (1979)
Raymond Viskanta
Purdue University
ISBN 0-915928-30-2

66. **Interior Ballistics of Guns** (1979)
Herman Krier
University of Illinois at Urbana-Champaign
Martin Summerfield
New York University
ISBN 0-915928-32-9

*67. **Remote Sensing of Earth from Space: Role of "Smart Sensors"** (1979)
Roger A. Breckenridge
NASA Langley Research Center
ISBN 0-915928-33-7

68. **Injection and Mixing in Turbulent Flow** (1980)
Joseph A. Schetz
Virginia Polytechnic Institute and State University
ISBN 0-915928-35-3

69. **Entry Heating and Thermal Protection** (1980)
Walter B. Olstad
NASA Headquarters
ISBN 0-915928-38-8

70. **Heat Transfer, Thermal Control, and Heat Pipes** (1980)
Walter B. Olstad
NASA Headquarters
ISBN 0-915928-39-6

*71. **Space Systems and Their Interactions with Earth's Space Environment** (1980)
Henry B. Garrett
Charles P. Pike
Hanscom Air Force Base
ISBN 0-915928-41-8

72. **Viscous Flow Drag Reduction** (1980)
Gary R. Hough
Vought Advanced Technology Center
ISBN 0-915928-44-2

73. **Combustion Experiments in a Zero-Gravity Laboratory** (1981)
Thomas H. Cochran
NASA Lewis Research Center
ISBN 0-915928-48-5

74. **Rarefied Gas Dynamics, Parts I and II** (two volumes) (1981)
Sam S. Fisher
University of Virginia
ISBN 0-915928-51-5

75. **Gasdynamics of Detonations and Explosions** (1981)
J.R. Bowen
University of Wisconsin at Madison
N. Manson
Université de Poitiers
A.K. Oppenheim
University of California at Berkeley
R.I. Soloukhin
Institute of Heat and Mass Transfer, BSSR Academy of Sciences
ISBN 0-915928-46-9

76. **Combustion in Reactive Systems** (1981)
J.R. Bowen
University of Wisconsin at Madison
N. Manson
Université de Poitiers
A.K. Oppenheim
University of California at Berkeley
R.I. Soloukhin
Institute of Heat and Mass Transfer, BSSR Academy of Sciences
ISBN 0-915928-47-7

77. **Aerothermodynamics and Planetary Entry** (1981)
A.L. Crosbie
University of Missouri-Rolla
ISBN 0-915928-52-3

78. **Heat Transfer and Thermal Control** (1981)
A.L. Crosbie
University of Missouri-Rolla
ISBN 0-915928-53-1

79. **Electric Propulsion and Its Applications to Space Missions** (1981)
Robert C. Finke
NASA Lewis Research Center
ISBN 0-915928-55-8

80. **Aero-Optical Phenomena** (1982)
Keith G. Gilbert
Leonard J. Otten
Air Force Weapons Laboratory
ISBN 0-915928-60-4

81. **Transonic Aerodynamics** (1982)
David Nixon
Nielsen Engineering & Research, Inc.
ISBN 0-915928-65-5

82. **Thermophysics of Atmospheric Entry** (1982)
T.E. Horton
University of Mississippi
ISBN 0-915928-66-3

83. **Spacecraft Radiative Transfer and Temperature Control** (1982)
T.E. Horton
University of Mississippi
ISBN 0-915928-67-1

84. **Liquid-Metal Flows and Magnetohydrodynamics** (1983)
H. Branover
Ben-Gurion University of the Negev
P.S. Lykoudis
Purdue University
A. Yakhot
Ben-Gurion University of the Negev
ISBN 0-915928-70-1

85. **Entry Vehicle Heating and Thermal Protection Systems: Space Shuttle, Solar Starprobe, Jupiter Galileo Probe** (1983)
Paul E. Bauer
McDonnell Douglas Astronautics Co.
Howard E. Collicott
The Boeing Co.
ISBN 0-915928-74-4

86. **Spacecraft Thermal Control, Design, and Operation** (1983)
Howard E. Collicott
The Boeing Co.
Paul E. Bauer
McDonnell Douglas Astronautics Co.
ISBN 0-915928-75-2

87. **Shock Waves, Explosions, and Detonations** (1983)
J.R. Bowen
University of Washington
N. Manson
Université de Poitiers
A.K. Oppenheim
University of California at Berkeley
R.I. Soloukhin
Institute of Heat and Mass Transfer, BSSR Academy of Sciences
ISBN 0-915928-76-0

88. **Flames, Lasers, and Reactive Systems** (1983)
J.R. Bowen
University of Washington
N. Manson
Université de Poitiers
A.K. Oppenheim
University of California at Berkeley
R.I. Soloukhin
Institute of Heat and Mass Transfer, BSSR Academy of Sciences
ISBN 0-915928-77-9

SERIES LISTING

89. **Orbit-Raising and Maneuvering Propulsion: Research Status and Needs** (1984)
Leonard H. Caveny
Air Force Office of Scientific Research
ISBN 0-915928-82-5

90. **Fundamentals of Solid-Propellant Combustion** (1984)
Kenneth K. Kuo
Pennsylvania State University
Martin Summerfield
Princeton Combustion Research Laboratories, Inc.
ISBN 0-915928-84-1

91. **Spacecraft Contamination: Sources and Prevention** (1984)
J.A. Roux
University of Mississippi
T.D. McCay
NASA Marshall Space Flight Center
ISBN 0-915928-85-X

92. **Combustion Diagnostics by Nonintrusive Methods** (1984)
T.D. McCay
NASA Marshall Space Flight Center
J.A. Roux
University of Mississippi
ISBN 0-915928-86-8

93. **The INTELSAT Global Satellite System** (1984)
Joel Alper
COMSAT Corp.
Joseph Pelton
INTELSAT
ISBN 0-915928-90-6

94. **Dynamics of Shock Waves, Explosions, and Detonations** (1984)
J.R. Bowen
University of Washington
N. Manson
Université de Poitiers
A.K. Oppenheim
University of California at Berkely
R.I. Soloukhin
Institute of Heat and Mass Transfer, BSSR Academy of Sciences
ISBN 0-915928-91-4

95. **Dynamics of Flames and Reactive Systems** (1984)
J.R. Bowen
University of Washington
N. Manson
Université de Poitiers
A.K. Oppenheim
University of California at Bereley
R.I. Soloukhin
Institute of Heat and Mass Transfer, BSSR Academy of Sciences
ISBN 0-915928-92-2

96. **Thermal Design of Aeroassisted Orbital Transfer Vehicles** (1985)
H.F. Nelson
University of Missouri-Rolla
ISBN 0-915928-94-9

97. **Monitoring Earth's Ocean, Land, and Atmosphere from Space — Sensors, Systems, and Applications** (1985)
Abraham Schnapf
Aerospace Systems Engineering
ISBN 0-915928-98-1

98. **Thrust and Drag: Its Prediction and Verification** (1985)
Eugene E. Covert
Massachusetts Institute of Technology
C.R. James
Vought Corp.
William F. Kimzey
Sverdrup Technology AEDC Group
George K. Richey
U.S. Air Force
Eugene C. Rooney
U.S. Navy Department of Defense
ISBN 0-930403-00-2

99. **Space Stations and Space Platforms — Concepts, Design, Infrastructure, and Uses** (1985)
Ivan Bekey
Daniel Herman
NASA Headquarters
ISBN 0-930403-01-0

100. **Single- and Multi-Phase Flows in an Electromagnetic Field: Energy, Metallurgical, and Solar Applications** (1985)
Herman Branover
Ben-Gurion University of the Negev
Paul S. Lykoudis
Purdue University
Michael Mond
Ben-Gurion University of the Negev
ISBN 0-930403-04-5

101. **MHD Energy Conversion: Physiotechnical Problems** (1986)
V.A. Kirillin
A.E. Sheyndlin
Soviet Academy of Sciences
ISBN 0-930403-05-3

102. **Numerical Methods for Engine-Airframe Integration** (1986)
S.N.B. Murthy
Purdue University
Gerald C. Paynter
Boeing Airplane Co.
ISBN 0-930403-09-6

103. **Thermophysical Aspects of Re-Entry Flows** (1986)
James N. Moss
NASA Langley Research Center
Carl D. Scott
NASA Johnson Space Center
ISBN 0-930403-10-X

104. **Tactical Missile Aerodynamics** (1986)
M.J. Hemsch
PRC Kentron, Inc.
J.N. Nielsen
NASA Ames Research Center
ISBN 0-930403-13-4

105. **Dynamics of Reactive Systems Part I: Flames and Configurations; Part II: Modeling and Heterogeneous Combustion** (1986)
J.R. Bowen
University of Washington
J.-C. Leyer
Université de Poitiers
R.I. Soloukhin
Institute of Heat and Mass Transfer, BSSR Academy of Sciences
ISBN 0-930403-14-2

106. **Dynamics of Explosions** (1986)
J.R. Bowen
University of Washington
J.-C. Leyer
Université de Poitiers
R.I. Soloukhin
Institute of Heat and Mass Transfer, BSSR Academy of Sciences
ISBN 0-930403-15-0

107. **Spacecraft Dielectric Material Properties and Spacecraft Charging** (1986)
A.R. Frederickson
U.S. Air Force Rome Air Development Center
D.B. Cotts
SRI International
J.A. Wall
U.S. Air Force Rome Air Development Center
F.L. Bouquet
Jet Propulsion Laboratory, California Institute of Technology
ISBN 0-930403-17-7

108. **Opportunities for Academic Research in a Low-Gravity Environment** (1986)
George A. Hazelrigg
National Science Foundation
Joseph M. Reynolds
Louisiana State University
ISBN 0-930403-18-5

109. **Gun Propulsion Technology** (1988)
Ludwig Stiefel
U.S. Army Armament Research, Development and Engineering Center
ISBN 0-930403-20-7

110. **Commercial Opportunities in Space** (1988)
F. Shahrokhi
K.E. Harwell
University of Tennessee Space Institute
C.C. Chao
National Cheng Kung University
ISBN 0-930403-39-8

111. **Liquid-Metal Flows: Magnetohydrodynamics and Applications** (1988)
Herman Branover,
Michael Mond, and
Yeshajahu Unger
Ben-Gurion University of the Negev
ISBN 0-930403-43-6

112. **Current Trends in Turbulence Research** (1988)
Herman Branover,
Michael Mond, and
Yeshajahu Unger
Ben-Gurion University of the Negev
ISBN 0-930403-44-4

113. **Dynamics of Reactive Systems Part I: Flames; Part II: Heterogeneous Combustion and Applications** (1988)
A.L. Kuhl
R & D Associates
J.R. Bowen
University of Washington
J.-C. Leyer
Université de Poitiers
A. Borisov
USSR Academy of Sciences
ISBN 0-930403-46-0

114. **Dynamics of Explosions** (1988)
A.L. Kuhl
R & D Associates
J.R. Bowen
University of Washington
J.-C. Leyer
Université de Poitiers
A. Borisov
USSR Academy of Sciences
ISBN 0-930403-47-9

115. **Machine Intelligence and Autonomy for Aerospace** (1988)
E. Heer
Heer Associates, Inc.
H. Lum
NASA Ames Research Center
ISBN 0-930403-48-7

116. **Rarefied Gas Dynamics: Space-Related Studies** (1989)
E.P. Muntz
University of Southern California
D.P. Weaver
U.S. Air Force Astronautics Laboratory (AFSC)
D.H. Campbell
University of Dayton Research Institute
ISBN 0-930403-53-3

117. **Rarefied Gas Dynamics: Physical Phenomena** (1989)
E.P. Muntz
University of Southern California
D.P. Weaver
U.S. Air Force Astronautics Laboratory (AFSC)
D. Campbell
University of Dayton Research Institute
ISBN 0-930403-54-1

118. **Rarefied Gas Dynamics: Theoretical and Computational Techniques** (1989)
E.P. Muntz
University of Southern California
D.P. Weaver
U.S. Air Force Astronautics Laboratory (AFSC)
D.H. Campbell
University of Dayton Research Institute
ISBN 0-930403-55-X

119. **Test and Evaluation of the Tactical Missile** (1989)
Emil J. Eichblatt Jr.
Pacific Missile Test Center
ISBN 0-930403-56-8

120. **Unsteady Transonic Aerodynamics** (1989)
David Nixon
Nielsen Engineering & Research, Inc.
ISBN 0-930403-52-5

121. **Orbital Debris from Upper-Stage Breakup** (1989)
Joseph P. Loftus Jr.
NASA Johnson Space Center
ISBN 0-930403-58-4

122. **Thermal-Hydraulics for Space Power, Propulsion and Thermal Management System Design** (1989)
William J. Krotiuk
General Electric Co.
ISBN 0-930403-64-9

123. **Viscous Drag Reduction in Boundary Layers** (1990)
Dennis M. Bushnell
Jerry N. Hefner
NASA Langley Research Center
ISBN 0-930403-66-5

124. **Tactical and Strategic Missile Guidance** (1990)
Paul Zarchan
Charles Stark Draper Laboratory, Inc.
ISBN 0-930403-68-1

125. **Applied Computational Aerodynamics** (1990)
P.A. Henne
Douglas Aircraft Company
ISBN 0-930403-69-X

126. **Space Commercialization: Launch Vehicles and Programs** (1990)
F. Shahrokhi
University of Tennessee Space Institute
J.S. Greenberg
Princeton Synergetics Inc.
T. Al-Saud
Ministry of Defense and Aviation Kingdom of Saudi Arabia
ISBN 0-930403-75-4

127. **Space Commercialization: Platforms and Processing** (1990)
F. Shahrokhi
University of Tennessee Space Institute
G. Hazelrigg
National Science Foundation
R. Bayuzick
Vanderbilt University
ISBN 0-930403-76-2

128. **Space Commercialization: Satellite Technology** (1990)
F. Shahrokhi
University of Tennessee Space Institute
N. Jasentuliyana
United Nations
N. Tarabzouni
King Abulaziz City for Science and Technology
ISBN 0-930403-77-0

129. **Mechanics and Control of Large Flexible Structures** (1990)
John L. Junkins
Texas A&M University
ISBN 0-930403-73-8

130. **Low-Gravity Fluid Dynamics and Transport Phenomena** (1990)
Jean N. Koster
Robert L. Sani
University of Colorado at Boulder
ISBN 0-930403-74-6

131. **Dynamics of Deflagrations and Reactive Systems: Flames** (1991)
A. L. Kuhl
Lawrence Livermore National Laboratory
J.-C. Leyer
Université de Poitiers
A. A. Borisov
USSR Academy of Sciences
W. A. Sirignano
University of California
ISBN 0-930403-95-9

132. **Dynamics of Deflagrations and Reactive Systems: Heterogeneous Combustion** (1991)
A. L. Kuhl
Lawrence Livermore National Laboratory
J.-C. Leyer
Université de Poitiers
A. A. Borisov
USSR Academy of Sciences
W. A. Sirignano
University of California
ISBN 0-930403-96-7

133. **Dynamics of Detonations and Explosions: Detonations** (1991)
A. L. Kuhl
Lawrence Livermore National Laboratory
J.-C. Leyer
Université de Poitiers
A. A. Borisov
USSR Academy of Sciences
W. A. Sirignano
University of California
ISBN 0-930403-97-5

134. **Dynamics of Detonations and Explosions: Explosion Phenomena** (1991)
A. L. Kuhl
Lawrence Livermore National Laboratory
J.-C. Leyer
Université de Poitiers
A. A. Borisov
USSR Academy of Sciences
W. A. Sirignano
University of California
ISBN 0-930403-98-3

135. **Numerical Approaches to Combustion Modeling** (1991)
Elaine S. Oran
Jay P. Boris
Naval Research Laboratory
ISBN 1-56347-004-7

136. **Aerospace Software Engineering** (1991)
Christine Anderson
U.S. Air Force Wright Laboratory
Merlin Dorfman
Lockheed Missiles & Space Company, Inc.
ISBN 1-56346-005-5

137. **High-Speed Flight Propulsion Systems** (1991)
S. N. B. Murthy
Purdue University
E. T. Curran
Wright Laboratory
ISBN 1-56347-011-X

138. **Propagation of Intensive Laser Radiation in Clouds** (1992)
O. A. Volkovitsky
Yu. S. Sedunov
L. P. Semenov
Institute of Experimental Meteorology
ISBN 1-56347-020-9

139. **Gun Muzzle Blast and Flash** (1992)
Günter Klingenberg
Fraunhofer-Institut für Kurzzeitdynamik, Ernst-Mach-Institut (EMI)
Joseph M. Heimerl
U.S. Army Ballistic Research Laboratory (BRL)
ISBN 1-56347-012-8

140. **Thermal Structures and Materials for High-Speed Flight** (1992)
Earl A. Thornton
University of Virginia
ISBN 1-56347-017-9

141. **Tactical Missile Aerodynamics: General Topics** (1992)
Michael J. Hemsch
Lockheed Engineering & Sciences Company
ISBN 1-56347-015-2

142. **Tactical Missile Aerodynamics: Prediction Methodology** (1992)
Michael R. Mendenhall
Nielsen Engineering & Research, Inc.
ISBN 1-56347-016-0

143. **Nonsteady Burning and Combustion Stability of Solid Propellants** (1992)
Luigi De Luca
Politecnico di Milano
Edward W. Price
Georgia Institute of Technology
Martin Summerfield
Princeton Combustion Research Laboratories, Inc.
ISBN 1-56347-014-4

144. **Space Economics**
Joel S. Greenberg
Princeton Synergetics, Inc.
Henry R. Hertzfeld
HRH Associates
ISBN 1-56347-042-X

145. **Mars: Past, Present, and Future**
E. Brian Pritchard
NASA Langley Research Center
ISBN 1-56347-043-8

(Other Volumes are planned.)